# Principles of Systematic Zoology

# Principles of Systematic Zoology

## Ernst Mayr

Alexander Agassiz Professor of Zoology, Harvard University
Director, Museum of Comparative Zoology
Member, International Commission on Zoological Nomenclature

**McGraw-Hill Book Company**

NEW YORK   ST. LOUIS   SAN FRANCISCO   TORONTO   LONDON   SYDNEY

**Principles of Systematic Zoology**

*Library of Congress Catalog Card Number* 68-54937

41143

1234567890 MAMM 7543210698

Dedicated to my friends

**E. Gorton Linsley**
**and**
**Robert L. Usinger**

without whose constant encouragement
and inspiration this work could not have
been undertaken nor completed.

# *Preface*

Systematics has had a remarkable renaissance during the last generation. The reasons for this are diverse. Taxonomists played a leading role in the new synthesis of evolutionary theory, and they have demonstrated that the study of organic diversity, the main concern of systematics, is a major integral branch of biology. Systematics has also been very important in initiating the entire field of population biology, including population genetics. The recognition of the importance of taxonomy has been helped by the realization that there are two great scientific methods, the experimental method and the comparative method (based on observations). Observational data are rather meaningless unless they are classified prior to comparison. Understanding this methodological necessity has created a new interest in methods and theory of classification in all comparative sciences. This interest has been enhanced by the coming of the computer and concurrent endeavors to automate classifying through electronic data processing. All this is superimposed on the increasing need for the applied aspects of taxonomy, such as the correct identification and classification of species in agriculture, public health, ecology, genetics, and behavioral biology.

In 1953 when *Methods and Principles of Systematic Zoology* was published, no other text on the systematic method was available except Hennig's specialized German volume. The current vigor of systematics is indicated by the number of recently published texts in this area (see Chap. 1), by the flourishing of the journal *Systematic Zoology* since its founding in 1952 and of similar periodicals in England and Germany, and by the annually increasing number of articles in journals throughout the world dealing with the methods and principles of systematics. The only way to cope with this massive accumulation of new information and new theory has been to prepare a radical revision of the 1953 volume.

Two of the three coauthors of this earlier volume (E. Gorton Linsley and Robert L. Usinger) were unfortunately too heavily committed by professional duties to participate in the time-consuming task of a revision. By mutual agreement Ernst Mayr, therefore, assumed full responsibility. In the course of the work it became apparent that more was needed than a mere revision. The outcome was a new work with a new title. Nevertheless it is a direct descendant of the 1953 book, and much that was valuable in the old volume is incorporated in the new one. A first draft of the manuscript was, therefore, critically read by the former coauthors. I deeply appreciate the help and encouragement I have had from my two friends, who continue to maintain an undiminished interest in this volume.

It has been a frequent complaint in the past that the training of the young taxonomist is too empirical, consisting too much in merely conditioning him to carry out the operations of experienced taxonomists. Such criticism is legitimate when this practical training is not supplemented by a study of the theory. Yet, at the same time, taxonomy is a subject which is so operational that it could not possibly be learned merely by reading a book. The main objective of this volume, therefore, is to serve as a guide and companion for those learning the subject and perhaps even more so for those who are teaching it. However, it cannot take the place of a laboratory in which the procedures of classification are actually demonstrated. The problems of taxonomy are different in every group of animals, and every teacher will want to use material and illustrations from the zoological groups with which he himself is most familiar. This is the reason why this volume does not cite more examples to illustrate basic principles and methods. The teacher himself will know best what examples will be most instructive in a given case. The problems of taxonomy are far too diverse to provide, as the beginner might love to have it, a set of recipes which can be applied to every situation. Actually, a clear presentation of the theory will be more helpful in the carrying out of practical tasks than such a recipe book.

It is for this reason that the theoretical aspects of taxonomy have received so much attention in this work. The object is not so much to make a contribution to theoretical biology as to establish a sound foundation for practical operations. It is important for the taxonomist at every stage of his work to understand exactly the meaning of such terms as species, taxon, category, classification, and type. The clearer these terms, and the concepts underlying them, are understood, the greater will be the agreement among taxonomists and the less time they will waste on sterile controversy.

The 1953 book, while presenting some original ideas of the authors, concentrated on presenting a balanced digest of the published literature of taxonomy. The new work goes somewhat further in offering original material, particularly on the theory of taxonomy (Chap. 4), on the procedures of classification (Chap. 10), and on the theory of nomenclature (Chap. 13). It is hoped that this will make the volume important even for the experienced worker without interfering with its utility for the beginner. As far as the theory of nomenclature and the procedure of classification are concerned, there are no other comprehensive treatments available in the literature. To add to the usefulness of the volume the main text of the Rules of Zoological Nomenclature is incorporated in Chap. 12.

In order to compensate for the considerable expansion of the volume through the addition of new material, an effort was made to shorten the presentation of subjects that are well covered in recent publications. The new Code of Zoological Nomenclature, for instance, gives so much detail on the formation of scientific names (in Appendix D) that it seemed advisable to omit this topic from the new edition. Likewise, the revised edition of *Quantitative Zoology* by Simpson, Roe, and Lewontin (1960) covers the aspects of statistics relevant to taxonomy so excellently that it was possible to shorten our own account drastically. The recent publication of several books devoted to scientific illustration and the techniques of drawing (Chap. 11) permitted shortening the section devoted to this subject.

Finally, the author's recent publication of a wide-ranging volume on evolution, *Animal Species and Evolution* (1963), made it unnecessary to cover the same subject matter in the present volume, as important as an understanding of the evolutionary processes is for the taxonomist. The exhaustive bibliography of that volume permits convenient access to the primary evolutionary literature, and the volume is therefore—specifically for this purpose—frequently quoted in the present work.

No author can ever adequately thank all those who by their constructive criticism have contributed to the improvement of the manuscript. The present work is one more example of such helpfulness. A first draft of the entire manuscript was read by E. G. Linsley, R. L. Usinger, Michael Ghiselin, Donn E. Rosen, W. F. Blair, and Richard D. Alexander. Of the next version Herbert Levi, K. Boss, and J. Lawrence read most chapters. Chapters 8 and 9 were read by S. Gould and R. A. Reyment and Chaps. 12 and 13 by J. Corliss, Eugene Eisenmann, W. J. Follett, R. P. Higgins, Myra Keen, Alfred and Helen Loeblich, Hobart M. Smith, and Ellis Yochelson. Every one of them, but particularly R. D. Alexander, made valuable suggestions. Each found a number of errors and inconsistencies in my treatment which I was able to correct. Those faults which still remain are entirely my own responsibility.

John Lawrence undertook the great labor of compiling the information on the number of known species presented in Table 1-1, much of it secured from the leading specialists in the respective groups.

I am greatly indebted to Susan Martin, Sabine Wespi, and Sally Loth for the careful typing of three consecutive versions of the manuscript and to William Jolly as well as to Vojislav Jovanovic for literature search, help with the bibliography, and other editorial assistance, which greatly lightened my own burden. More than anyone else it was my wife who has assisted me in numerous ways and has truly made the completion of this volume possible.

*Ernst Mayr*

# Contents

*Principles of Systematic Zoology*

# Chapter 1 The Science of Taxonomy

*T*he amount of diversity in the living world is staggering. About 1 million species of animals and half a million species of plants have already been described, and estimates on the number of still undescribed living species range from 3 to 10 million. An estimate of half a billion for the extinct species is consistent with the known facts. Each species may exist in numerous different forms (sexes, age classes, seasonal forms, morphs, and other phena). It would be impossible to deal with this enormous diversity if it were not ordered and classified. Systematic zoology endeavors to order the rich diversity of the animal world and to develop methods and principles to make this task possible.

## 1.1 TAXONOMY AND SYSTEMATICS

The study of organic diversity has changed its objectives and enlarged its scope in the course of history, as happens in any branch of science. A detailed history will be presented in Chap. 4, but some of it must be anticipated here in order to make clear subtle changes in the meanings of commonly used terms (e.g., systematics).

The ancients looked for a natural order (*kosmos*) which would explain the bewildering diversity of phenomena. They attempted to discover the true "nature" of things and approached the classification of inanimate objects and living beings by the procedures of logic. The major purpose of a classification was to serve as an identification key, and the philosophy of the early taxonomists was well suited for the utilitarian purposes of taxonomy. The objective shifted and the interests of the taxonomist broadened when, after 1859, organic diversity was interpreted as the result of evolutionary divergence. No longer was he interested merely in producing identification keys; he now interpreted groups of organisms as descendants of common ancestors, and inevitably he became interested in the pathways and causations responsible for evolutionary changes. Also, since this force deals with the living organism, the taxonomist became increasingly a student of living organisms, particularly in the field. These field studies, in turn, showed that behavior and ecology often supply far more important taxonomic characters of species than the morphological differences of preserved specimens. Imperceptibly a new branch of biology began to emerge, *the study of the diversity of organisms.*

The ultimate result of these developments has been the recognition that the universe of the taxonomist is far greater than was previously envisioned. This had an effect on the definition of the terms taxonomy and systematics. Until quite recently these terms were generally considered to be synonymous. Now it has become advantageous to restrict the term taxonomy to its conventional meaning, but to define the term systematics more broadly as the study of organic diversity.

The term taxonomy is derived from the Greek words *taxis,* arrangement, and *nomos,* law, and was first proposed, in its French form by de Candolle (1813) for the theory of plant classification. Analogous to astronomy, agronomy, economy, etc., it is correctly formed and need not be amended (Mayr, 1966). It agrees best with current thinking to define it as follows: *Taxonomy is the theory and practice of classifying organisms.*

The term systematics stems from the latinized Greek word *systema,* as applied to the systems of classification developed by the early naturalists, notably Linnaeus (*Systema naturae,* 1st ed., 1735). We follow Simpson's (1961) modern redefinition of this term: *"Systematics is the scientific study of the kinds and diversity of organisms and of any and all relationships among them,"* or more simply, *systematics is the science of the diversity of organisms.* The word "relationship" is not used in a narrow phylogenetic sense, but is broadly conceived to include all biological relationships among organisms. This explains why such a broad area of common interest has developed between systematics, evolutionary biology, ecology, and behavioral biology.

*Place of Systematics in Biology.* Systematics is unique among the biological sciences in its dominant concern with diversity. In all subdivisions of functional biology the main concern is with basic processes and mechanisms shared by all or most organisms. Hence the reductionist tendency at the cellular and molecular levels—the endeavor to reduce everything to common denominators, i.e., to penetrate to the universal building stones. If all biologists were to share this objective, biology would become very one-sided. It is the student of systematics who helps to restore the balance by his interest in and insistence on uniqueness, whole organisms, and systems. One of the major preoccupations of systematics is to determine, by comparison, what the unique properties of every species and higher taxon are. Another is to determine what properties certain taxa have in common with each other, and what the biological causes for the differences or shared characters are. Finally, it concerns itself with variation within taxa. In all these concerns systematics holds a unique and indispensable position among the biological sciences. Classification makes the organic diversity accessible to the other biological disciplines. Without it most of them would be unable to give meaning to their findings.

Systematics deals with populations, species, and higher taxa. No other branch of biology occupies itself in a similar manner with this level of integration in the organic world. It not only supplies urgently needed information about these levels but, more important, it cultivates a way of thinking, a way of approaching biological problems which is tremendously important for the balance and well-being of biology as a whole (Mayr, 1968).

As in every generation there is also at present a questioning of the basic objectives of taxonomy. Have recent developments in other branches of biology made taxonomy superfluous? Is it futile to search for a classification based on evolution? Is the method of binominal nomenclature in conflict with its function in information retrieval? Although these questions are sometimes posed in a manner to produce more heat than light, the taxonomist would ignore his obligations if he did not face them squarely. Stimulating—if not provocative—reading is provided by Michener (1963), Ehrlich (1964), Constance (1964), and Rollins (1964). The impression the taxonomist receives from a discriminating reading of the current biological literature is that there was never a greater need for a strong science of systematics than there is at present. There is a need for someone to stress the diversity of the living world, the most truly biological quality of organisms, in the face of current reductionist tendencies. However legitimate the study is of that which all organisms have in common (much of it being the physics and chemistry of organisms), it is equally legitimate to study the unique characteristics of taxa at all levels down to the species. And this is precisely what the taxonomist is doing.

## 1.2  TERMS AND DEFINITIONS

In systematic zoology, as in all branches of science, the danger of misunderstandings is greatly reduced by the precise definition of terms. Terms that are frequently used in this volume, such as species, type, polytypic, etc., will be carefully defined in the relevant chapters and in the glossary. At this point some terms will be considered that relate to all the chapters.

The terms *taxonomy* and *systematics* were defined above. This leaves the term *classification,* which partly overlaps with taxonomy. The word is used with two different meanings. Most commonly it designates the product of the activity of the taxonomist, the classification of the primates or of the bees. But it is also used as a term for the activity of classifying: "Zoological classification is the ordering of animals into groups (or sets) on the basis of their relationships" (Simpson, 1961). In this sense classification coincides largely with what is sometimes designated as *beta taxonomy.* Both usages are so well established and so easily distinguished by their context that it would seem futile and impractical to try to restrict the term classification to only one of the two meanings. A certain amount of overlap between the terms systematics, taxonomy, and classification is perhaps unavoidable and not necessarily harmful.

The process of *classification* is totally different from that of *identification.* In classification we undertake the ordering of populations and groups of populations at all levels by inductive procedures; in identification we place individuals by deductive procedures into previously established classes (taxa). (See 4.3.1 for the theory and 6.3 for the practices of identification.)

*Zoological nomenclature* is the application of distinctive names to each of the groups recognized in the zoological classification. The rules governing it and the interpretation of these rules will be presented in Chaps. 12 and 13

The most important aspects of classification are the grouping and ranking of organisms (Chap. 10). Precision and unambiguity of terms are of the utmost importance in these operations. Simpson (1961, pp. 16–21; 1963) gives excellent analyses of the relevant terms.

In the earlier taxonomic literature there was frequent confusion between the zoological objects which groups of populations represent and their rank in the hierarchy of taxonomic categories. These are two very different phenomena. We have somewhat analogous situations in our daily affairs: Fred Smith is a concrete person, but "captain" (or "associate professor") is his rank in a hierarchy of levels.

*Taxon.* The words bluebirds, thrushes, songbirds, or vertebrates refer to groups of organisms. These are the concrete objects of zoological classification. Any such group of populations is called a *taxon* if the zoologist considers it sufficiently distinct to be worthy of being formally assigned to a definite category in the hierarchic classification. As Simpson defines it, "A taxon is a group of real organisms recognized as a formal unit at any level of a hierarchic classification." The same thought can also be expressed as follows: *A taxon is a taxonomic group of any rank that is sufficiently distinct to be worthy of being assigned to a definite category.*

This definition calls attention to the fact that the delimitation of a taxon against other taxa of the same rank is often subject to the judgment of the taxonomist.

Two aspects must be stressed. A taxon always refers to concrete zoological objects. Thus *the* species is not a taxon, but a given species such as the robin (*Turdus migratorius*) is. Secondly, the taxon must be formally recognized by the taxonomist. Within any large genus, for instance, groupings of species can be recognized. They are taxa only if and when they are formally distinguished, as, for instance, by being recognized as separate subgenera. Likewise, demes and geographical isolates become taxa only when formally recognized as subspecies.

We speak of higher taxa, such as thrushes, birds, or vertebrates, and lower taxa, such as bluebirds or robins. Taxa of species rank are what the taxonomist ordinarily classifies. Yet there is a great deal of variation within most taxa, as will be discussed in later chapters. The recognition of what belongs to a given taxon of species rank is often the most difficult step in classification owing to sexual dimorphism, seasonal changes, age variation, and genetic polymorphism.

*Phenon.* The first step in classification is the separation of reasonably uniform samples and their assortment into taxa at the species level. There is no generally accepted technical term for *a phenotypically reasonably uniform sample,* but it may be designated as a *phenon,* a term introduced by Camp and Gilly (1943) for phenotypically homogeneous samples at the species level. Males and females often belong to different phena, while in the case of sibling species it is possible that several species belong to a single phenon. The term *morphospecies* has sometimes been applied confusingly to what is here designated as a phenon. Recognition of a technical term for the phenotypically uniform sample greatly facilitates the description of the taxonomic procedure. Its recognition is of particular importance in the procedures of computer taxonomy. Sokal and Sneath (1963) use the term phenon in a very different sense. Chapters 8 and 9 deal with the taxonomic treatment of phena.

A *category* designates rank or level in a hierarchic classification. *It is a class, the members of which are all the taxa assigned a given rank.* For instance, the species category is a class the members of which are the species taxa.

A full understanding of the meaning of category depends on an understanding of hierarchical classification, which will be discussed in Chap. 5. Such terms as species, genus, family, and order designate categories. A category, thus, is an abstract term, a class name, while the taxa placed in these categories are concrete zoological objects. Until the word taxon was introduced into the literature, the term category was often confusingly used both for group and for rank.

Ghiselin (1966*a*) suggests that confusion will be further reduced by exercising more precision in the use of the verbs associated with the terms category and taxon. Since we define words, we may also legitimately define categories such as *the* species, or *the* genus. Taxa, however, are things, and we can only describe (or delimit) but not define things. When this distinction is clearly kept in mind, many of the arguments about "defining species" are automatically eliminated.

For a fuller discussion of the theory of classification and the meaning of the stated terms see Gregg (1954), Beckner (1959), Cain (1958, 1962), Simpson (1961, 1963), and Buck and Hull (1966).

## 1.3  THE CONTRIBUTION OF SYSTEMATICS TO BIOLOGY

A consideration of the contribution of systematics to other branches of biology and to mankind as a whole adds to an appreciation of its scope.

Leaders in many fields of biology have acknowledged their total dependence on taxonomy. Elton (1947) made this statement regarding ecology:

> The extent to which progress in ecology depends upon accurate identification, and upon the existence of a sound systematic groundwork for all groups of animals, cannot be too much impressed upon the beginner in ecology. This is the essential basis of the whole thing; without it the ecologist is helpless, and the whole of his work may be rendered useless.

No scientific ecological survey can be carried out without the most painstaking identification of all the species of ecological significance. A similar dependence on taxonomy is true for other areas of science. The entire geological chronology and stratigraphy hinges on the correct identification of fossil key species. Even the experimental biologist has learned to appreciate the necessity for sound taxonomy. There are many genera with two, three, or more very similar species. Such species often differ more conspicuously in their physiological traits or in their cytology than in their external morphological characters. Every biologist will recall examples in which two workers came to very different conclusions concerning the physiological properties of a certain "species" because, in fact, one of them had been working with species *a* and the other with species *b*. Comparative biochemistry is vitally interested in sound classification. The evolution of molecules, an increasingly important area of research in molecular biology, can be understood only against the background of a sound classification. It is solely in consultation with the taxonomist that the biochemist can determine what organisms might supply the key to important steps in the evolution of molecules.

One of the greatest assets of a sound classification is its predictive value. It permits extrapolation from known to previously unstudied characters. An analysis of a few species strategically scattered through the natural system may provide us with much of the needed information on the distribution of a new enzyme, hormone, or metabolic pathway. Many animals cannot be kept in the laboratory, and others will not reproduce in captivity. Again, a sound system will permit all sorts of inferences from the genetically well-known types. (The validity of these inferences, of course, is limited by the fact that every species is a unique system.) The systematist can fill many gaps in our knowledge that are inaccessible to the specialists in the experimental branches of biology. There has been increasingly close collaboration in recent years among taxonomists and workers in immunology, in comparative biochemistry, in comparative physiology, and in the study of animal behavior.

It may be proper to single out some specific areas to which taxonomy has made noteworthy contributions.

**1.3.1  Applied Biology.** The contribution of taxonomy to the applied sciences has been both direct and indirect. It relates to medicine, public health, agriculture, conservation, management of natural resources, etc. Taxonomic breakthroughs have often supplied the key to the solution of previously perplexing problems in economic entomology. The famous case of the epidemiology of malaria is a good example. The supposed vector in Europe, the malaria mosquito *Anopheles maculipennis* Meigen, was reported throughout the continent, and yet malaria was restricted to local districts. Large amounts of money were wasted because no one understood the connection between the distribution of the mosquito and of malaria. Careful taxonomic studies, summarized by Hackett (1937) and Bates (1940), finally provided the key to the situation. The *maculipennis* complex was found to consist of several sibling species with different habitat preferences and breeding habits, only some of which are responsible for the transmission of malaria in a given area. This new information enabled control measures to be directed to the exact spots where they would be most effective.

With biological control of insect pests again receiving increased attention, the determination of the exact country of origin of insect pests and of their total fauna of parasites and parasitoids is restored to the great importance it had prior to the short period when at least some applied entomologists thought that they could completely control insects with pesticides.

Pemberton (1941) cites an outstanding instance of the value of insect collections, assembled for taxonomic study, in the solution of a biological control problem. Some twenty years ago the fern weevil, *Syagrius fulvitarsis* Pascoe, became very destructive to Sadleria ferns in a forest reserve on the island of Hawaii, and control measures became necessary. Entomological literature failed to reveal its occurrence anywhere outside Hawaii except in greenhouses in Australia and Ireland. These records, of course, gave no clue as to the country of origin. However, while engaged on other problems in Australia in 1921, Pemberton had the opportunity of examining an old private insect collection at Sydney, and among the beetle specimens was a single *Syagrius fulvitarsis* bearing the date of collection, 1857, and the name of the locality in Australia from which it was obtained. This provided the key to the solution of the problem, for a search of the forest areas indicated on the label revealed a small population of the beetles and, better still, a braconid parasite attacking the larvae. Collections were made immediately for shipment to Hawaii, and the establishment of the parasite was quickly followed by satisfactory control of the pest. The data borne on a label attached to a single insect specimen in 1857, in Australia, thus contributed directly to the successful biological control of the pest in Hawaii sixty-five years later.

Clausen (1942) and Sabrosky (1955) have presented in detail some of the achievements of taxonomy in biological control as well as reports on very expensive failures due to incorrect identifications.

**1.3.2  Systematics and Theoretical Biology.**  The service functions of taxonomy are often stressed to such a degree that the important contributions of systematics to the conceptual structure of biology are overlooked. Population thinking, for instance, has come into biology through taxonomy (Mayr, 1963), and indeed one of the two roots of population genetics is taxonomy (Chap. 3). The problem of the multiplication of species was solved by taxonomists, and they have made the greatest contributions to our understanding of the structure of species and of the evolutionary role of peripheral populations. It was taxonomists who continued to uphold the importance of natural selection at the time when the early Mendelians thought that mutation had eliminated the role of natural selection as an evolutionary factor. Taxonomists like H. W. Bates and F. Müller made significant contributions to the understanding of mimicry and related evolutionary phenomena. Taxonomists and naturalists in close contact with taxonomy were instrumental in the development of ethology and the study of the phylogeny of behavior. Taxonomists have consistently played an important role in counteracting the reductionist tendencies dominant in so much of functional biology. They have contributed thereby to a healthy balance in biological science. (See also Simpson, 1962*a,* on the status of taxonomy.)

*The Role of Taxonomy.* The multiple role of taxonomy in biology can be summarized as follows:

1. It works out for us a vivid picture of the existing organic diversity of our earth and is the only science that does so.

2. It provides much of the information permitting a reconstruction of the phylogeny of life.
3. It reveals numerous interesting evolutionary phenomena and thus makes them available for causal study by other branches of biology.
4. It supplies, almost exclusively, the information needed for entire branches of biology (e.g. biogeography).
5. It supplies classifications which are of great heuristic and explanatory value in most branches of biology, e.g., evolutionary biochemistry, immunology, ecology, genetics, ethology, historical geology.
6. It is indispensable in the study of economically or medically important organisms.
7. In the hands of its foremost exponents it makes important conceptual contributions (such as population thinking), not otherwise so easily accessible to experimental biologists. Thus it contributes significantly to a broadening of biology and to a better balance within biological science as a whole.

## 1.4   THE TASK OF THE TAXONOMIST

There is much uncertainty in the minds of some zoologists as to the function of the taxonomist. Some laboratory workers and ecologists limit his role to service. He should content himself with identifying material and devising keys. Beyond that he should keep his collections in good order, describe new species, and have every specimen properly labeled. According to this view, systematics is the mere pigeonholing of specimens.

In actuality systematics is one of the major subdivisions of biology, broader in base than genetics or biochemistry. It includes not only the service functions of identifying and classifying but the comparative study of all aspects of organisms, as well as interpretation of the role of lower and higher taxa in the economy of nature and in evolutionary history. It is a synthesis of many kinds of knowledge, theory, and method, applied to all aspects of classification. The ultimate task of the systematist is not only to describe the diversity of the living world but also to contribute to its understanding.

The modern taxonomist is far more than the caretaker of a collection. He is a well-trained field naturalist who studies the ecology and the behavior of his species in their native environment. Most younger systematists have had a thorough training in various branches of biology, including genetics. This experience in both field and laboratory gives them an excellent background for more fundamental studies.

The practice of classifying consists of two principal procedures, each in turn consisting of two separate steps (see also Chap. 10).

1.4.1   **The Discrimination of Entities.** The systematist's first task is to sort that portion of the diversity of individuals which he encounters into easily recognizable and internally homogeneous groups, and to find constant differences between such groups. Each such aggregate is a *phenon* (see 1.2). A phenon is not necessarily a population in the biological sense but may also be either a biased sample from a population (males, juveniles, morphs, etc.) or else (in the case of sibling species) a mixture of several populations, and finally, in the case of geographically heterogeneous material, possibly a mixture of several subspecies.

The second step is the assignment of phena to species, the lowest taxon routinely used in classification (Chap. 9). In most zoological groups instances are known in which several phena within a single biological species have been described and named as species. In certain groups the specialists have not succeeded completely even today in matching the sexes (e.g. mutillid wasps), the castes (e.g. ants), or the generations (e.g. trematodes). For some groups we have separate identification schemes for larvae and for adults. These should not be called biological classifications because we as biologists classify species, while larvae and adults are merely different phenotypic expressions of the same genotypes.

As soon as the basic units are discriminated, it becomes necessary to supply identifying symbols, "names," to facilitate communication among zoologists. Universality and stability are among the two most important qualities of such names, as in all other communication systems (Chap. 12).

*The Inventory of Species.* Even this first task of the taxonomist, the discrimination and description of species, becomes the more formidable the greater the increase in our knowledge. Linnaeus in 1758 recorded 4,162 species of animals; Möbius in 1898 listed 415,600 Recent species. At present the figure presumably exceeds 1 million (Table 1-1).

It must be remembered that all estimates or counts contain two sources of error, which fortunately tend to cancel one another out. First, a lesser or major proportion of the existing species is still undescribed in all groups of animals. On the other hand, in the more poorly known groups the so-called number of species is the number of recorded names. This includes many synonyms and the names of geographic races still listed as full species.

Estimates of the total number of living species (including nematodes, mites, and protozoans) range as high as 5 or even 10 million. The number of fossil species has been estimated to be 50 or 100 times as high (Cailleux, 1954). Thus only about one in 5,000 fossil species has so far been described. Of the one single group of protozoans, the Foraminifera, 28,000 species had been described by 1964. 685,000 species of insects were counted by Sabrosky in 1953, and allowing for an annual increment of about 6,000 species, the figure would now be around 750,000.

It is of considerable significance that some strikingly new types are still being discovered. This includes the only surviving coelacanth, *Latimeria,* discovered in 1938; the primitive mollusk, *Neopilina,* described in

TABLE 1-1.  NUMBER OF SPECIES IN MAJOR ANIMAL TAXA

ANIMALIA (Recent)................................... 1,071,000
  PROTOZOA........................................ 28,350
    Sarcomastigophora............................... 17,650
      Mastigophora.................................. 6,000
      Opalinata..................................... 200
      Sarcodina..................................... 11,450
    Sporozoa...................................... 3,600
    Cnidospora.................................... 1,100
    Ciliophora.................................... 6,000
  MESOZOA........................................ 50
  PORIFERA....................................... 4,800
  COELENTERATA (CNIDARIA)...................... 5,300
  CTENOPHORA..................................... 80
  PLATYHELMINTHES................................ 12,700
    Turbellaria.................................... 3,000
    Trematoda..................................... 6,300
    Cestoda....................................... 3,400
    Gnathostomulida............................... 45
  ENTOPROCTA (KAMPTOZOA)...................... 75
  NEMERTINEA..................................... 800
  ASCHELMINTHES (NEMATHELMINTHES)............ 12,500
    Gastrotricha................................... 170
    Rotatoria..................................... 1,500
    Nematoda...................................... 10,000
    Nematomorpha................................. 230
    Kinorhyncha................................... 100
    Acanthocephala................................ 500
  PRIAPULIDA..................................... 8
  MOLLUSCA....................................... 107,250
    Polyplacophora (Loricata)...................... 1,000
    Aplacophora (Solenogastres).................... 150
    Monoplacophora............................... 3
    Gastropoda.................................... 80,000
    Scaphopoda.................................... 350
    Bivalvia (Lamellibranchia)..................... 25,000
    Cephalopoda................................... 750
  SIPUNCULIDA.................................... 250
  ECHIURIDA...................................... 150
  ANNELIDA....................................... 8,500
  ONYCHOPHORA................................... 70
  TARDIGRADA..................................... 350
  PENTASTOMIDA (LINGUATULIDA).................. 65
  ARTHROPODA.................................... 838,000
    Chelicerata.................................... 57,500
      Merostomata (Xiphosura)...................... 4

| | |
|---|---:|
| Arachnida | 57,000 |
| Pantopoda (Pycnogonida) | 500 |
| Mandibulata | 780,500 |
| Crustacea | 20,000 |
| Chilopoda | 2,800 |
| Diplopoda | 7,200 |
| Pauropoda | 380 |
| Symphyla | 120 |
| Insecta | 750,000 |
| LOPHOPHORATA (TENTACULATA) | 3,750 |
| Phoronidea | 18 |
| Bryozoa | 3,500 |
| Brachiopoda | 230 |
| HEMICHORDATA (BRANCHIOTREMATA) | 80 |
| ECHINODERMATA | 6,000 |
| Echinozoa | 1,750 |
| Holothuroidea | 900 |
| Echinoidea | 850 |
| Crinozoa | 650 |
| Asterozoa | 3,600 |
| Somasteroidea | 1 |
| Asteroidea | 1,700 |
| Ophiuroidea | 1,900 |
| POGONOPHORA | 100 |
| CHAETOGNATHA | 50 |
| CHORDATA | 43,000 |
| Tunicata | 1,300 |
| Cephalochordata | 25 |
| Vertebrata | 41,700 |
| Agnatha | 50 |
| Chondrichthyes | −550 |
| Osteichthyes | 20,000 |
| Amphibia | 2,500 |
| Reptilia | 6,300 |
| Aves | 8,600 |
| Mammalia | 3,700 |
| Total about | 1,071,000 |

1956; the phylum Pogonophora, with over 100 species mostly described since 1950; the Cephalocarida, discovered in 1955; and the Gnathostomulida, discovered in 1956. Several of these discoveries have had a strong impact on our concepts of phylogeny.

The degree to which the inventory taking has been completed differs from group to group. In birds 99 percent of all living species have surely been described, and in mammals and reptiles it may well be more than

90 percent. Yet there are poorly studied groups of arthropods, protozoans, and marine invertebrates in which less than 10 percent of the species existing in the world have so far been described. These are merely estimates, as no one really knows.

Only a few non-taxonomists appreciate how poorly most animal groups are known taxonomically. A striking illustration of this is presented by Remane's work on the microscopic marine fauna of the Kieler Bucht, an area previously considered to be well known. By thorough search and with the application of new methods, Remane found 300 new species in ten years, including representatives of 15 new families. Sabrosky (1950) has pointed out how poorly much of the North American insect fauna is known. Many so-called "common species" actually represent whole complexes of previously overlooked good species. This is the situation in the temperate zone. In the tropics the discrimination of species is even farther behind; often in a generic revision every species is found to be new.

To complete this part of the task of the taxonomist will require the labors of several more generations. Considering the limited number of specialists, we cannot expect, for instance, to settle all the problems of mite taxonomy in the next thirty years. We must take it for granted that a large part of the mite fauna of the world will remain unsampled, unnamed, and unclassified for decades to come. The same is probably true for the majority of the kinds of animals. It is regrettable that literally thousands

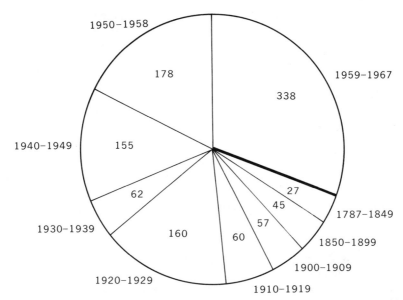

Fig. 1-1. Number of new species of Drosophila described in stated periods. Almost as many (516) new species were described in the last 18 years as in the preceding 163 years (566 species). [Data from M. R. Wheeler.]

of species will become extinct in the next generation in all parts of the world owing to habitat destruction before they were ever collected and scientifically described.

**1.4.2  Grouping and Ranking.** The recognition and accurate description of species is only the first step in classification. Should the taxonomist stop here, he would soon be confronted with a chaotic accumulation of species names. The second task, then, of the taxonomist is to put the species in order. He must group them into smaller and larger arrays of related species, taxa, and higher taxa, and must place these in a hierarchy of categories. In other words, he must make a classification (see Chaps. 4 and 10).

*Information Storage.* There is perhaps no other branch of science in which information storage and retrieval is as formidable and as crucially important a task as in taxonomy. It is impossible to prepare a reliable revision or monograph of a higher taxon if one does not first compile a listing of the described species. *Mutatis mutandis* this is true for any operation in beta and gamma taxonomy. To make matters worse there are few fields in which the literature is more scattered. The more species are described and the more people work in taxonomy, the worse the problem gets.

Hardly anyone doubts that computerized information storage is the inevitable solution. It is probably correct to say that every taxonomist spends more time on searching and extracting the literature than on original work. This is so traditional and taken so much for granted that it is virtually never pointed out how appallingly inefficient this research method is. Coding all taxonomic information for computer storage will be very expensive, but in the long run perhaps considerably less so than the uneconomic waste of time now required of the working taxonomist. Whatever new kinds of information become important to the taxonomist—behavioral, cytological, biochemical, ecological—he will soon be totally overwhelmed by it if new ways for storing and retrieving it are not soon found.

**1.4.3  Biological Systematics.** The making of a classification is not the end of the taxonomist's concern. Consistent with Simpson's definition of systematics as "the scientific study of the kinds and diversity of organisms and of any and all relationships among them," the systematist studies all aspects of living organisms. He is interested in more than description; as a scientist he is concerned with meanings and the study of causation. Species formation, the factors of evolution, the structure of natural populations, biogeography—all these are of concern to the taxonomist, and he has made important contributions to all these branches of biological science. Indeed it may be claimed that they all are part of systematics. The entire broad field of comparative zoology is included in his concern. Activities in this area are sometimes referred to as gamma taxonomy (see next paragraph).

*Stages in Classification.* The classification of a given taxonomic group goes through various stages of maturation. In the first stage, often called *alpha taxonomy,* emphasis is on the description of new species and their preliminary arrangement in comprehensive genera. In *beta taxonomy* relationships are worked out more carefully on the species level and on that of the higher categories; emphasis is placed on the development of a sound classification. At the level of *gamma taxonomy* much attention is paid to intraspecific variation, to various sorts of evolutionary studies, and to a causal interpretation of organic diversity. The three levels overlap and intergrade, but the trend from alpha to gamma is unmistakable in the taxonomic history of any group. Though there is need for more work on the alpha and beta levels even in the taxonomically best-known groups, it is the ultimate endeavor of the biologically minded taxonomist to achieve an understanding of the group in which he specializes.

The reason for gamma taxonomy is the fact that causal explanation is part of any science, and systematic zoology is no exception. If systematics were satisfied with merely pigeonholing the diversity of the living world, it would be a technology but not a science.

**1.4.4 Strategy of Taxonomic Research.** Like most other scientists, the taxonomist must frequently make the decision what to do next. Should he concentrate on work in alpha taxonomy, describe new species, and prepare catalogs of the available names? Should he concentrate on beta taxonomy and prepare a basic revision or monograph, based on an examination of all included species and preferably all type-specimens, and culminating in a well-balanced, carefully reasoned reclassification? Or finally, should he devote himself to some aspects of gamma taxonomy, such as making a detailed study of individual and geographical variation in a single species, based on numerous large population samples, or studying the behavioral or chemical characteristics of a set of species, etc.? The answer to these questions is twofold. First, depending on the level of maturity reached in the group of which he is a specialist, either one or the other of these levels will permit the most productive work. Secondly, most specialists avoid mental fatigue by alternating different approaches. When one gets tired of the tedious compilation of data necessary for the preparation of a revision, it is relaxing to do a study of geographic variation in a single species with rich material, or of living material, or of certain distributional aspects. Field systematics offers countless challenges which are particularly stimulating when alternated with museum research.

Finally, as pointed out by Gosline (1965), the geographic location of a taxonomist will considerably influence his research strategy. Some problems are most advantageously pursued in a small institution, others in a large national museum. The specialist working in Washington, London, or Leningrad, in spite of his many other advantages, will be unable to settle certain questions, particularly those dealing with life cycles, sexual dimorphism, seasonal occurrence, niche occupation, concealing or warning coloration, relative frequency, etc., that are easily accessible to the specialist in Trinidad, Belem, Honolulu, Lae, Kuala Lumpur, or Entebbe. To devote one's life to the compilation of a nomenclator of generic names, when one happens to be at a tropical station with a poor library, is about as misguided a strategy as can be imagined (and yet this has actually happened).

Some nonsystematists have proposed in recent years to abandon most of taxonomy as "old-fashioned," "an exhausted mine," or "useless," and to concentrate instead on some specialized area, such as comparative protein chemistry, the taxonomy of behavioral traits, or the principles of geographic variation. Advocates of such specialized approaches forget a number of things: (1) that the various approaches are not mutually exclusive, (2) that in many groups there is still abundant need for the most basic alpha taxonomy, and (3) that man is one of the most polymorphic species. Some workers are interested in using electronic computers, others in watching behavior; some are interested in insects, others in fish; some like to work with books, others with test tubes, some with preserved specimens, others with living material. Human polymorphism cannot be controlled by regimentation. The breadth of systematics has room for students with the most diversified interests and talents.

A good biological education will reveal to the student where the problems are that are most exciting to him. His natural inclination will lead him to select one area or another. And it will be only for the good of biological science if not too many students of a given year class jump on the same bandwagon.

Questions of strategy also arise in connection with publication. These are dealt with in Chap. 11.

## 1.5    SYSTEMATICS AS A PROFESSION

### 1.5.1    Opportunities and Difficulties.

Positions for professional taxonomists exist in museums, at universities, and in various agencies of the government. In the United States there are perhaps at present more taxonomists in teaching positions in various colleges and universities than in government or museum positions. A well-trained taxonomist is particularly well qualified to teach a general course in zoology or biology, because he has the broad background in zoology, morphology, physiology, genetics, and ecology which some specialists, for lack of time, are unable to acquire. At the present time (1968) there are numerous unfilled positions for taxonomists, particularly for specialists in the Arachnida and marine invertebrates. Indeed, the professional opportunities in taxonomy are far greater than is generally realized. There are, however, drawbacks. The situation in applied taxonomy often changes rapidly. When the importance of Foraminifera for the stratigraphic determination of oil-bearing geological formations was discovered, there was such a demand for micropaleontologists that it took decades to fill it. With geophysical methods now dominating oil exploration, the demand for micropaleontologists has decreased sharply. During the height of chemical insect control, the demand for well-trained insect taxonomists lagged. With the renewed recognition of the importance of biological control, there is a greatly increased need for qualified insect taxonomists. The recent stepping-up of oceanographic research revealed an extraordinary shortage of marine zoologists. New positions for specialists in marine invertebrates were soon created in numerous museums and universities. No one can predict where the next need will arise. It is conceivable, for instance, that institutes in molecular biology may begin to employ taxonomists for consultation on the numerous evolutionary and taxonomic problems which comparative biochemistry encounters at every step. As in all fields of biology, there is always room for superior workers, regardless of temporary fluctuations in the employment situation.

To the same extent to which the field of taxonomy as a whole has become increasingly diversified, the need for different kinds of taxonomists has sharply increased. As Rozeboom (1962) has said rightly: "The person trained to program data for analysis with electronic computers is not likely to be willing to spend three or five years in the jungles of Africa collecting ecological data on *Anopheles gambiae*. I question also whether he would be willing to spend an equal amount of time searching for minute morphological characters suitable for some kind of measurement." Nor is the person who is painstaking enough to compile a checklist or nomenclator likely to be equally interested in comparative behavior and biochemical studies, or vice versa. There are some taxonomists who do their best work when undertaking population analyses, while others are virtual geniuses in the weighting of characters preparatory to construction of a classification of higher taxa. All these talents are needed, and someone can usually be found to fill each niche, thanks to the extraordinary variation in man's talents.

The number of pure research positions is as limited in taxonomy as in most branches of science. Most taxonomists earn their living as teachers, curators, members of identification services, or in other branches of applied biology. Another group of taxonomists are "amateurs," that is, they earn their living as civil servants, businessmen, lawyers, or doctors, and conduct their taxonomic research as a hobby. The role of the amateur in systematics has been changing over the years. When collecting and "naming" was believed to be the essence of taxonomy, almost any untrained person could become a specialist in a group of which knowledge was still at the level of alpha taxonomy. The stamp-collector type of amateur has been losing ground as the quality of taxonomic research has been improving. However, when field systematics became an increasingly important branch of taxonomy, a new niche opened up for the amateur naturalist. He supplies information on behavior and ecology which may be very important taxonomically. The work of the better amateurs is sometimes of the same high level of quality as that of the good professional. Considering the enormous size of the taxonomic task still to be done, taxonomy would make slow progress if it were not for the dedicated amateur.

One of the most interesting current developments is the blurring of the borderline between "taxonomists" and "other biologists." Population geneticists such as Sturtevant, Dobzhansky, Carson, and Wheeler have been most active in elucidating the classification of the genus *Drosophila* and in the describing of new species. Herpetologists have discovered new sibling species of frogs by analyzing their vocalization. Entomologists have done the same for cicadas and grasshoppers. Ecologists have sometimes taken up taxonomic studies in order to give a new dimension to their ecological research. Virtually all taxonomists have a broad interest in biogeography, and most major contributions to this branch of biology were made by taxonomists. Now that "taxonomic character" means not only aspects of morphology, but every kind of manifestation of the genotype, there is continuous transgression of the borders between taxonomy, physiology, behavior, genetics, biochemistry, and so forth. This active interchange is evident from recent symposia (Leone, 1964; Handler, 1965; Bryson and Vogel, 1965). What this means is that there is room in taxonomy for biologists of the most diverse interests. Even the mathematically inclined student can make contributions by applying computer methods to taxonomy.

**1.5.2    The Training of the Taxonomist.** Regular courses in methods and principles of taxonomy are a comparatively new phenomenon. Formerly the young taxonomist learned as apprentice to a master. Perhaps this is in part responsible for the highly uneven quality of taxonomic work done in the past. A competence of such breadth is demanded from the professional taxonomist of today that the method of apprenticeship is no longer sufficient. One expects from the well-trained taxonomist a broad knowledge of zoology, a thorough knowledge of the comparative morphology of the group in which he is specializing, and an understanding of genetics and evolutionary biology. He must be versed in statistical methods and, if possible, in the use of computers. The discussion of taxonomic characters in Chap. 7 indicates some of the other areas of biology in which the taxonomist should be knowledgeable. One might add, incidentally, that a broad knowledge of the classification of animals should be not merely useful, but an indispensable part of the equipment of every biologist, regardless of field of specialization.

1.5.3   **Professional Organizations.** Systematic zoologists of most countries have their own professional organizations. Some of these are devoted to systematics as a whole, others are specialized societies, like the American Ornithologists' Union, the Entomological Society of America, or the Lepidopterists' Society. Blackwelder (1967) gives a listing of such societies.

All aspects of animal taxonomy are represented in the Society of Systematic Zoology, the only broad systematics society in this country and publisher of *Systematic Zoology*. A comparable organization in Great Britain is the Systematics Association, publisher of symposia.

As far as general systematics is concerned, much relevant material is also published in the journals *Evolution, American Naturalist,* and *Ecology;* in Great Britain in the *Journal of the Linnaean Society of London;* and in Germany in the *Zeitschrift für zoologische Systematik und Evolutionsforschung* (from 1963 onward).

1.5.4   **The Future of Systematics.** Judging from the developments of recent decades, the systematist has every reason to be optimistic about the future of his field. Simpson (1945) put it very well when he said that systematics "is at the same time the most elementary and most inclusive part of zoology, most elementary because animals cannot be discussed or treated in a scientific way until some taxonomy has been achieved, and most inclusive because [systematics] in its various branches gathers together, utilizes, summarizes, and implements, everything that is known about animals, whether morphological, physiological, psychological, or ecological."

There will always be a need for the ordering and classifying activities of the taxonomist, even if society should, unexpectedly, stress increasingly the applied aspects of science. There will always be a need, and as it seems to us, a growing need, to study organic diversity and its meaning, for there is no other subject that teaches us more about the world we live in than systematics. And is it not equally important to reveal the unknown life on our own planet as it is to map additional specks on the sky?

This age of leisure and automation offers unparalleled opportunities for the cultivation of avocational interests. There are few hobbies as rewarding as natural history, and the greatest joy always comes from deep knowledge and true understanding. We agree with Crowson (1958) that the pursuit of taxonomy not only gives great pleasure to its devotees, but that it conveys a reverence for the wonders of living nature which should be part of the outlook of every truly human being.

# Part I Principles of Zoological Classification

# Chapter 2  The Species Category

*T*he twenty or more categories which the taxonomist uses in classification are of unequal value and of different significance. They fall quite naturally into three groups:

1. The species category (Chap. 2)
2. Categories for distinguishable populations within species ( = infraspecific categories, Chap. 3)
3. Categories for higher taxa, that is, for groupings of species (collective categories = higher categories, Chap. 5)

The species, in a number of different ways, occupies a unique position in the taxonomic hierarchy.

## 2.1  THE SPECIES PROBLEM

It seems to be one of man's most elementary urges to want to identify things and name them. Even the most primitive native peoples have names for kinds of birds, fishes, flowers, or trees. If only individuals existed, and the diversity of nature were continuous, it would be difficult to sort them into groups and distinguish "kinds." Fortunately, at least in the animal world, the diversity of nature is discontinuous, consisting in any local fauna

23

of more or less well-defined "kinds" of animals which we call species. Around New York City, for instance, there are about 150 "kinds" of breeding birds. These are the species of the taxonomist. Primitive natives in the mountains of New Guinea will distinguish the same kinds of organisms as, quite independently, does the specialist in the big national museums (Diamond, 1966).

The concept of species seems so absurdly simple that it always comes as something of a shock to a beginning taxonomist to learn how voluminous and seemingly endless the debate about the species problem has been. In zoology there is now fair agreement on the species concept, although heterodox views are still vigorously defended. For recent summaries see Mayr (1957a, 1963) and Simpson (1961). For statements of the botanical point of view see Heslop-Harrison (1963) and Löve (1964).

The species problem has been made to appear more difficult than it is by a confusion of the concepts underlying the terms phenon, taxon, and category (see 1.2). The working taxonomist sorts specimens (individuals) into phena and decides which of these are members of populations each of which belongs to a single taxon of the species category. To be able to undertake the ranking of taxa, the taxonomist must have a clear conception of the category species. If he defines it (as a morphospecies) in such a way that it coincides with the phenon, he may facilitate his task of sorting specimens, but his activity will result in species that are biologically, and hence scientifically, meaningless. The objective of a scientifically sound concept of the species category is to facilitate the assembling of phena into meaningful taxa on the species level. A short survey of the history of species concepts will show how different the species taxa are which one is forced to distinguish if one adopts different species concepts.

## 2.2  SPECIES CONCEPTS

Taxonomic literature reports innumerable species concepts (Mayr, 1957b; Heslop-Harrison, 1963). By their philosophical basis, all these concepts fall into three groups. The first two have mainly historical significance, but are still upheld by a few contemporary authors.

**2.2.1  Typological Species Concept.** According to this concept the observed diversity of the universe reflects the existence of a limited number of underlying "universals" or types (*eidos* of Plato). Individuals do not stand in any special relation to each other, being merely expressions of the same type. Variation is the result of imperfect manifestations of the idea implicit in each species. This species concept, going back to the philosophies of Plato and Aristotle, was the species concept of Linnaeus and his followers (Cain, 1958). Since this philosophical tradition is sometimes referred to as essentialism, the typological definition is also sometimes called the essentialist species definition. (For a discussion of essentialism see 4.3.2.) Various attempts at a purely numerical or mathematical species definition (e.g. Ginsburg, 1938) are logical equivalents of this species concept. It must be emphasized that there is a complete difference between basing one's species concept on morphology and using morphological evidence as inference for the application of a biological species concept (Simpson, 1961).

Two practical reasons exist for the now quite universal rejection of the typological species concept: (1) Individuals are frequently found in nature that are clearly conspecific with other individuals in spite of striking differences in structure owing to sexual dimorphism, age differences, polymorphism, and other forms of individual variation. Although often described originally as different species they are deprived of their species status, regardless of the degree of morphological difference, as soon as they are found to be members of the same breeding population. Different phena that belong to a single population can not be considered different species. (2) Sibling species differ hardly at all morphologically, yet are good biological species. Degree of difference is not the decisive criterion in the ranking of taxa as species.

Its own adherents abandon the typological species concept whenever they discover that they have named as a separate species something that is nothing but a conspecific phenon. At present the typological species concept is still defended by some writers adhering to Thomistic philosophy.

**2.2.2   Nominalistic Species Concept.** The nominalists (Occam and his followers) deny the existence of "real" universals (4.3.3). For them only individuals exist, while species are man-made abstractions. (When they have to deal with a species, they treat it as an individual on a higher plane.) The nominalistic species concept was popular in France in the eighteenth century (Buffon in his early writings, Robinet, Lamarck) and has adherents to the present day. Bessey (1908) expressed this viewpoint particularly well: "Nature produces individuals and nothing more . . . species have no actual existence in nature. They are mental concepts and nothing more . . . species have been invented in order that we may refer to great numbers of individuals collectively."

Any naturalist, whether a primitive native or a trained population geneticist, knows that this is simply not true. Species of animals are not human constructs, nor are they types in the sense of Plato and Aristotle, but they are something for which there is no equivalent in the realm of inanimate objects.

**2.2.3   The Biological Species Concept.** In the late eighteenth century it began to be realized that neither of the medieval species concepts discussed in the two preceding sections was applicable to biological species. An entirely new species concept began to emerge after about 1750. It is augured by statements made by Buffon (in his later writings), Merrem, Voigt, Walsh (1864) and many other naturalists and taxonomists of the nineteenth century. K. Jordan (1905), however, was the first who clearly formulated the concept in all of its consequences. It combines elements of the typological and nominalistic concepts by stating that species have independent reality and are typified by the statistics of populations of individuals. It differs from both by stressing the populational aspect and genetic cohesion of the species, and by pointing out that it receives its reality from the historically evolved, shared information content of its gene pool.

As a result, the members of a species form (1) *a reproductive community*. The individuals of a species of animals recognize each other as potential mates and seek each other for the purpose of reproduction. A multitude of devices ensures intraspecific reproduction in all organisms. The species is also (2) *an ecological unit* which, regardless of the individuals composing it interacts as a unit with other species with which it shares the environment. The species, finally, is (3) *a genetic unit* consisting of a large, intercommunicating gene pool, whereas the individual is merely a temporary vessel holding a small portion of the contents of the gene pool for a short period of time. These three properties raise the species above the typological interpretation of a "class of objects" (Mayr, 1963, p. 21). The species definition which results from this theoretical species concept is:

*Species are groups of interbreeding natural populations that are reproductively isolated from other such groups.*

The development of the biological concept of the species is one of the earliest manifestations of the emancipation of biology from an inappropriate philosophy based on the phenomena of inanimate nature. This species concept is called biological not because it deals with biological taxa, but because the definition is biological. It utilizes criteria that are meaningless as far as the inanimate world is concerned.

When encountering difficulties (see also 2.5) it is important to focus on the basic biological meaning of the species: A species is a protected gene pool. It is a Mendelian population which has its own devices (called isolating mechanisms) which protect it against harmful gene flow from other gene pools. Genes of the same gene pool form harmonious combinations because they have become coadapted by natural selection. Mixing the genes of two different species leads to a high frequency of disharmonious gene combinations; mechanisms that prevent this are therefore favored by selection. This makes it quite clear that the word species in biology is a relational term: *A* is a species in relation to *B* and *C* because it is reproductively isolated from them. It has its primary significance with respect to sympatric and synchronic populations, and these are precisely the situations where the application of the concept faces the fewest difficulties ("the nondimensional species"). The more distant two populations are in space and time, the more difficult it becomes to test their species status in relation to each other, but the more irrelevant biologically this also becomes.

The biological species concept also solves the paradox caused by the conflict between the fixity of the species of the naturalist and the fluidity of the species of the evolutionist. It was this conflict which made Linnaeus deny evolution and Darwin the reality of species (Mayr, 1957b). The biological species combines the discreteness of the local species at a given time with an evolutionary potential for continuing change.

The unique position of species in the hierarchy of taxonomic categories has been pointed out by many authors. Taxa of the species category can be delimited against each other by operationally defined criteria ("interbreeding versus noninterbreeding of populations"). It is the only taxonomic category for which the boundaries between the taxa at that level are defined objectively.

Logicians (e.g. Gregg, 1954; Buck and Hull, 1966) do not fully understand that the terminology of "class" and "member" fails to bring out the complete difference between, on one hand, the relation of individuals to the species and, on the other hand, of species taxa to higher taxa. The statement that something is "a member of a class" has an entirely different meaning for an individual which through its genotype is a member of a species, and a species taxon which is included in a higher taxon. A category is not a class in the same sense as a higher taxon, but a designation of rank. Logicians do not appreciate that the "higher" and "lower" rank of taxa is a relative, not an absolute property. One can compare taxa in a single phyletic line, but one cannot say that the genus is the same thing in birds, ammonites, bivalve mollusks, protozoans, and weevils, and that they are all equal. Again, the species is an exception because (at least in sexual species) the species is an equivalent phenomenon in all groups of animals.

Intraspecific categories designate groupings of populations within species. Normally, however, the species is the lowest category used in routine taxonomy. The higher categories are groupings of species. In view of this key position of the species and the fact that in nature one encounters individuals and phena, the assigning of individuals and phena to species taxa is one of the key problems of taxonomy.

## 2.3   FROM PHENON TO TAXON TO CATEGORY

A failure to understand the meaning of these three terms and their theoretical foundation has led taxonomists into much confusion. It has been the cause of most attacks on the biological species concept. When an author says, "As a paleontologist I cannot employ the biological species concept because I cannot test the reproductive isolation of fossils," he reveals his lack of understanding. What the taxonomist observes directly are individuals which he sorts into phena. On the basis of certain biological concepts and information, such as an awareness of the possibility of sexual dimorphism, growth, alternation of generations, nongenetic modifications of the phenotype, etc., he assigns the phena to populations, which in turn he classifies into taxa. The ranking of a taxon as subspecies, species, or genus by a taxonomist is based on inferences drawn from the available data. This methodology of basing inferences on evidence and its justification has been perceptively discussed by Simpson (1961, pp. 68–69):

> "Evolutionary classification uses, for the most part, concepts and definitions for which the data are not directly observable. This is not a feature peculiar to taxonomy. It is shared in greater or less degree by most of the inductive sciences. . . . Here it is necessary again to emphasize the distinction between definition and the evidence that the definition is met. We propose to define taxonomic categories in evolutionary and to the largest extent phylogenetic terms, but to use evidence that is almost entirely non-phylogenetic when taken as individual observations. In spite of considerable confusion about this distinction, even among some taxonomists, it is really not particularly difficult or esoteric. The well-known example of monozygotic ("identical") twins is explanatory and is something more than an analogy. We define such twins as two individuals developed from one zygote. No one has ever seen this occur in humans, but we recognize when the definition is met by evidence of similarities sufficient to sustain the inference. The individuals in question are not twins because they are similar but, quite the contrary, are similar because they are twins. Precisely so, individuals do not belong in the same taxon because they are similar, but they are similar because they belong to the same taxon. (Linnaeus was quite right when he said that the genus makes the characters, not vice versa, even though he did not know what makes the genus.) That statement is a central element in evolutionary taxonomy, and the alternative clearly distinguishes it from non-evolutionary taxonomy. Another way to put the matter is to say that categories are defined in phylogenetic terms but that taxa are defined by somatic relationships that result from phylogeny and are evidence that the categorical definition is met.

The reproductive isolation of a biological species, the protection of its collective gene pool against pollution by genes from other species, results in a discontinuity not only of the genotype of the species, but also of its morphology and other aspects of the phenotype produced by this genotype. This is the fact on which taxonomic practice is based. Reproductive isolation cannot, of course, be directly observed in samples of preserved specimens. However, it can be inferred on the basis of various types of evidence, as for instance the presence of a discontinuity, a bridgeless gap, between two correlated character complexes. In living species, of course, such inferences can be tested by observation and experiment.

The crucial difference between the reasoning of the typologist and the adherent of the biological species concept is as follows: The typologist says, "There is a clear-cut morphological difference between samples *a* and *b,* therefore they are, by definition, two morphospecies, that is, two species." Any list of synonymies will quickly reveal how often this philosophy has led to the description of phena as species. The biological taxonomist asks, "Is the morphological difference between samples *a* and *b* of the kind one would expect to find between two reproductively isolated populations, that is, between two biological species?" In other words, he uses the amount and kind of morphological difference only as an indication of reproductive isolation, only as evidence to draw an inference. This is a legitimate and reliable technique. Where the typologist would recognize phena as (morpho-) species, the biologist will draw the right inferences from largely morphological evidence, and his species are usually confirmed by subsequent researches. When competent taxonomic work based on morphological evidence is re-examined in the light of the findings of behavior or biochemistry, it is usually confirmed in its entirety.

It is not always realized that the classification of phena is based on entirely different evidence than the classification of species. The classification of species is based on weighted similarity, evaluating all sorts of comparative data, be they morphological, physiological, behavioral or what not. The classification of phena is based on their relation to the gene pool of the population to which they belong. Ultimately this can be established only by breeding behavior. This in turn can be either observed in nature or studied experimentally. It does not matter whether one deals with strikingly different sexes in birds, insects, or marine invertebrates, or with larval forms, or alternating generations of parasites; breeding alone (or the piecing together of growth stages) will establish what phena together form a population. The experienced taxonomist knows what variation to expect within a biological species. No computer method has yet been found that would empirically assign phena to species. The taxonomist does this rapidly and with a high degree of precision on the basis of his accumulated knowledge of the biology of the species concerned. In this taxonomic operation the classical methods still reign supreme "because they are enormously faster than the numerical methods" (Michener, 1963).

## 2.4  SPECIES NAMES

The scientific name (binomen) of a species consists of two words, the generic name and the specific name. The rules concerning the coining of species names and all other aspects of the nomenclature of such names are discussed in Chap. 13 (particularly 13.31, 13.40, and 13.43), where cross references to the appropriate articles of the Code (Chap. 12) are also provided.

Species names, to become available, must be accompanied by a description. Advice concerning the preparation of descriptions is given in Chap. 11.

## 2.5    DIFFICULTIES IN THE APPLICATION OF THE BIOLOGICAL SPECIES CONCEPT

The fact that difficulties sometimes arise when the biological species concept is applied to natural taxa does not mean that the concept as such is invalid. This has been shown by Simpson (1961, p. 150) and Mayr (1963, pp. 21–22).

Many generally accepted concepts face similar difficulties when they have to be applied in a particular situation or to a specific sample. The concept *tree,* for instance, is not invalidated by the existence of spreading junipers, dwarf willows, giant cacti, and strangler figs. One must make a clear distinction between a concept and its application to a particular case.

The more ordinary problems of taxonomic discrimination at the species level, in particular the criteria for ranking a taxon as a species rather than a subspecies, will be dealt with in Chap. 9B.2.

The three most serious difficulties in the application of the biological species concept, discussed in this chapter, are those caused by the lack of pertinent information, those caused by uniparental reproduction, and those caused by evolutionary intermediacy.

2.5.1    **Insufficient Information.** Individual variation in all of its forms often raises doubt as to whether a certain morphotype is a separate species or only a phenon within a variable population. Sexual dimorphism, age differences, polymorphism, and other such types of variation can be unmasked as individual variation through a study of life histories and through population analysis. This is dealt with more fully in Chap. 8A and 8B. The neontologist who normally works with preserved material is confronted by the same problems as the paleontologist, who likewise must assign phena (morphotypes) to species.

2.5.2    **Uniparental Reproduction.** In many organisms systems of reproduction are found that are not based on the principle of an obligatory recombination of genetic material between individuals prior to the formation of a new individual. Self-fertilization, parthenogenesis, pseudogamy, and vegetative reproduction (budding or fission) are some of these forms of uniparental reproduction.

A population, as defined in evolutionary biology, is an interbreeding group. By definition, therefore, an asexual biological population is a contradiction, even though the word population has other usages in which a combination with asexual is not contradictory. The biological species concept based on the presence or absence of interbreeding between populations is therefore inappropriate for uniparentally reproducing organisms.

How to solve this dilemma has been discussed by Simpson (1961, pp. 161–163) and by Mayr (1963, pp. 27–29). Attempts to define agamospecies or asexual species, with or without using the word population, have not been particularly successful. Fortunately, there are usually well-defined morphological discontinuities among kinds of uniparentally reproducing organisms. These discontinuities are apparently produced by natural selection among the various mutants which occur in the asexual clones. It is customary to utilize the existence of such discontinuities, and the amount of morphological difference among them, to delimit species among uniparentally reproducing types.

Species recognition among asexual organisms is based not merely on analogy but also on the fact that each of the morphological entities, separated by a gap from other similar entities, seems to occupy an ecological niche of its own; it plays its own evolutionary role. In groups like the bdelloid rotifers, all of which reproduce by obligatory parthenogenesis, there is evidence for a definite biological meaning to the recognized morphological species. The treatment of asexual entities that do not qualify as species is discussed in Chap. 3 (3.3.8).

Examples are known in which a form that is as distinct as a good species reproduces strictly parthenogenetically, and no biparental species is known from which it might have branched off. Nomenclatural recognition is justified in such cases. Whenever several reproductively isolated chromosome types occur within such a "species," as in various crustaceans (e.g. *Artemia salina* Linnaeus) (White, 1954), it may be convenient to distinguish them nomenclaturally. Although conventionally referred to as races, reproductively isolated chromosomal populations are more logically designated (micro-)species.

**2.5.3 Evolutionary Intermediacy.** The species, as manifested by a reproductive gap between populations, exists in full classical distinctness only in a local fauna. As soon as the dimensions of space (longitude and latitude) and time are added, the stage is set for incipient speciation. Populations will be found under these circumstances which are in the process of becoming separate species and have acquired some but not yet all of the attributes of distinct species. In particular, the acquisition of morphological distinctness is not always closely correlated with the acquisition of reproductive isolation. The various difficulties for the taxonomist which may result from evolutionary intermediacy may be tabulated as follows (see also Mayr, 1957*a*, p. 375):

1. *Acquisition of reproductive isolation without equivalent morphological change.* Reproductively isolated species without (or with very slight) morphological difference are called *sibling species*. Their taxonomic treatment will be discussed in Chap. 9A.2.

2. *Acquisition of strong morphological difference without reproductive isolation.* A number of genera of animals and plants are known in which morphologically very different populations interbreed at random wherever they come in contact. The typological solution of calling every morphologically distinct population a species is clearly inappropriate in such situations. Conversely, there are genera in which the isolating mechanisms between any two species may break down occasionally. To consider such species conspecific would be going to the opposite extreme. No generalized solution is possible where morphological divergence and acquisition of reproductive isolation do not coincide. The only recommendation to the specialist is that he delimit his species in such a way that they form biologically meaningful, natural entities. The difficulty posed by the rapid morphological divergence of populations without acquisition of reproductive isolation is well-illustrated by the West Indian snail genus *Cerion* (see Fig. 2-1).

3. *The occasional breakdown of isolating mechanisms (hybridization).* Reproductive isolation may break down occasionally even between good species. Most frequently this will lead only to the production of occasional hybrids that are either sterile or of lowered viability, and this will not cause any taxonomic difficulty. More rarely, there is a complete local breakdown of isolation resulting in the production of extensive hybrid swarms and more or less complete introgression (Mayr, 1963, pp. 110–135).

Hybrid individuals are sometimes described as species before their hybrid nature is discovered. Such names lose their validity as soon as the hybridism is established (see 13.21). Only populations are recognized as taxa, and hybrid individuals are not populations.

Taxonomically more difficult are situations where new populations are formed as a result of hybridization. We can recognize several types of natural populations that owe their origin to hybridization. The taxonomic treatment of secondary intergradation, the result of the fusion of previously isolated populations, is discussed in Chap. 9B.2b. Two other kinds of hybridism concern us here.

3a. *Sympatric hybridization.* In all instances in which the two parental species maintain their genetic integrity over a more or less wide area in which they occur together, it is advisable to uphold their species status even though in a portion of their ranges there is a breakdown of the isolation. The example of the two Mexican towhees (*Pipilo erythrophthalmus* and *P. ocai*) is an excellent illustration of this situation (Sibley, 1954) (Fig. 2-2). No taxonomic recognition is given to hybrid populations that result from such a local breakdown of reproductive isolation. The only possible exception is such a complete breakdown of the isolation that the two parental species fuse into a single new species. The taxonomic literature records a number of instances that have been interpreted in this manner, but we are not aware of a single thorough analysis that would have established this unequivocally.

3b. *Amphiploidy.* Hybridism in plants may lead to the instantaneous production of an allopolyploid, an individual that combines the chromosome sets of two parental species. Such hybrids may give rise by uniparental reproduction to a

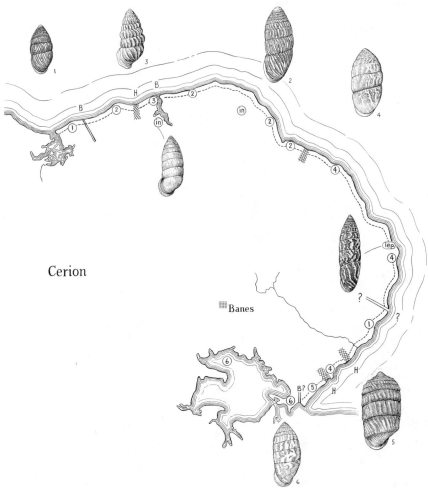

Cerion

Banes

Fig. 2–1. Irregular distribution of populations of the halophilous land snail *Cerion* in eastern Cuba. Numbers refer to different races or species. Where two populations come in contact (with the exception of *lepida*) they hybridize (*H*), regardless of difference. In other cases contact is prevented by a barrier (*B*). *In* = isolated population (*from Mayr*, 1963).

new population that is reproductively isolated from both parents and behaves like a new species if it is able to reproduce and, by occupying a new ecological niche, is able to compete with other species (including the parents). It is doubtful that such speciation by amphiploidy has ever been incontrovertibly established in sexually reproducing animals. Polyploidy among parthenogenetically reproducing insects, annelids, and turbellarians is not infrequent, but it is not certain that any of these parthenogenetic lines owes its origin to hybridization.

4. *Semispecies.* Geographical isolates occasionally have an intermediate status between subspecies and species. On the basis of some criteria they would be considered species; on the basis of others, they would not. It is usually more con-

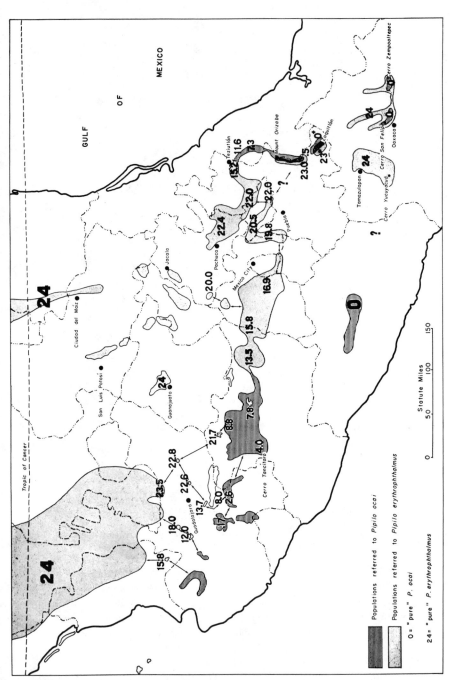

Fig. 2-2. Sympatry and hybridization of two species of towhees (*Pipilo*) in Mexico. Pure *erythrophthalmus* (24) in the north and southeast, pure *ocai* (0) in the south and southeast. The numbers (from 0–24) designate the mean character indices of various hybrid populations (*from Sibley*, 1954).

venient for the taxonomist (see 9B.2) to attach such doubtful populations to the species with which they are most nearly allied. Circular overlaps and other borderline cases (Mayr, 1963, pp. 496–512) are other instances of evolutionary intermediacy that will have to be decided from case to case on the basis of convenience and degree of evolutionary intermediacy.

## 2.6  SPECIAL SITUATIONS

**2.6.1    Paleontology.** The paleontologist has to cope with two special kinds of difficulties in species delimitation in addition to those which the neontologist encounters.

These difficulties have been discussed frequently and with considerable detail in the paleontological literature. We refer to Imbrie (1957), Sylvester-Bradley (1956, 1958), Simpson (1960; 1961, pp. 152–155 and 163–171), and various articles in paleontological journals. As these authors point out, the difficulties encountered by the paleontologist are often exaggerated. Many paleontologists have confused phenon and species, taxon and category, and evidence and inference. A clear understanding of these terms and the underlying concepts removes much of the difficulty.

*Evolutionary Continuity.* Species are evolving systems, and the vertical delimitation of species in the time dimension should in theory be impossible. Unbroken sequences of fossil populations are, however, extremely rare. Where they exist, the named morphospecies are usually so similar that they are better not recognized at all or, at most, ranked as subspecies.

In most fossil sequences there are convenient breaks between horizons to permit a nonarbitrary delimitation of species. This is true even for Brinkmann's (1929) remarkably complete sequence of Jurassic ammonites (*Kosmoceras*). Since much speciation occurs in peripheral isolates, the discovery of strata with intermediate populations (incipient species) is highly improbable and will occur only rarely. Range fluctuations contribute to the appearance of breaks even in cases of speciation owing to uncomplicated phyletic evolution in a single vertical column (cf. Fig. 3-8). An evolutionary species may be absent from a given locality for shorter or longer periods, and when it reoccurs, it may have changed sufficiently to be classified as a different species.

*Amount of Evidence Limited.* Taxonomic evidence supplied by behavior, chromosomes, properties of proteins and other chemical constituents, and by other attributes of living populations, is not available to the paleontologist. However, most classification of living species is also made without access to such information, and when it finally becomes available it usually confirms the existing classification. As Imbrie (1957) and others have pointed out, the paleontologist has far more information available than that of mere morphology. The variation of his samples, the associated faunas, the paleoecology, the geography, horizon and time level, etc., contribute abundant information that facilitates inferences on the delimitation of taxa and their categorical ranking. The analysis, as in much neontological work, often has to begin with a separation of the samples into phena, but a consideration of all the collateral evidence usually allows an unequivocal assignment of these phena to species or subspecies, as was well demonstrated by Sylvester-Bradley (1958) for the two chronological species *Ostrea knorri knorri* and *O.k. lotharingica* (Table 2-1).

TABLE 2-1.  FREQUENCY OF 12 PHENA IN TWO SUBSPECIES OF *Ostrea knorri (from Sylvester-Bradley, 1958)*

| 12 Phena | Schönmatt (*knorri*) | Geisingen (*lotharingica*) |
|---|---|---|
| D + B + M | 49 | 0 |
| A + G | 41 | 5 |
| K + J + E | 10 | 5 |
| F + H | 3 | 14 |
| C + L | 0 | 45 |
| Total | 103 | 69 |

**2.6.2   Parasites.** Populations of parasites on different hosts are often slightly different. There are three possibilities for interpreting these morphological differences.

1. The differences are caused by nongenetic modification resulting from the different physiological environments of the different hosts. Many trematodes may mature in a large number of possible hosts and show great differences. Specimens of the liver fluke *Fasciola hepatica* from a cow, a rabbit, and a guinea pig manifest differences far greater than those usually employed to distinguish species (Stunkard, 1957). It is possible that large numbers of nominal species in certain genera of trematodes such as *Hymenolepis* may be shown eventually to be nothing but host-induced, nongenetic modifications.

2. The differences are indicative of subspecific rank. The differences between host populations of Mallophaga, although constant, are often so slight that they are best treated as subspecies (Clay, 1958). Host separation in this case corresponds to geographic isolation in free-living species, and the amount of gene flow between Mallophaga occurring on different host species is presumably very slight.

3. The differences are indicative of rank as full species. Numerous cases have been described in the parasitological literature where exceedingly similar parasites could not be transferred from one host to another. There is no opportunity for gene interchange in the case of species without intermediate host. In such a case, even though the morphological difference is comparatively minor, one must assume that the genetic barrier has reached species level.

# Chapter 3

## The Polytypic Species Population Systematics and Infraspecific Categories

$A$t a given locality a species of animal is usually separated from other sympatric species by a complete gap. This is the species of the local naturalist, the species of Ray and Linnaeus. It may also be called the *nondimensional species* because it lacks the dimensions of space and time. Combining the properties of a species and of a single local population, the nondimensional species can usually be delimited unequivocally.

Every species consists, however, of numerous local populations, and some of these are visibly different from each other. If a taxonomist finds a population which he considers sufficiently distinct from the population of the original type locality of the species (or of other previously named subspecies), he calls it a new *subspecies* (see below). Species that contain two or more subspecies are called *polytypic species*. Species that are not subdivided into subspecies are called *monotypic species*. Recognition of the fact that many species, particularly widely distributed species, are polytypic was one of the most important developments in taxonomy. For a full treatment of this development and of various aspects of the polytypic species, see Mayr (1963, chap. 12).

## 3.1  IMPORTANCE OF POLYTYPIC SPECIES

The greatest benefit derived from the recognition of polytypic species is that in well-known groups of animals such as birds, mammals, butterflies, or snails it has led to a considerable simplification of the classification. The reclassification into polytypic species of geographically representative forms that had originally been separately described as monotypic species led to a great clarification of the system. This reorganization of classification on the species level is virtually completed in birds. It is in full swing for mammals and some groups of insects and land mollusks, but has hardly begun in most other groups of animals. The 19,000 monotypic species of birds listed in 1910 (together with numerous species discovered since then) have now been reduced to about 8,600 species. A similar simplification has been reported for many other groups of vertebrates and invertebrates. Of much greater significance is the restoration to the species category of a definite biological meaning and homogeneity. Awarding species rank to every local population, no matter how slight its difference, completely destroyed the biological significance of the species category.

The task of assembling local populations into polytypic species, or, more broadly, of sorting large numbers of "nominal species" and "varieties" into polytypic species, reveals many taxonomically and biologically interesting situations (Mayr, 1963, p. 343). The consistent application of the polytypic species concept to all groups of animals is one of the major tasks of taxonomy.

**3.1.1  Difficulties.** When establishing polytypic species the taxonomist encounters two difficulties. Polytypic species are composed of allopatric or allochronic populations that differ from each other. However, all populations of sexually reproducing organisms differ slightly from each other, and certain standards must be met before subspecies can be recognized (9B.1). The other difficulty is that, occasionally, closely related species with similar ecological requirements replace each other geographically and yet are full species and not subspecies. How to choose between these two alternatives will be discussed in 9B.2.

## 3.2  OCCURRENCE OF POLYTYPIC SPECIES IN
## THE ANIMAL KINGDOM

The frequency of polytypic species differs from animal group to animal group. They are most frequent where species tend to form geographic isolates. In most well-studied groups between 40 and 80 percent of the species are polytypic, but some highly specialized species, particularly host-plant

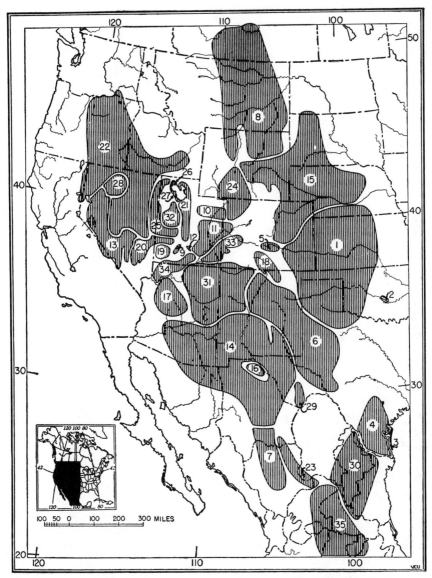

Fig. 3–1. The distribution of 35 subspecies of the kangaroo rat *Dipodomys ordii* Woodhouse, an example of a range map of a polytypic species (*after Setzer*). Numbers designate the ranges of the various subspecies.

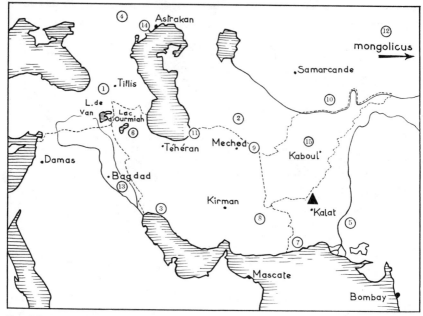

Fig. 3–2. Type localities (1–15) of fifteen subspecies of the scorpion *Mesobuthus eupeus* in the middle east (*from Vachon*, 1958).

specific insects, do not readily form polytypic species. Polytypic species are also scarce or absent in groups with slight species differences (e.g., groups of sibling species).

A number of other technical terms, such as *Formenkreis* (Lorenz, Kleinschmidt) and *Rassenkreis* (Rensch), have been applied to polytypic species but have not become established.

The polytypic species is, in a sense, the lowest of the higher categories. Being multidimensional it lacks the simplicity and objectivity of the nondimensional species. Most of the difficulties of delimiting species of animals concern situations where there is doubt as to whether two allopatric populations belong to the same polytypic species or not. Among birds, even such borderline cases are in the minority. The claim that such difficulties are more frequent in other groups of animals still awaits verification.

**3.2.1 Nomenclatural Problems.** A polytypic species is often a compound of several "species" originally proposed as monotypic. It differs thus from the Linnaean species by no longer being the lowest category (which is now the subspecies) and by being a collective category. What scientific name should be given to this new collective taxon, and who should be the author? When Linnaeus named the white wagtail *Motacilla alba,* to cite one example, he had in mind the Swedish population with the diagnostic characters described by him. The *M. alba* of Linnaeus is now called the *nominate* subspecies *M. alba alba* Linnaeus. When *M. alba* of Linnaeus was combined with eight or more other taxa, all originally described as separate species (*lugubris* Temminck, *dukhunensis* Sykes, *baicalensis* Swinhoe, *leucopsis* Gould, *personata* Gould, *hodgsoni* Blyth, *ocularis* Swinhoe, *lugens* Kittlitz, etc.), the newly-formed polytypic species was something very different from the original *M. alba* of Linnaeus. If we still associate the name Linnaeus with the new polytypic species *M. alba,* it is to indicate Linnaeus as the author of the name *alba,* and not of the drastically reconstituted taxon to which we now attach the name.

## 3.3   INFRASPECIFIC CATEGORIES AND TERMS

**3.3.1   The Variety.** This, as *varietas,* was the only subdivision of the species recognized by Linnaeus. It designated any deviation from the type of the species. As a consequence the varieties of the early taxonomist were a heterogeneous potpourri of individual variants (see Chap. 8) and various kinds of races. As a result of this confusion the term variety came into discredit and is hardly used any longer by animal taxonomists. For a further discussion see Simpson (1961, p. 177) and Mayr (1963, p. 346).

**3.3.2   The Subspecies.** The term subspecies when it came into general usage during the nineteenth century was a replacement for the term variety in its meaning of "geographic race." It was considered a taxonomic unit like the morphological species, but at a lower taxonomic level. Many early authors used the term subspecies indiscriminately, almost like the term variety, for distinguishable entities that were not as distinct as species. Ant specialists, for instance, employed the term for sibling species and for individual variants. When an author reports several subspecies of one species from the same locality, it strongly indicates a wrong usage of the term. Subspecies are normally allopatric and allochronic, exceptions occurring, however, in migratory species and in parasites with sympatric host subspecies. A purely morphological definition of the subspecies, as attempted by typologists, often results in a sympatry of entities, thus defined. In view of the many misuses of the term it must be emphasized that the subspecies is a category quite different from the species. No nonarbitrary criterion is available to define the category subspecies. Nor is the subspecies a unit of evolution, except where it happens to coincide with a geographical or other genetic isolate.

The subspecies may be defined as follows: *A subspecies is an aggregate of phenotypically similar populations of a species, inhabiting a geographic subdivision of the range of a species, and differing taxonomically from other populations of the species.*

The reasons for the particular wording of the definition are as follows:

1. A subspecies may consist of many local populations all of which, though very similar, are slightly different from each other genetically and phenotypically; a subspecies is therefore a collective category.
2. Every local population is slightly different from every other local population, and the presence of these differences can be established through sufficiently sensitive measurements and statistics. It would be absurd and would lead to nomenclatural chaos if each such population were given the formal trinominal name that is customary for subspecies. Therefore, subspecies are to be named only if they differ "taxonomically," that is, by sufficient diagnostic morphological characters (see 9B.1).
3. Even when it is possible to assign populations to subspecies, an assignment on the basis of phenotype alone is not necessarily possible for every individual, owing to an overlap of the ranges of variation of neighboring populations.

The term *overlap* is often misused. The breeding ranges of two species may overlap geographically but not those of two subspecies of the same species. If two discrete breeding populations coexist at the same locality, they are full species (except in the rare cases of "circular overlap"). Where two subspecies meet, intermediate or hybrid populations may occur which combine the characters of both subspecies. It would be misleading in such a case to say that the two subspecies overlap in this area, since the species is represented in this area only by a single population, no matter how variable.

*Difficulties in the Application of the Subspecies Category.* Recognition of the polytypic species requires the use of the subspecies category, with all the concomitant benefits described above. However, various aspects of geographic variation cause difficulties. Indeed, the subspecies has been misused in many ways. Some authors applied the term to individual variants and sibling species, many authors named insignificant local populations as subspecies, and finally some authors considered the subspecies as a unit of evolution rather than as an arbitrary device to facilitate intraspecific classification. As a result the practice of describing subspecies was criticized by numerous authors, most cogently by Wilson and Brown (1953) (see also Inger, 1961). They pointed out four aspects of the subspecies which reduce its usefulness:

1. The tendency of different characters to show independent trends of geographic variation
2. The independent reoccurrence of similar or phenotypically indistinguishable populations in geographically separated areas ("polytopic subspecies"—see Sec. 9B.1.4)
3. The occurrence of microgeographic races within formally recognized subspecies

4. The arbitrariness of the degree of distinction considered by different specialists as justifying subspecific separation of slightly differentiated local populations

Numerous articles on the pros and cons of the subspecies question can be found in volumes 3 to 5 of *Systematic Zoology* (1954–1956). For general reviews of the subspecies question see also Simpson (1961, pp. 171–176) and Mayr (1963, pp. 347–350).

Recent arguments have, we hope, led to a more critical attitude toward subspecies. It has also shown that sensible use of the category subspecies is still a convenient device for classifying population samples in geographically variable species.

Practical problems which the population taxonomist frequently faces will be dealt with in Chap. 9. They concern particularly the following questions: How different must a population be in order to justify subspecific recognition? How shall one treat intermediate populations? How does one delimit subspecies against adjacent subspecies? Shall one recognize polytopic subspecies for indistinguishable but geographically separated populations? When should geographical isolates be called species, and when subspecies?

*The Nomenclature of Subspecies.* A trinomen is used to designate a subspecies taxon (Arts. 5 and 45 of the Code). For instance, the British Red Deer is called *Cervus elaphus scoticus,* while the continental Red Deer, originally named by Linnaeus binominally, becomes the nominate subspecies *Cervus elaphus elaphus.* For further nomenclatural aspects see 13.40 and 13.41; see also 13.25 (combined and divided taxa), 13.36 (subordinate taxa), 13.46 (homonymy among subspecies), and 13.57 (type-localities).

**3.3.3 Temporal Subspecies.** In paleontology slightly different populations separated in time are increasingly often assigned to the subspecies category. It does not seem advisable to make a terminological distinction between geographical and temporal subspecies because it is usually impossible, when different subspecies of a fossil species are found at different localities, to determine whether or not they are precisely contemporary. Even when there is a sequence of subspecies at a single locality, it need not necessarily be purely temporal. Subspecies found in succeeding strata may actually be geographic races that replaced each other owing to climatic changes (Fig. 3-3).

The paleontologist, when applying the subspecies category, faces certain difficulties not encountered by the neontologist. There may be a differential deposition of various age classes and sexes in different horizons, as well as the occurrence of nongenetic habitat forms. Yet treating fossils as the remains of formerly existing populations usually leads to a deeper analysis as well as to a better understanding of relationships and of the meaning of evolutionary trends. As with living species, it must be kept

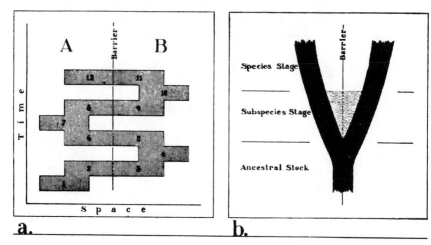

Fig. 3–3. (a) Geography and time in subspeciation and speciation. Diagram illustrating how geographical fragmentation of successive populations (the numbered rectangles) may accompany vertical differentiation of a phyletic line. The populations rarely remain in one locality for long but migrate. Some migrants become isolated from the parent stock by barriers, becoming ultimately differentiated into geographical races. The faunal succession in any locality (A or B) is never absolutely continuous, even though gaps may be obscure. The gaps may be produced by migrations, by depositional hiatus, and by local extermination. (b) A population becomes divided by a barrier causing partial isolation with limited gene flow for a time—the subspecies stage in speciation. After sufficient genetic differentiation has been reached, interbreeding ceases, gene flow is stopped, and the two branches become separate species (*Newell*, 1947).

in mind at all times that the subspecies is merely a classificatory device. For further discussions on the subspecies in paleontology see Newell (1947, 1956a), Sylvester-Bradley (1951, 1956), and Simpson (1961, pp. 175–176).

   **3.3.4    Race.**  A race that is not formally designated as a subspecies is not recognized in the taxonomic hierarchy. However, the terms subspecies and geographical race are frequently used interchangeably by taxonomists of mammals, birds, and insects. Other taxonomists apply the word race to local populations within subspecies.

   The nature of ecological races among animals is still controversial (Mayr, 1963, p. 355). Since no two localities are exactly identical with respect to their environment, every subspecies is at least theoretically also an ecological race. However some populations differ in their ecological requirements without acquiring taxonomically significant differences. More important from the taxonomic as well as the evolutionary point of view are host races among parasites and species-specific plant feeders. If gene flow between populations on different hosts is drastically reduced, such host races are the equivalent of geographic races in free-living animals. Also, such host races often develop subspecific characters.

   **3.3.5    Cline.**  This term was coined by J. S. Huxley (1939) for a character gradient. It is not a taxonomic category. A single population may belong to as many different clines as it has characters. A cline is formed by a series of contiguous populations in which a given character changes gradually. At right angles to the cline are the lines of equal expression of the character (points of identical phenotype); such a line is called an *isophene*. For instance, if in the range of a species of butterfly the percentage of white specimens varies from north to south, the corresponding isophenes may be indicated on a map (Fig. 3-4).

Any character, be it a morphological, physiological, or other genetically determined character, may vary clinally. Clines may be smooth, or

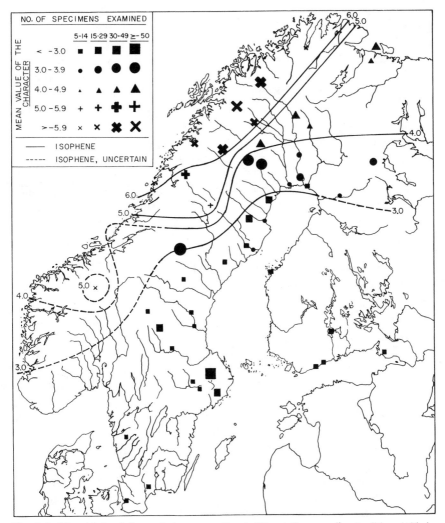

Fig. 3–4. Cline in the darkness of the upper side at different Fennoscandian localities of *Pieris napi* females first generation. Isophenes of various darkness values indicated on the map (*from Petersen*, 1949).

they may be "step clines" with rather sudden changes of values. Clines do not receive nomenclatural recognition. Indeed, when the geographic variation of a species is clinal, it is usually inadvisable to recognize subspecies, except possibly for the two opposite ends of the cline when they are very different or separated by a pronounced step. For a further discussion of clines see Simpson (1961, p. 178) and Mayr (1963, pp. 361–366).

**3.3.6  Infrasubspecific Categories.** The subspecies is the lowest taxonomic category recognized in the Code of Nomenclature (Art. 45c) (see 13.41).

In the days when the subspecies was still defined typologically, many proposals were made to subdivide heterogeneous subspecies into still smaller, hopefully uniform taxa, and terms were proposed for such taxa, e.g. natio. Now that it is being realized that every local population is different from every other one, even if they live only a few miles apart or less, and that these populations are not sharply separated from each other (except where separated by barriers), there is no longer any excuse for a formal recognition of innumerable local subdivisions of subspecies. The term *deme* adopted by zoologists for the evolutionary unit corresponding to a local population (Mayr, 1963, p. 137) is not the name of a taxonomic category.

**3.3.7  Intrapopulation Variants.** Taxa are populations, and populations are the material of classification. Phena, composed of intrapopulation variants, have no taxonomic status and deserve no formal recognition in nomenclature. They can be referred to by a vernacular terminology, e.g. "albinos," as discussed in 13.41. The *morphs* found in polymorphic populations are such variants.

**3.3.8  Asexual Entities.** Uniparental reproduction through parthenogenesis, vegetative budding, or fission is not infrequent among the lower invertebrates, with parthenogenesis occurring even among insects and lower vertebrates up to reptiles. Since interbreeding is the ultimate test of conspecificity in animals, and since this criterion is available only in sexually reproducing organisms, determination of categorical rank is difficult for taxa of uniparentally reproducing organisms. How should the taxonomist treat clones, pure lines, biotypes, and so-called "strains" or "stocks" of such organisms?

Parthenogenesis is usually only a temporary condition in animals. In aphids, cladocerans, rotifers, and various others, females of many species are parthenogenetic during part of the year but return to sexual reproduction when environmental conditions change. Nomenclatural recognition is not given to such temporary clones. They terminate sooner or later, either by extinction or by returning to the common gene pool of the parental sexually reproducing species.

In the case of permanently uniparentally reproducing lines, the species category is applied on the basis of degree of morphological difference. Morphological difference between clones can be used as an indication of the underlying genetic difference and this in turn for an inference on probable species status. To group uniparentally reproducing organisms into distinct taxa is often not difficult. The ranking of these taxa, that is their assignment to a definite category, is almost invariably somewhat arbitrary. Nature, unfortunately, is not always as tidy as the taxonomist would wish.

**3.3.9  Neutral Terms.** It is very convenient in taxonomic work to have some terms that can be given informally to phena, particularly in incompletely analyzed cases. These are the so-called "neutral terms." The ones that are most frequently used in taxonomy are *form,* for a single unit, and *group* or *complex* for a number of units. We may speak of a "form" when we do not know whether the phenon in question is a full species or a subspecies, or whether it is a subspecies or an individual variant. Seasonal and polymorphic variants are often referred to as forms. The term is also used in the plural when two unequal units are referred to. For instance, when describing attributes common to a species and a subspecies of another species, one may refer jointly to the species and the subspecies as "these two forms."

The term *group* is more commonly applied to an assemblage of closely related taxa which one does not want to place in a separate category. In the large genus *Drosophila,* for example, numerous species groups are recognized, such as the *melanogaster* group, the *virilis* group, the *obscura* group, and so forth. A species group is a group of closely related and presumably recently evolved species. The use of the species group in formal taxonomy has been spreading in recent years, because it reduces the need for the recognition of subgenera. In large polytypic species the term group is also applied to subspecies groups. The common Palearctic jay, *Garrulus glandarius,* has a total of 28 subspecies which can be arranged in seven subspecies groups, the *garrulus* group, the *bispecularis* group, and others. The term group is used more rarely for aggregates of genera and other higher categories. The word complex is sometimes used synonymously with the word group.

Terms like *section, series,* and *division,* are generally used for the higher categories. Their use is not standardized, however, and in different branches of systematic zoology they may be used above or below the family, the order, or the class.

## 3.4  POPULATION TAXONOMY

The increasing preoccupation with the description of new subspecies and the establishment of polytypic species, beginning in the second half of the nineteenth century and continuing into the present, resulted in a subtle change of emphasis and outlook. The emerging understanding of the species taxon as a geographically variable aggregate of populations accelerated the replacement of the typological species concept, and its taxonomic equivalent, the morphospecies, by the biological species concept. The taxonomist was no longer satisfied to separate his collections into types and duplicates. He began to sample the species at many localities, and tried to assemble large series from every locality. This type of study was initiated in the second half of the nineteenth century almost simultaneously by ornithologists, entomologists, and malacologists.

Although the study of populations reached its dominant position in systematics only within recent generations, its roots go back to the pre-Darwinian period. A short history is found in Mayr, 1963, chaps. 11 and 12.

Populations are variable, and consequently the description, measurement, and evaluation of variation has become one of the principal preoccupations of the taxonomist who studies taxa in the lower categories. A typologist needed only one or two "typical" specimens of a species; when he had more, he disposed of them as "duplicates." The modern taxonomist attempts to collect large series at many localities throughout the range of a species. Subsequently he evaluates this material with the methods of population analysis and statistics (see Chap. 8c).

The work of population taxonomy not only led to a simplification of taxonomy through the introduction of polytypic species, but it also led to a new approach in the study of evolution. Systematics has made many important conceptual contributions, and one of its greatest achievements is to have assisted in the introduction of the population concept into biology (1.3.2).

## 3.5  POPULATION STRUCTURE

The study of the population structure of species shows that the conventional division of species into subspecies is a very inadequate and sometimes even misleading representation of the actual phenomena. A species does not consist of a number of "little species" called subspecies. Rather, a species consists of innumerable local populations or demes which stand in a certain relationship to each other. When species are studied strictly from the viewpoint of their population structure, it is found that they can best be described in terms of three major population phenomena (Mayr, 1963, chap. 13):

**3.5.1  The Population Continuum.** A large part of the range of many species, particularly the central part, is occupied by a series of essentially contiguous populations. Even when there are minor breaks in distribution, owing to the unsuitability of the habitat, such breaks are bridged by steady dispersal, resulting in copious gene exchange among populations. Variation in such a population continuum is essentially clinal. Terminal populations at the opposite ends of a continuum may be very different phenotypically and may deserve subspecific recognition (3.3.5).

3.5.2    **The Geographical Isolate.** This term designates all geographically isolated populations, or groups of populations, which have only limited or no gene exchange with other populations of the species. Any insular population is normally such an isolate, and isolates are therefore particularly common near the periphery of the species range. Isolates are frequently of sufficient difference to merit subspecies rank. The biological importance of the geographical isolate is that every isolate, regardless of its taxonomic rank, is an incipient species; it is an important unit of evolution.

3.5.3    **The Zone of Secondary Intergradation.** Whenever a geographical isolate reestablishes contact with the main body of the species, the two will interbreed, if the isolate has not yet acquired an effective set of isolating mechanisms. Depending on the degree of genetic and phenotypic difference achieved by the previously isolated populations, a more or less well-defined hybrid belt or zone of secondary intergradation will develop. Fusion lines between ex-isolates can be found in many species.

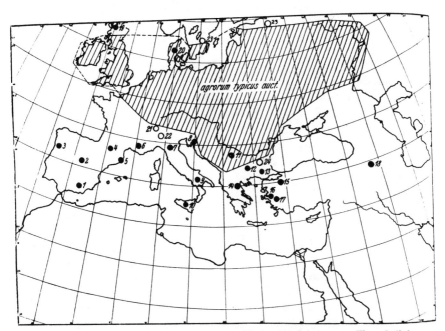

**Fig. 3-5.** Pattern of geographic variation in the bumble bee *Bombus agrorum*. There is little geographic variation in the continuous range of the nominate subspecies *agrorum*, while each peripherally isolated population (nos. 1–24) is distinct and mostly recognized as a separate subspecies (*after Reinig*).

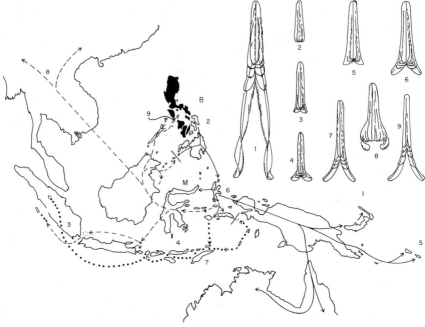

Fig. 3–6. Peripheral isolates at the ends of various lines of expansion in the polytypic bird species *Dicrurus hottentottus*. The figures indicate the ranges of the nine forms, the tails of which are shown in the insert. The tails of 4 and 6 are typical for most populations of the species; the tails of the peripheral forms 1–3, 5, and 7–9 are aberrant and specialized in various directions (*from Mayr and Vaurie, 1948*).

There are only a few groups of animals in which the population structure of species is sufficiently well known to permit such a detailed analysis. This has been done by Keast (1961) for all the species in a number of families of Australian birds. The study of the population structure of species does not replace classical taxonomy but is a superimposed refinement of the classical methods. It is possible only in groups where taxonomic analysis and population sampling has reached a degree of maturity that permits such detailed analysis.

## 3.6  THE NEW SYSTEMATICS

The approach of the population taxonomist differs rather drastically from the simple pigeonholing of classical Linnaean taxonomy. To emphasize the difference, Huxley (1940) introduced the term *new systematics* for the newer approach, but its roots actually go back to the first half of the nineteenth century, some traces of the new systematics being found in the writings of taxonomists as far back as 125 years ago. Every generation

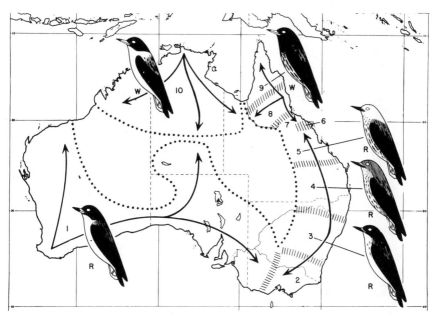

Fig. 3-7. Secondary hybrid belts among tree runners (*Neositta*) of Australia. The arrows indicate expansion from post-Pleistocene aridity refuges. Wherever former isolates have met, they have formed hybrid belts (indicated by hatching). *R*, subspecies with red wing bar; *W*, subspecies with white wing bar (*from Mayr*, 1963).

has its own new systematics, and what we consider as new systematics in the year 1968 may indeed be very old systematics 50 years hence. Some writers have placed the new systematics in opposition to alpha taxonomy. This is a misleading antithesis. Mayr (1964c) wrote:

> What then is the new systematics? Perhaps it is best described as a viewpoint, an attitude, a general philosophy. It started primarily as a rebellion against the nominalistic-typological and thoroughly non-biological approach of certain, alas all too many, taxonomists of the preceding period.
>
> The worker in the new systematics considers himself a biologist rather than a filing clerk. This has a number of well-defined consequences in his attitude toward his material and toward various techniques:
>
> 1. He is conscious at all times that he is classifying organisms, not the remains of organisms or merely names.
> 2. As a consequence, he places considerable emphasis on so-called biological characteristics, that is on non-morphological information derived from behavior, physiology, biochemistry, ecology, and so forth.
> 3. He appreciates that all organisms occur in nature as members of populations and that specimens cannot be understood and properly classified unless they are treated as samples of natural populations.
> 4. As a consequence, he attempts to collect statistically adequate samples, in the case of variable species often amounting to hundreds and thousands

of specimens, in order to be able to undertake a study of individual and geographic variation with the help of the best biometric and statistical tools.

If we would attempt to describe the current model of the new systematics, we would see at once that every single item is merely the continuation of a trend which in most cases had started more than one hundred years ago. Let me mention some of these points:

1. The utilization of an ever-increasing number of kinds of characters and a continued depreciation of key characters and of single character classifications in contrast to the typological approach.
2. A ready acceptance of new tools and techniques, such as
   (a) the visual analysis (by sonagrams, etc.) of sounds in insects (cicadas, orthopterans), frogs, and birds;
   (b) the analysis of courtship displays and other behavior;
   (c) the utilization of biochemical characters, particularly those yielded by various methods of protein analysis; and
   (d) the utilization of computers, to reduce the danger of subjectivity in character evaluation.
3. A further clarification of concepts, for instance,
   (a) a clearer separation of taxon from category;
   (b) the recognition of the subspecies as a category, not an evolutionary unit; and
   (c) a clearer understanding of the causes for similarities and differences between taxa.

It is evident that the new systematics is neither a special technique nor a special method, but rather, as I have said, a viewpoint or attitude which can be applied at every taxonomic level.

For a somewhat different statement on the characteristic trends in modern taxonomy see Simpson (1961, pp. 63–66).


## 3.7   THE SUPERSPECIES

Allopatric populations are often so distinct from each other that there is little doubt about their having reached species level. Rensch (1929) proposed for such groups of allopatric species the German term *Artenkreis*. Since the literal translation "circle of species" was frequently misunderstood, Mayr (1931) introduced the term superspecies as a convenient international equivalent.

*A superspecies is a monophyletic group of closely related and largely or entirely allopatric species.*

When the ranges of the component species are plotted on a map, the superspecies usually presents the picture of a polytypic species. Yet,

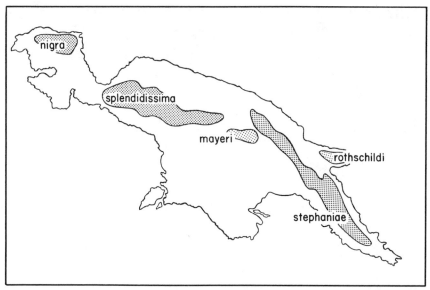

Fig. 3-8. A superspecies of paradise magpies (*Astrapia*) in the mountains of New Guinea. Some hybridization has been recorded in the zone of contact between *mayeri* and *stephaniae* (*from Mayr, 1963*).

there are three kinds of evidence to indicate that the component species have attained reproductive isolation. Either these species, although completely isolated from each other, are morphologically as different as are sympatric species in the respective genus, or they are in some areas in geographical contact (*parapatry*) without interbreeding, or there is actually a slight distributional overlap. Superspecies are not given nomenclatural recognition, but are listed as such in monographs and catalogs by an appropriate use of headings or symbols. They are important chiefly in zoogeographical and speciation studies. See Mayr (1963, pp. 499–501) for examples of superspecies and further discussion (also Cain, 1955, and Amadon, 1967).

The component species of a superspecies were originally designated *semispecies*. However, various authors have suggested a broadening of the term semispecies to include not only members of superspecies but all borderline cases in speciation (Mayr, 1963, p. 501).

# Chapter 4 Theories of Biological Classification and Their History

## 4.1 INTRODUCTION

If the taxonomist were satisfied merely with the describing and naming of species, he would be left with total chaos, considering the several million species of animals in existence. To convert this chaos into order is the task of classification. The ability to classify components of the environment did not originate with man. Indeed, classifying of objects of the environment occurs well below the human level, although not codified in language. Animals prove by their reactions that they classify objects of their environment as "food" or "non-food," as "competitors" or "potential mates," as "enemies" or "prey." Man has classified things by his use of generic or collective terms ever since he began to communicate by speech. Classification by man of animals encountered in his environment is as old as man himself, even though early classifications may have been broad and vague as indicated by terms like animals, bugs, and worms. A good classification makes an enormous amount of diverse information readily and conveniently available to biologists and, indeed, to anyone dealing with organisms. It gives meaning by association. It should therefore be the objective of the taxonomist to produce a system which has high predictive value and will allow maximum information retrieval.

The meaning and the principles of biological classification, as well as all the associated difficulties, have been excellently discussed by Simpson (1961) and by numerous other authors—for instance, Michener (1957), Beckner (1959), Cain and Harrison (1960), Cain (1962), Gisin (1964), and Mayr (1965a, 1965c). The present chapter is devoted to a discussion of the history and theory of classification. The application of this theory, the procedure of constructing a classification, will be dealt with in Chap. 10.

To define classification without being circular and without including terms that bias the definition in favor of a particular philosophy of classification is difficult. Most zoologists would perhaps be willing to accept a tentative definition, such as:

*Zoological classification is the ordering of animals into groups on the basis of their similarity and relationship* (in the broadest, not necessarily biological, meaning of the latter word). The two terms, similarity and relationship, used in this definition, are the reason for controversies that have raged for hundreds of years.

## 4.2  HISTORY OF THEORIES OF CLASSIFICATION

The history of taxonomy goes back to the beginnings of mankind. Since that time we can distinguish a number of periods or levels of increasing knowledge and understanding. Simpson (1961) in his second chapter gives a valuable survey of the history of taxonomy and the development of modern theory. His treatment goes much deeper into the philosophical aspects of the subject than the present short one, and it should be read in conjunction. Simpson also cites bibliographic references to other papers on the history of taxonomy, such as his own earlier one (Simpson, 1959) and, for example, Cain (1958ff) and Hopwood (in Sprague, 1950).

We shall distinguish six historical periods, a division which is admittedly somewhat arbitrary. They are not sharply separated from each other, and sometimes several trends are concurrent during a single period. Progress in the classification of various animal groups (and in the study of animals from different regions) has been uneven. Activity in some groups of arthropods and other invertebrates still concentrates on identification and description, while the methods of population systematics have been prominent in some better-known groups, for instance birds, mammals, and butterflies.

An authoritative history of taxonomy has not yet been written; the treatments in the standard histories of zoology and botany are superficial and purely descriptive. Indeed, it remained impossible to attempt a penetrating history as long as the roots of taxonomic method and principle in essentialism (4.3.2), in Aristotelian logic, and in nominalism (4.3.3) were not understood.[1] The following short treatment attempts to incorporate some of the recent findings.

**4.2.1 First Period: The Study of Local Faunas.** Natives of even the most primitive tribes are often excellent naturalists, and attach specific names to the more prominent plants and animals of their home country. Seashore tribes usually have names for all species of shoreline fishes and for all edible or poisonous inshore invertebrates. Schemes to classify the distinguished entities are usually rudimentary (Conklin, 1962), but binominal nomenclatures are found in native tribes of Asia and America—they are not the sole invention of Linnaeus.

Several early Greek scholars, notably Hippocrates (460–377 B.C.), enumerated kinds of animals, but there is no indication of a useful classification in the remaining fragments of their work. There is no doubt that Aristotle (384–322 B.C.) was the father of biological classification. He lived for some years on the island of Lesbos, where he seems to have devoted himself almost entirely to the study of zoology, in particular to the study of marine organisms. He not only studied comparative morphology but also paid much attention to embryology, habits, and ecology. Emphasizing that all attributes must be taken into consideration, he said, "Animals may be characterized according to their way of living, their actions, their habits, and their bodily parts." He referred to such major groups of animals as birds, fishes, whales, and insects, distinguishing among the last both mandibulate and haustellate types and winged and wingless conditions; he used certain terms for lesser groups, such as Coleoptera and Diptera, which persist today. He established numerous collective categories, or genera, using as differentiating characters blooded or bloodless, two-footed or four-footed, hairy or feathered, with or without an outer shell, and so forth. All this was a tremendous advance over anything that had previously existed, and it completely dominated animal classification for the next 2,000 years. Nevertheless, Aristotle did not supply an orderly, fully consistent classification of animals.

Aristotle, however, was not only the father of biological taxonomy, he was also the proponent of a well-rounded philosophy (including metaphysics) and the founder of logic, and his impact on taxonomy by means of these was far greater than through the actual classification he proposed. Among other things it was he who chiefly suggested to naturalists and philosophers the idea of arranging all animals in a single graded *scala naturae* according to their degree of "perfection" (Lovejoy, 1936, pp. 58–59). This led to the endeavor to classify animals into those that are "lower" and those that are "higher," an endeavor translated after 1859 into evolutionary terms.

[1] Consult the glossary and index for a definition of terms left unexplained in this historical survey.

Likewise, it was through Aristotle and his followers (including Linnaeus) that "typological" or essentialist thinking became entrenched in taxonomy (4.3.2). Eventually typological thinking was replaced by population thinking (1.3.2 and 3.4). Typology should not be confused with the type method of modern nomenclature (13.48).

The energetic revival of natural history after the Dark Ages, together with the vigorous worldwide explorations and discoveries since the fifteenth century, led to a precipitous increase in the number of known kinds of animals and plants. Yet the great encyclopedias of Gesner (1551–1558) and Aldrovandi (1599) were either alphabetical or followed Aristotle's rudimentary classification. Of all the pre-Linnaean authors the one who arrived at the most natural higher classification was John Ray (1627–1705). On the whole, throughout this period the botanists were far ahead of the zoologists and were the first to look for new methods and new principles. These endeavors reached their climax in the next period.

**4.2.2   Second Period: Linnaeus and His Contemporaries.** The great Swedish naturalist Linnaeus (1707–1778) exerted such an important influence on the entire subsequent development that he has been called the father of taxonomy. The binominal method of nomenclature (13.11, 13.12)

CAROLI LINNÆI
EQUITIS DE STELLA POLARI,
ARCHIATRI REGII, MED. & BOTAN. PROFESS. UPSAL.;
ACAD. UPSAL. HOLMEN. PETROPOL. BEROL. IMPER.
LOND. MONSPEL. TOLOS. FLORENT. SOC.

# SYSTEMA NATURÆ

PER

*REGNA TRIA NATURÆ,*

SECUNDUM

CLASSES, ORDINES,
GENERA, SPECIES,

CUM

*CHARACTERIBUS, DIFFERENTIIS, SYNONYMIS, LOCIS.*

TOMUS I.

EDITIO DECIMA, REFORMATA.

*Cum Privilegio S:æ R:æ M:tis Sveciæ.*

*HOLMIÆ,*
IMPENSIS DIRECT. LAURENTII SALVII,
1758.

Fig. 4-1. Carolus Linnaeus (1707–1778) and title page of the foundation work in zoological nomenclature.

was for the first time consistently applied by him to animals in the tenth edition of his *Systema naturae* (1758). Linnaeus himself considered this a rather inconsequential achievement, and was far prouder of his consistent application of the Aristotelian system of logic to classification. The world as we see it was to him the work of the Lord, and classification was the presentation of the plan of creation. The words "nature" and "natural," as in natural system, had a highly specific technical meaning to the Aristotelians (Cain, 1958). They referred to the operational manifestations of the "essence" of a being (in terms of the Thomistic philosophy) (4.3.2). Great innovator that Linnaeus was in his binominal method and in his extraordinarily useful catalogs of names and diagnoses, in his basic philosophy he was looking backward to the scholastic philosophy of the Middle Ages and to Aristotle. Yet the conveniences of a rigid hierarchy of categories and an unambiguous binominal nomenclature proved themselves so great that succeeding generations had little trouble getting rid of the Linnaean philosophy and yet retaining the best of his method. In more recent years even the value of binominalism has been questioned (13.13, 13.14).

Most early classifications were quite frankly identification schemes, based on single characters. They were usually presented as simple dichotomous keys, using single key characters. These often related to broad or special adaptations, as when Pliny classified animals into those of land, water, and air. Water birds with webbed feet were classed together, and so were all wading birds with long legs. Superimposed on this pragmatic approach were various taxonomic theories (4.3.2) culminating in the *Systema naturae* of Linnaeus. The more perceptive naturalists soon realized that these identification schemes often resulted in very heterogeneous groupings, that is, in "artificial classifications." The botanist Hieronymus Bock was the first (in 1546) to endeavor explicitly to place together those plants which resembled one another in external characters, or, as he put it, those which Nature seemed to have joined together by similarity of form.

Even these guiding principles led to improvement only slowly. Early classifications were a curious mixture of "natural" and artificial groups. Plants, for instance, might be divided into mosses, ferns, grasses, herbs, shrubs, and trees. The former three correspond to natural taxa, the latter three do not. A division of animals into Vermes, Insecta, and Vertebrata was equally mixed because Vermes is an artificial conglomerate of worm-shaped animals. Continuing improvement resulted from a study of an increasing number of characteristics which led to the delimitation of groups that had many characters in common, not merely a key character. Linnaeus himself adopted this empirical approach for the groups he was most familiar with, for instance the insects, producing a classification that is still largely acceptable. His classifications of other groups, like birds, amphibians, and lower invertebrates, were very poor.

The Aristotelian philosophy of Linnaeus was heavily attacked by contemporary naturalists with empirical or nominalistic tendencies, particularly by M. Adanson.

**4.2.3  Third Period: The Empirical Approach.** The hundred years between the tenth edition of the *Systema naturae* and the publication of Darwin's *Origin of Species* was a period of subtle but steady transition. Deductive, a priori principles were increasingly rejected, and taxonomists more and more delimited taxa empirically, on the basis of the totality of characters, not just a few "essential" ones. The term "natural system" lost the Linnaean meaning of a system based on the inherent "natures" of organisms (as defined in scholastic philosophy); "natural" came to signify unbiased by a priori considerations, and based on a consideration of the totality of characteristics. Eventually what the Aristotelians had considered as "natural" came to be considered arbitrary and artificial.

Lamarck (1744–1829), who lived during this period, had no visible influence on these developments except for some purely practical contributions he made to the classification of the invertebrates. Cuvier (1769–1832) was far more influential (Coleman, 1964). His taxonomic theory was an extraordinary mixture of antiquated concepts (such as his a priori weighting of characters on the basis of their physiological importance) and some very sound practical taxonomy. His insistence on the total independence of four major types (embranchements) of animals—vertebrates, mollusks, arthropods, and radiates—was the death knell of the *scala naturae* which had still dominated Lamarck's classification. Now that the invertebrates, the "bloodless" animals, were broken up, it became easy to continue the dismemberment of unnatural groupings, and this was indeed a major preoccupation of zoologists during the first half of the nineteenth century.

A steady and enormous increase in the number of known animals characterized this period. Voyages all over the globe acquainted zoologists with the animals of Africa, Australia, and the Americas. The local naturalists were being replaced by specialists who studied birds, reptiles, mollusks, or insects, or indeed only one particular group of insects, be it butterflies, beetles, or ants.

The systematic work of this period confirmed the conviction of the early naturalists that the endless variety of organic life is organized into natural groups. There are bluebirds (*Sialia*), which are one of the subdivisions of the thrushes (Turdidae), which together with many other similar subdivisions form the songbirds (Oscines), one of the subdivisions of the birds, belonging to the vertebrates, and so on. The empirical taxonomists of this period did a magnificent job in developing a "natural system," in the new meaning of this term. However, disappointed by the philosophical speculations of the Aristotelians, of the nominalists, and of Lamarck, and also kept far too busy by the avalanche of new species descending upon the museums, they did not try to give meaning to the orderliness of nature discovered by them.

The best empiricists were not satisfied merely to search for "characters in common," they also established all the major methods and principles of classification that are still recognized as leading most efficiently to the establishment of sound taxa. In particular, they developed principles of a posteriori weighting of characters, for instance through the study of their correlation with other characters (10.4). Even though Ray and others before him had stressed the fact that some characters correlate better with natural groups than others, it is only in this post-Linnaean period, and in part as a conscious rebellion against Aristotelian apriorism, that the system of empirical a posteriori weighting developed. The empiricists also undertook the evaluation of gaps between taxa, and they supported hierarchical arrangements of categories (Chap. 5) on the basis of degree of similarity.

**4.2.4    Fourth Period: Darwin and Phylogeny.** Prior to 1859 the taxonomist had to choose between two alternatives to explain the naturalness of the system. He could side with the nominalists and claim that natural

ON

# THE ORIGIN OF SPECIES

### BY MEANS OF NATURAL SELECTION,

OR THE

PRESERVATION OF FAVOURED RACES IN THE STRUGGLE FOR LIFE.

By CHARLES DARWIN, M.A.,

FELLOW OF THE ROYAL, GEOLOGICAL, LINNÆAN, ETC., SOCIETIES; AUTHOR OF 'JOURNAL OF RESEARCHES DURING H. M. S. BEAGLE'S VOYAGE ROUND THE WORLD.'

LONDON: JOHN MURRAY, ALBEMARLE STREET. 1859.

**Fig. 4-2.** Charles Robert Darwin (1809–1882) and title page of the foundation work in evolutionary biology.

groups do not exist and that taxa are merely the arbitrary products of the ordering human mind. This conclusion was so clearly contradicted by the empirically found naturalness of most taxa that it had hardly any adherents by 1859. The alternative was to believe that the order of nature was due to the plan of the Creator, and that each taxon consisted of variants of an underlying type, all of them containing, however, the essence of this type. It is Darwin's everlasting merit to have proposed a third alternative in the *Origin of Species*.

When Charles Darwin (1809–1882) in 1831 joined the *Beagle* as a naturalist, he still accepted the creationist dogma. During this voyage, however, he encountered so many phenomena of distribution, variation, structure, and adaptation that were quite improbable on a creationist interpretation that he adopted the evolutionary interpretation. All at once the enigma of the natural system was solved. "Natural" groups exist because the members of such a group had descended from a common ancestor. Fortunately, accepting evolution did not necessitate any change in the taxonomic technique. No longer did the taxonomist have to "make" taxa, evolution had done this for him. All he needed to do was to discover these groups.

It is not surprising in the least that the adoption of the evolutionary theory had virtually no impact on the established classifications. All it did was to give intellectual justification to what had already been standard practice among the best empirical taxonomists. What the evolutionary theory supplied was the explanation for the fact that variation in nature is not continuous but consists, in Darwin's words, of "groups within groups."

But Darwin did more than provide the theoretical basis for a natural system. He also gave some clear practical rules on how to avoid the circular reasoning of Linnaeus, Cuvier, and other predecessors. It is here that his many years of concentrated work on the classification of barnacles repaid him abundantly. Far more clearly than most of his successors, Darwin realized that two processes occur during phylogeny—branching and subsequent divergence. Accordingly Darwin stressed (1859, p. 420) that the separation of taxa must be based on branching ("propinquity of descent"), but that in the ranking of these taxa into various categories due consideration must be given "to the different degrees of modification which they have undergone."

Darwin made yet one other fundamental contribution to taxonomic theory. He, like the empiricists, rejected both the a priori weighting of taxonomic characters, as practiced by Linnaeus and Cuvier, as well as the disapproval of all weighting; he proposed instead (1859, pp. 425–426) a number of empirical rules on how to discover taxonomically useful characters, that is, how to undertake a posteriori weighting. They include the constant presence of the character in related forms, "especially those having very different habits of life," and particularly a constant association of several characters ("we know that such correlated or aggregated characters have a special value in classification"). Simpson (1959, 1961) and Cain (1959a) have given us lucid interpretations of Darwin's contribution to taxonomic theory.

The natural system was studied in the generations following Darwin primarily as important evidence in favor of the evolutionary theory. It is only rather recently that the theory of biological classification has been studied as a branch of methodology and philosophy of science as such. Only then was it recognized that the a priori principles that are useful in the classification of inanimate objects are largely inapplicable, if not decidedly misleading, when applied to organisms with an evolved information content (4.3.6).

The empirical taxonomists were greatly encouraged to learn that Darwin and his evolution theory gave meaning to their classifying activities and continued with increased vigor. One aspect in particular now came to the fore, namely, the search for missing links between seemingly unconnected taxa and the reconstruction of "primitive ancestors." Ernst Haeckel's (1834–1919) phylogenetic trees and speculations greatly stimulated this type of activity. The search for facts to substantiate these phylogenetic trees and to improve their design dominated biology during the second half of the nineteenth century and led to a boom in the fields of comparative systematics, comparative morphology, and comparative embryology. Even though the results were far less permanent than was then believed, this concern with phylogeny resulted in much sound research and in an interest in aberrant groups of organisms that might have otherwise been totally ignored. More importantly, these descriptive studies laid the foundation for the functional and experimental branches of biology whose flowering started in the 1870s and 1880s.

The period of the discovery of major new types of animals was essentially over well before the end of the nineteenth century. By then the need to prove the fact of evolution had ceased to exist. Taxonomy no longer was an exciting bandwagon, and taxonomists were forced to concentrate on the necessary though tedious job of describing, diagnosing, and classifying the seemingly endless number of species. A minority of the species describers were dilettantes who brought discredit to the field by the creation of numerous synonyms and by an excessive splitting of families and genera. Others concentrated on the unearthing of long-forgotten synonyms, thus arousing the ire of general biologists, who complained quite rightly that this defeated the basic objective of nomenclature as an information retrieval system. There is little question that taxonomy fell into some disrepute during the latter part of the nineteenth and the early twentieth century. The fact that the taxonomists sided against the immensely popular and powerful

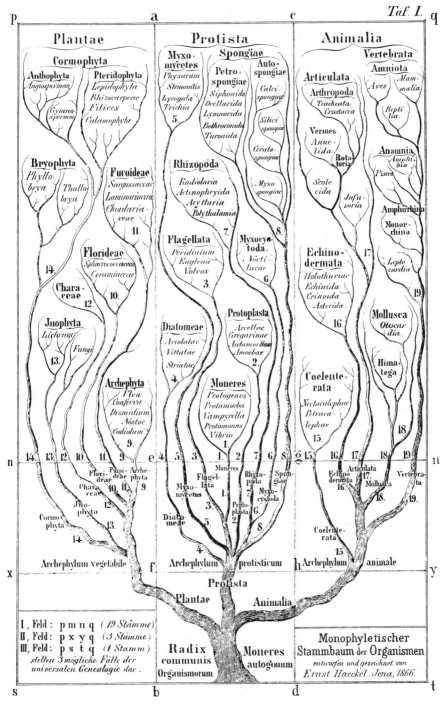

Fig. 4-3. The phylogeny of living beings as conceived by Haeckel (1866) and expressed in a formal treelike diagram.

early Mendelians with their antiselectionist and saltationist interpretation of evolution aggravated the situation. However, a turn for the better began to develop in the 1920s.

**4.2.5    Fifth Period: Population Systematics.** For the sake of convenience, taxonomists had continued to treat species, in a typological manner, as invariant units until long after the invalidity of the typological dogma had become apparent. Yet, whenever population samples from different portions of the geographic range of a species were compared, smaller or greater differences were found. This resulted eventually in the replacement of the typologically defined species by the polytypic species composed of different populations in the dimensions of space and time. The study and comparison of intraspecific populations became the objective of population systematics (3.4). The history of this development, beginning in the first half of the nineteenth century and reaching its climax in the 1930s and 1940s, has been described in detail by Mayr (1942, 1963). Replacement of typological thinking by population thinking has had important consequences in many areas of taxonomy. Considering taxa as populations or aggregates of populations greatly facilitated the study of variation and the definition of lower taxa and categories. Labeled by J. S. Huxley (1940) as the *"new systematics,"* it led to a reevaluation of the species concept and to a more biological approach in taxonomy. The population systematist understands that all organisms occur in nature as members of populations and that specimens cannot be understood and properly classified unless they are treated as samples of natural populations.

The same period saw two additional aspects come to the fore. One is what might be called the biological approach to taxonomy. The taxonomist moved more and more from the museum into the field, he increasingly supplemented the morphological characters with characteristics of the living animal, such as behavior, voice, ecological requirements, physiology, and biochemistry. Taxonomy truly became biological taxonomy. The other development was the introduction of the experiment into taxonomy. Although this has been far more characteristic for botany than for zoology, nevertheless the experimental analysis of isolating mechanisms—particularly in vertebrates, in Drosophila, and in protozoans—and the application of other experimental methods have been very helpful. At no time, however, did a separate field, experimental taxonomy, develop, because there was little resistance against the use of experimental methods.

Population systematics is not an alternative to classical taxonomy, but only an extension of it. In those groups in which the inventory-taking of species is still in full progress, and where too few local populations have been sampled, one cannot easily apply the methods of population systematics. Being centered on the population level, the new systematics naturally had little impact on the theory of classification at the level of the higher taxa. The population thinking of the new systematics was one of the major sources of the new science of population genetics, which in turn influenced the further development of population systematics. Together they greatly helped to clarify our thinking about evolution on the species level and were instrumental in producing the great synthesis in evolutionary biology.

**4.2.6    Sixth Period: Current Trends.** The current period is characterized by three developments. One is renewed examination of the whole theory of taxonomy, as indicated by the publications of Hennig (1950, 1966), Remane (1952), Gregg (1954), Bloch (1956), Cain (1958ff), Beckner (1959), Simpson (1961), Günther (1962), and Mayr (1965b). Articles published in *Systematic Zoology* and in the *Publications* (nos. 1–6) of the Systematics Association are indicative of the same interest. The use of electronic computers and associated endeavors to revive a nominalistic approach to taxonomy (Sokal and Sneath, 1963) form the second significant development of this period. The third is the vigorous introduction of biochemical techniques and, more importantly, a growing realization among molecular biologists of the importance of understanding the phylogeny of organisms as a basis for the understanding of the evolution of macromolecules. Comparative ethology likewise has stimulated a deeper interest in taxonomy.

## 4.3    THEORIES OF CLASSIFICATION

The time since about 1930 has been a period of unprecedented activity in systematics. After an early and almost total preoccupation with population systematics a concern began to develop over the meaning of higher taxa and the hierarchy of categories, and indeed over the theory of classification itself. Previously, when asked to explain their theory of classification, taxonomists had been singularly inarticulate. They would have said that they wanted to establish "natural groups" or that they wanted to group together species agreeing in their "essential properties" or showing "natural affinity." Such vague and seemingly noncommittal phrases are actually the expression of succinct philosophies or taxonomic theories. The publications listed in 4.2.6, among others, have helped to disentangle the manifold, often strangely interwoven lines of thought. It is not yet possible to write a balanced history of taxonomic theory, but it seems that from the beginning there have been only five theories of classification. Either pure or mixed they seem to supply the theoretical foundation of the work of all practicing taxonomists:

1. Essentialism (Aristotle to Linnaeus)
2. Nominalism
3. Empiricism
4. Cladism
5. Evolutionary classification

The first three are pre-Darwinian, the latter two post-Darwinian. The emergence of these theories is closely correlated with the general history of taxonomy (4.2).

Much of the recent literature is devoted to criticism. Curiously, taxonomic theory is being attacked for aspects that are common to all scientific theory and that are taken for granted when found in other sciences, for instance in physics. The provisional nature of theories is one of them. The fact that most findings of science cannot be observed directly but must be inferred is another. The necessity for unproven models and working hypotheses is a third. Evolutionary taxonomy is sometimes accused of circular reasoning, but without foundation (4.4). Actually, the theoretical foundations of taxonomic science are far more solid, secure, and elaborate than realized by most nontaxonomists.

Before these theories can be discussed, the place of identification in the conceptual framework of taxonomy must be clearly understood.

**4.3.1    Identification.** Many classifications attempt to be both identification schemes and classifications. This, as we now know, creates conflicts. The procedure of identification is based on deductive reasoning. One starts out with a given set of taxa (classes) and attempts to fit the investigated specimen into one of them. If one succeeds, one has identified it. Identification deals with individuals. The procedure of classification is inductive. Unlike identification, which deals with a few characters (ideally a single one) that throw a given specimen into one or the other line of the key, classification deals with and evaluates a multitude of characters, ideally all of them. Classification deals with populations and aggregates of populations. Much of the development of taxonomy has been toward clearer and clearer separation of these two entirely different operations. He who tries to include classification and identification in a single operation is bound to become confused and thwarted in both endeavors. For details on the procedure of identification see 7.3.

**4.3.2    Essentialism (Aristotle's Natural System).** The dominant theory of classification for many centuries was based on Aristotelian logic. It was adopted and elaborated by the Thomists and later by Linnaeus; its major tenets are still defended by some taxonomists, e.g., Thompson (1952, 1962), Borgmeier (1957), Blackwelder and Boyden (1952).

The philosopher Karl Popper (1950, p. 34) describes this approach in these words:

> I use the name methodological *essentialism* to characterize the view, held by Plato and many of his followers, that it is the task of pure knowledge or "science" to discover and to describe the true nature of things, i.e. their hidden reality or essence . . . [all essentialists, including Aristotle] agreed with Plato in determining the task of pure knowledge as the discovery of the hidden nature or form or essence of things. All these methodological

essentialists also agreed with Plato in holding that these essences may be discovered and discussed with the help of intellectual intuition; that every essence has a name proper to it, the name after which the sensible thing is called; and that it may be described in words. A description of the essence of a thing they called a "definition."

This philosophy, when applied to the classification of organic diversity, attempts to assign the variability of nature to a fixed number of basic types at various levels. It postulates that all members of a taxon reflect the same essential nature, or in other words that they conform to the same type. This is why the essentialist ideology is also referred to as typology. Variation, consequently, is considered by the typologist as trivial and irrelevant. The constancy of taxa and the sharpness of the gaps separating them tend to be exaggerated by him. The fatal flaw of essentialism is that there is no way of determining what the essential properties of an organism are and why these and no other properties of an organism are essential. Simpson (1961) and Hull (1965) have adequately refuted the theoretical arguments of the essentialist school. It also encounters numerous purely practical difficulties. The first is the polythetic nature of most taxa (4.7). Virtually every higher taxon contains certain species which, on the basis of the total evidence, are clearly members of the taxon even though they lack some of the "essential natures" of that taxon. A second reason is that it does not distinguish between "characters in common" resulting from descent and those resulting from convergence. Indeed its entire emphasis on "characters in common" is misleading (Simpson, 1961). Finally the essentialist approach is singularly defenseless against conflicting evidence from different groups of correlated characters (Crowson, 1965). Essentialists attempt to combine classification and a system of logic in a single scheme. Everything has to be classifiable into *a* and non-*a*, *b* and non-*b*. The essential characters in this theory of classification have two great virtues: they are exclusive ("winged" versus "nonwinged," "six legs" versus "four legs"), and they serve as diagnostic ( = key) characters. Indeed many early "classifications" were published in the form of dichotomous keys, and there are still taxonomists who find it difficult to distinguish between a classification and an identification key.

Classification continues to attract the attention of logicians right to the present—see Gregg (1954), Beckner (1959), Simpson (1961), or Buck and Hull (1966). Perhaps these attempts have helped to make the language of the taxonomist more precise, but they do not appear to have added to the theory of classification.

The ideal of essentialist classification was the discovery (rather than establishment!) of the Natural System. Their natural system was nothing more or less than the plan of creation. Among the empiricists and evolutionists the term natural system acquired very different meanings, yet the concept is so permeated with essentialist-creationist ideology that its use invariably evokes a misconception among nontaxonomists. Since there is no such thing as *the* Natural System, the term is best not used at all.

One curious by-product of the search for order in the natural system was the attempt to express the order in numbers, attempts that can be traced all the way back to the Pythagorean school. Linnaeus had a great fondness for numbers, and for very specific numbers at that (for instance, six). Vigors, Swainson, Oken, Kaup, and others attempted to devise systems based on the numbers three, four, or five. All these misguided plans to discover in nature a numerical blueprint collapsed in 1859 (Stresemann, 1950).

**4.3.3  Nominalism.** According to this philosophy only individuals exist. All groupings, all classes, all universals are artifacts of the human mind. This is equally true for species and for higher taxa. According to this philosophy (Gilmour, 1940):

> The process of classification is as follows: The classifier experiences a vast number of sense data which he clips together into classes . . . thus a class of blue things may be made for sense data exhibiting a certain range of color, and so on . . . the important point to emphasize is that the construction of these classes is an activity of reason, and hence, provided they are based on experienced data, such classes can be manipulated at will to serve the purpose of the classifier. . . . The classification of animals and plants . . . is essentially similar in principle to the classification of inanimate objects.

Thus there are no such things as birds or snakes, but only names invented by man and attached to groups of individuals considered by him to be similar. Bessey's statement on the species category (2.2.2) describes the nominalist theory particularly well.

This philosophy ignores the fact that there is indeed a difference in principle between classifying inanimate objects (including human artifacts) and organisms (Darwin, 1859, p. 411). It ignores the fact that groups of organisms, related by descent, possess a unity by the shared portion of their DNA heritage, a causation for shared characteristics for which there is no equivalent among inanimate objects. Birds are not an arbitrary aggregate of organisms resulting merely from "an activity of reason," as claimed by the nominalists, but they are a natural group because of the common heritage they share.

The basic fallacy of the nominalists is their misinterpretation of the causal relation between similarity and relationship. As Simpson (1961) has emphasized correctly, members of a taxon are similar because they share in a common heritage; they do not, as the nominalist would have it, belong to the taxon because they are similar. It is exactly as with identical twins; two brothers are not identical twins because they are similar, but they are similar because they are both derived from a single zygote, that is, because they are identical twins (2.3). The fatal weakness of nominalist thinking, when applied to the classification of organisms, is the reversal of an existing causal relation between "similarity and affinity." [*]

The numerical pheneticists (Sokal and Sneath, 1963) have, in principle, adopted the nominalist philosophy. A purely phenetic approach, an approach that "makes taxa" on the basis of the degree of observed similarity, usually leads to a classification similar to one based on the evolutionary approach. The reason is that, by and large, two organisms will be the more similar the more closely related they are by descent (10.2). Nevertheless, the phenetic approach is exposed to the risk of reaching unsound classifications, because in giving equal weight to all characters it does not allow for mosaic evolution, special adaptation, convergence, parallelism, developmental and genetic homeostasis, and other evolutionary, genetic, and developmental phenomena that disturb the expected close correlation between phenetic similarity and phylogeny. Worst of all, the theoretical basis of its nominalistic approach is unsound.

For recent critiques of the philosophy of numerical taxonomy see Mayr (1965b), Simpson (1961), Gisin (1964), and Rollins (1965).

A direct consequence of the assumption that natural groupings do not exist, but that all "species" or "classes" are the product of the human mind, is the postulate that definitions should be "operational." This might be a legitimate request for arbitrary classes of inanimate objects, and it is therefore not surprising that a physicist (Bridgman) was the original proponent of the operational approach. It works best for the definition of units of measurements but breaks down, even in physics, for more complex concepts. Operational definitions are certainly not only inapplicable but altogether inappropriate for evolved phenomena. A species or the species concept is not made nor tested by *my* operations. The fact that it is possible to call delimitations of higher taxa "operational" when they are based on calculated degrees of difference of a set of arbitrarily selected characters proves the inappropriateness of operationalism when applied to the products of evolution. Birds, bats, and other higher taxa are not *made* by the arbitrary operations of the taxonomist, rather they are the products of evolution. Operationalism is an altogether invalid approach in most areas of evolutionary biology, when based on the phenetic method.

**4.3.4    Empiricism.** According to this approach to taxonomy (4.2.3), there is no need for a theory of classification. Provided enough characters are intelligently evaluated, a natural system (the meaning of "natural" being very different from the Aristotelian one) will emerge automatically. Even though the working taxonomist usually proceeds on the basis of these empiricist principles, he feels that the resulting classifications would be biologically meaningless if not supplied with a theoretical foundation. This was provided by Darwin. Two new theories of classification were proposed after 1858, cladism (4.3.5) and Darwin's own evolutionary taxonomy (4.3.6), at present adopted by the majority of animal taxonomists.

**4.3.5    Cladism.** This term refers to a taxonomic theory by which organisms are ranked and classified exclusively according to "recency of common descent." Categorical status according to this theory depends on the position of the branching points on the phylogenetic tree. Hennig (1950, 1966), the most consistent proponent of this thesis, and others have designated themselves misleadingly as the phylogenetic school, and this has confused all arguments since 1950. The heated controversy of that period, allegedly concerned with the defense or criticism of "phylogenetic classification," actually deals with the validity of basing classification exclusively on the position of branching points (Mayr, 1965b). Since the splitting of branches is only one of several phylogenetic processes, it will avoid misunderstandings to refer to this viewpoint as cladism, according to the terminology of Rensch (1947) and Cain and Harrison (1960). Users of the recent literature are warned to look out for the misleading use of the term phylogeny by the cladists. Cladism has also been designated the genealogical approach (Gisin, 1964).

The basic fallacy of cladism is to overlook the fact that "relationship" in the evolutionary sense is determined by both processes of phylogeny, namely, branching and subsequent divergence (Darwin, 1859, p. 420).

> The argument of the cladist fails to recognize that the term relationship has two distinct meanings, genetic relationship and genealogical relationship. The two happen to coincide for all practical purposes as long as we deal with close relatives. . . . In phylogeny, where thousands and millions of generations are involved, thousands and millions of occasions for changes in gene frequencies owing to mutation, recombination, and selection, it is no longer legitimate to express relationship in terms of genealogy. The amount of genetic similarity now becomes the dominant consideration for a biologist . . . if one of the lines is exposed to severe selection pressures and as a result diverges dramatically from its genealogically nearest relatives, it may become genetically so different that it would be a biological absurdity to continue calling them near relatives. Even though the crocodilians are cladistically nearest to the birds (both having descended from the pseudosuchians) [Fig. 4-4], the crocodilians are still closer to many of the other reptiles, as far as the total gene composition is concerned, than they are to the birds, which have so drastically altered their genetic composition [as a result of their adaptation to life in the air]. (Mayr, 1965b, p. 79.)

To rank taxa according to branching points is nearly always misleading. It might necessitate, for instance, the inclusion of the African apes (*Pan*) in the family Hominidae and their exclusion from the family Pongidae.

Their exclusive concern with branching causes cladists to ignore and

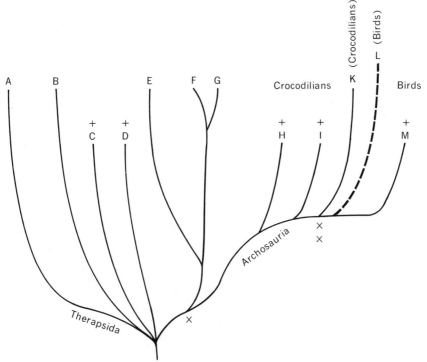

**Fig 4-4.** Inferred phylogeny of the reptilian branch of the vertebrates. The branching point between birds and crocodilians (xx) is much more recent than that between crocodilians and other surviving classes of reptilians.

A = mammals,  B = turtles,  C = ichthyosaurs,  D = plesiosaurs,  E = rhynchocephalians, F = lizards,  G = snakes,  H = ornithischian dinosaurs,  I = pterosaurians,  K = crocodilians, L = birds, M = saurischian dinosaurs; + = extinct.

even deny the existence or relevance of different rates of evolutionary change in different phyletic lines. Indeed, Hennig ignores rate of evolution so completely that he asserts (1966, p. 235): "Decisive is the fact that processes of species cleavage are the characteristic feature of evolution; they are the only positively demonstrable historical processes that take place in supra-individual organism groups in nature." This elimination from consideration of different evolutionary rates would be defensible only if evolutionary rates were the same in all lines. By equating genealogical distance with genetic distance, Hennig does indeed make this assumption.

This, in turn, leads him to postulate that one can determine the categorical rank of a taxon by fixing the branching point (from its sister taxon) in the geological time scale. The taxa in the hierarchy "are subordinated to one another according to the temporal distance between their origins and the present; the sequence of subordination corresponds to the 'recency of common ancestry'" [of the taxa] (p. 83). . . . . "In the phy-

logenetic [= cladistic] system the absolute rank order cannot be independent of the age of the group since . . . the subordination of groups is
by definition set by their relative age of origin" (p. 160). Consistently
with these principles Hennig decides that the position of the branching
point on the geological time scale determines categorical rank. Taxa that
originate from a split in the Precambrian are to be ranked as phyla; between Cambrian and Devonian as classes; between Mississippian and Per-
mian as orders; between Triassic and Lower Cretaceous as families; between upper Cretaceous and Oligocene as tribes; and in the Miocene as
genera (his fig. 58, p. 186). "Then the mammals would have to be called
an order . . . the Marsupialia and Placentalia would have to be downgraded to families, and the 'orders' of the Placentalia would be tribes"
(p. 187). The absurdity of the proposal is self-evident.

Throckmorton (1965) shows that the invalidity of the assumptions
of the cladists is singularly well demonstrated by evolution in the family
Drosophilidae. "Most of the diversification in this family has occurred by
divergence from a single lineage that was itself changing slowly in time"
(p. 233). Lines diverging from the same point diverge and diversify to
different degrees. The end points of many side lines are formed by different
genera. "Those that have diverged in their external and traditionally diagnostic features are classified in other genera. Where these same features
have remained unchanged and in spite of other changes, the forms are
classified as *Drosophila*" (p. 233). "In most instances in Drosophila closely
related species are complex mosaics of the characteristics of their nearest

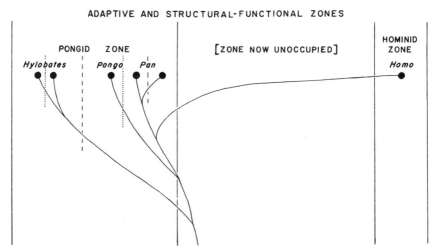

ADAPTIVE AND STRUCTURAL-FUNCTIONAL ZONES

Fig. 4-5. Recency of descent versus degree of adaptive divergence. Dendrogram of probable affinities
of recent hominoids in relationship to their radiation into adaptive-structural-functional zones. The two
major occupied adaptive zones are bordered by solid lines. Pongid radiation into sub- and sub-subzones
is schematically suggested by broken and dotted lines (*from Simpson*, 1963).

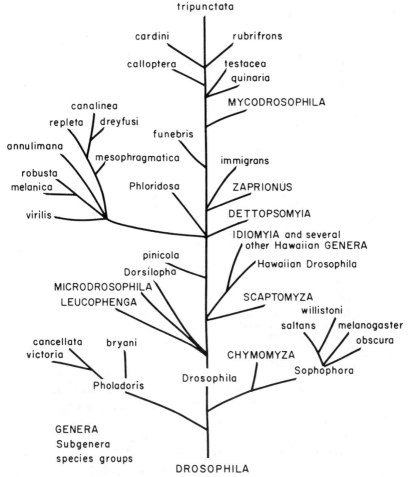

Fig. 4-6. Phylogeny of Drosophila and related genera. Many of these genera are specialized end points of certain species groups of Drosophila (*from Throckmorton, 1965*).

relatives. They show individually very little that is unique to themselves. They show instead unique combinations of the characters found among other close relatives" (p. 227).

As Sharov (1965) points out, cladists make the further assumption (which in most instances is not true) that a parental taxon expires when it gives rise, by splitting, to two daughter taxa ( = sister groups of Hennig). In reality, this is what rarely happens in phylogeny. A new group almost invariably buds off from a parental taxon which continues to exist with· very little change, sometimes for more than 100 million years (Fig. 4-7). A new group in these cases is the "sister group" of the parental taxon.

The fallacy of the cladistic approach consists not only in the equation

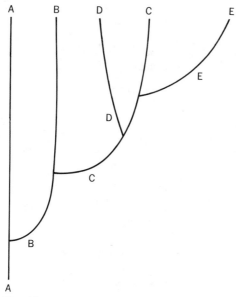

Fig. 4-7. Tetrapod phylogeny. New classes originate by
the branching off of daughter lines while the parental
line continues, essentially unaffected. A = Crossopterygia,
B = Amphibia,   C = Reptilia,   D = Mammalia,   E =
Aves. B is a daughter line of A, C of B, and D and E of C.

of genealogical and genetic distance but also in the ignoring of mosaic
evolution (Chap. 10) and in the fact that a classification is an information
retrieval system and not just an intellectual exercise. It places highly dissimilar organisms (like crocodilians and birds) in a single taxon while separating extremely similar ones of slowly evolving lines simply because the branching points which separate them occurred early on the geological time scale.

The ranking of taxa on the basis of the cladistic method is quite
meaningless. The method is difficult to apply in all groups with a poor
fossil record. In such groups the branching points are reconstructed on
the basis of evaluated similarity, and thus the method itself is negated.

Cladists tend to forget that the ordering of the diversity of nature
is the foremost task of classification. To use only one of the two processes
leading to the diversification of groups—namely, branching—as evidence
in the ordering procedure is self-defeating. The evolutionary taxonomist
agrees with the cladist in assuming that, usually, the more recently their
phyletic lines separated, the more similar two taxa are. However, the evolutionary taxonomist also gives due weight in his classifications to any unequal
divergence of the descendant lines. By deliberately ignoring these differences,
the cladist is often forced to recognize taxa of very unequal value.

Cladists sometimes claim that their theory of classification is the only one that can lead to the establishment of monophyletic taxa. This claim is based on a misinterpretation of the concept monophyly.

*Monophyly.* Once we accept the basic principle of biological classification, that organisms are to be classified according to the information content of their genetic program, it is evident that monophyly must be required of all taxa. This is therefore one of the postulates of evolutionary taxonomy. Artificial taxa, containing the descendants of different ancestors, would be unable to fulfill the demands one places on a scientific theory (see 4.5), owing to the heterogeneity of the included genetic programs.

The issue of monophyly has been clouded by various confusions. Some authors have referred to a "polyphyly" of a taxon when only a polyphyly of the diagnostic character of the taxon was involved, the taxon itself being monophyletic. It sometimes happens that a certain *grade* of morphological change is reached independently in several lines derived from a single ancestral group. The group showing this grade of development is of course monophyletic. As always in evolution, one must distinguish between what happens to the phenotype and what happens to the genotype. The diagnostic mammalian structure of the jaw-ear region, for instance, is believed to have evolved several times from ancestral therapsid reptiles, which had the needed genetic program to predispose them toward evolving the mammalian grade when exposed to the same selection pressures. This is not polyphyly, because the genotype permitting these parallel evolutionary changes goes back to the same ancestral program. We classify taxa ( = genotypes) and not characters ( = phenotypes). The usual phrasing of the principle of monophyly ("a taxon is monophyletic if its members are descendants of a common ancestor") is too vague to be helpful in more complicated cases, such as that of the mammals. Simpson (1961) has therefore given a more concrete definition: "Monophyly is the derivation of a taxon through one or more lineages, from one immediately ancestral taxon of the same or lower rank." The class Mammalia is monophyletic because all mammalian lines were derived from the immediately ancestral taxon of therapsid reptiles. Most cases of alleged polyphyly reported in the literature do not stand up under critical analysis.

The other misconception about monophyly is to consider it not only a retrospective but also a prospective principle. "The species included in each higher taxon must be derivable from a common stem species [ = retrospective postulate], and no species having arisen from this stem species can be placed outside this taxon [ = prospective postulate]" (Hennig, 1966, p. 71, also pp. 72–73). The latter postulate, of course, is completely contradicted by common sense and is in opposition to the phenomena of evolutionary divergence. If a descendant group, such as the birds among the archosaurian reptiles, evolves more rapidly than the other collateral lines, it not only can but it *must* be ranked in a higher category than its sister groups. This does not violate the principle of monophyly, retrospectively defined. The taxon Aves is monophyletic, and so are the taxa Crocodilia and Reptilia.

The concept of monophyly is important only at the level of the higher categories. Taxa that are still crossable, like subspecies, may produce hybrids which by definition would not be monophyletic. Yet the zoologist is not concerned with monophyly at the species level.

**4.3.6    Evolutionary Classification.** Like the empirical schemes of classification, evolutionary classification is based on the simple fact that readily delimitable groups of species, such as birds, penguins, bats, beetles, and the like, occur in nature. Evolutionary taxonomy differs from empiricism by demanding an explanation for the existence of such groupings and by using the answer to this question for the improvement of classification. The answer, of course, was given by Darwin (1859, p. 413): "I believe that something more is included [in our classification, than mere resemblance]; and that propinquity of descent,—the only known cause of the similarity of organic beings,—is the bond, hidden as it is by various degrees of modification, which is partially revealed to us by our classifications." By finding the reason for the existence of natural groups, Darwin changed the whole basis of classification. The taxonomist no longer "makes" taxa, he becomes a "discoverer" of groups made by evolution. Now he classifies not characters but organisms, and characters are downgraded to serve merely as *evidence* for something known by the biologist to have independent existence in nature (Simpson, 1961). The consequences of this difference in approach are by no means fully understood by all taxonomists. This is the reason why we may define taxa polythetically (4.7) and are not concerned when a species lacks some of the diagnostic characters of the higher taxon to which it belongs or when an individual lacks the diagnostic characters of its species. Stenzel (1963) has perceptively called attention to such cases.

A failure to understand the fact that we classify taxa and not characters has been the reason for most oversplitting of genera and higher taxa. The fact that a genus of spiders differs in eye structure from all other genera of its family is not sufficient cause for the creation of a new subfamily or family, if the genus in question agrees with the remaining genera in all other characters. *Mutatis mutandis* this principle is true for all kinds of organisms and all taxonomic levels. Classifying taxa means looking at the totality of characters as a single integrated ensemble, not at single, disconnected characters in an atomistic manner.

The difference between the logician—who applies the same rules to the classification of organisms and of artifacts—and the evolutionary taxonomist is now clear. First of all, the biologist classifies populations, not individuals or phena. The lower taxa are not arbitrary aggregates, but reproductive communities tied together by courtship responses and separated from other similar units not by arbitrary decisions of the classifier but by isolating mechanisms encoded in the genetic program of the organism (2.2.3). The higher taxa, likewise, are characterized by the joint possession of components of an ancestral genetic program.

Organisms have another unique property which distinguishes them from inanimate objects: they have a phenotype and a genotype (7.2). The classification of an inanimate object is completed once its "phenotype" has been classified. When we classify organisms, classification by phenotype is only the first step. As the second step we attempt to infer the genotype, the evolved genetic program, which has a far greater explanatory and predictive value than the phenotype. The phenotype is susceptible to all sorts of irrelevant similarities, and it is only the analysis of the inferred genotype which permits us to determine what similarities in the phenotype are due to convergence and what others are an expression of the ancestral genotype. The various kinds of similarities, and how to evaluate them when classifying taxa, are treated in Chap. 10.

The biological significance of discontinuities between taxa reveals another distinction between organisms and inanimate objects (Inger, 1958). Adaptive radiation, extinction, unequal rates of evolution, and many other purely biological phenomena discussed in the evolutionary literature (Simpson, 1953; Rensch, 1960; Mayr, 1963) are responsible for the observed inequalities in the gaps between species and groups of species. The fact that the observed clustering of species can be made the basis for many alternate classifications does not contradict the fact that species in nature are clustered and that there are real gaps between taxa at any given time level. Nor is this conclusion weakened by the fact that "missing links" must have existed at one time and that throughout the animal system a few species and species groups exist which connect major higher taxa. How to deal with clusters and gaps is explained in Chap. 10.

## 4.4  CLASSIFICATION AND PHYLOGENY

Classifications and the reconstruction of phylogenies are derived from the same basic evidence, a comparison of more closely or more distantly related species and an evaluation of similarities (or differences) in individual characters. One might call this method comparative character analysis. Its heuristic value has long been known among biologists and has led not only to the flourishing science of comparative anatomy but also to comparative physiology and ethology.

The raw data permit (1) the reconstruction of phylogenies and (2) the establishment of classifications. Yet, neither is "phylogeny based on classification" nor "classification based on phylogeny." Both are based on a study of "natural groups" found in nature, groups having character combinations one would expect in the descendants of a common ancestor. Both sciences are based on the same comparisons of organisms and their characteristics and on a careful evaluation of the established similarities and differences. This method is not circular (Hull, 1967).

The importance of phylogeny for classification is that a meaningful interpretation and evaluation of the characters of evolving and evolved organisms is not possible without carefully considering their probable evolution. What are primitive (ancestral) characters? Which characters form a single character complex that changed in response to a single adaptive shift? What similarities are the result of convergence (including the independent loss of structures)? We must ask these searching questions concerning characters because we use characters merely as evidence for the determination of genetic relationships among taxa. An answer cannot be given until several phylogenetic models have been tested. Classification then is not based on phylogeny, but phylogenetic considerations are important in the weighting of characters that are used in the construction of classifications. The practice of such weighting is discussed in Chap. 10.

Many misleading statements have been made about evolutionary taxonomy. For instance, it is not true that it is "the sole aim of evolutionary classification to reflect as accurately as possible the facts of evolution." Actually the most important aim of evolutionary classification is exactly the same as of all genuine classifications (in contrast to identification schemes): to combine maximal information content with maximal ease of retrieval of this information. The evolutionist believes that a classification consistent with our reconstruction of phylogeny has a better chance of meeting these objectives than any other method of classification. Taxa delimited in such a way as to coincide with phylogenetic groups (lineages) are apt not only to share the greatest number of joint attributes, but at the same time to have an explanatory basis for their existence. Since classifications are not based on established phylogenies, the objection that phylogeny is still largely unknown is irrelevant. What we do base our classification on is the universal fact that all organisms are the product of evolution. This permits us to set up models of inferred phylogeny, to test them against various alternate schemes of classification, and to undertake a taxonomic weighting of characters. There is no excuse for abandoning the evolutionary approach to classification because no man-made, hierarchical system of categories is capable of expressing precisely all the known or inferred facts of evolution. In spite of all its admitted inadequacies, the evolutionary approach produces a sounder basis for the classification of organisms than any other.

Evolutionism, as a philosophical basis for classification, is a valid approach only if and when natural groups of organisms are the result of divergent evolution. If reticulate evolution were common owing to the frequent fusion of previously separated evolutionary lines, or if convergence were frequently so complete as to lead to groupings that could not be unmasked as having a polyphyletic origin, then the claims of evolutionism as a proper theoretical basis for biological classification would indeed be questionable. No one has so far presented convincing evidence that these two potential difficulties are at all frequent in the evolution of animals. Unnatural taxa in animals almost invariably result from the use or availability of too few characters, and particularly from reliance on a few poorly chosen key characters.

## 4.5  THE OBJECTIVES OF A CLASSIFICATION

There has been much argument about the objectives of a classification. Some taxonomists claim that the only purpose of a classification is to create a reliable, easy-to-use filing system for the bewildering diversity of nature. All the early "identification systems" (see 4.3.1) had this as their objective, and even some modern authors contend that this is the only purpose for which a classification is constructed. This claim is mistaken. As important as this function of classification is, it is not the only one. To reduce the taxonomist to a filing clerk is to misunderstand his role. This would be even more true if the filing clerk–taxonomist were asked to file the items by superficial resemblance rather than on the basis of a thorough understanding of the contents. It would mean not only reducing taxonomy to a service function for other branches of biology, but also causing it to do this service badly.

4.5.1  **Scientific Theory.** For the scientist-taxonomist the most important meaning of a classification is that it is a scientific theory, with all the qualities of a scientific theory. First of all, it has an *explanatory* value, elucidating the reasons for the joint attributes of taxa, for the gaps separating taxa, and for the hierarchy of categories. It is precisely this explanatory property of evolutionary classifications which led to their rapid and almost universal acceptance after 1859. It is sometimes argued that the descriptive and the explanatory aspects of classifications should be neatly separated. This is impossible. A good classification of organisms is automatically explanatory.

The second property of a good classification is that it has a high *predictive* value. The common genetic program characteristic for the members of a natural taxon guarantees with a high probability that all the members of this taxon share certain characteristics. If I identify an individual as a thrush, I can make precise statements concerning its skeleton,

heart, physiology, and reproduction without ever testing it. A good classification predicts future experiences for the taxonomist. Indeed, one can test
the soundness of a classification by the ease with which it can accommodate
the findings derived from new characters and newly found species (Mayr,
1965b).

In turn the closeness of correlation between characters and classification permits conclusions on the genetic basis and biological significance
of characters. Characters controlled by one or few genes are usually irregularly distributed or limited to lower taxa. Characters that are shared by
most species of a higher taxon are believed to be the expression of a complex
genotype, particularly when the given character is not directly correlated
with utilization of the adaptive zone of the taxon. Like any scientific theory
a classification has a strong *heuristic* aspect. An evolutionary classification
stimulates efforts to recognize homologous structures and to test the concordance of various types of characters (e.g. gross morphological, cytological, biochemical, and behavioral). The advantage of this approach in protozoology has been well described by Corliss (1962a).

Finally, like any theory, a classification is *provisional*. The discovery
of new species and the availability of new character complexes is likely
to lead to a modification of the theory, that is, to an improvement of
the classification. Single-character identification schemes inevitably lead to
artificial groupings which have to be abandoned sooner or later. Yet even
the most sophisticated multi-character approach is provisional and subject
to future improvement. The demand for "final classifications" for the convenience of computer programmers can rarely if ever be fulfilled. In this
respect classification is not different from any other theory. All scientific
theories are provisional, subject to continuous testing, and rejected when
found to be wanting.

Recognizing a classification as a scientific theory also answers the
questions as to how important it is to have a classification, and whether
it makes any difference what kind of classification one adopts. Our discussions must have made it clear that the predictive worth of a classification
depends on the genetic homogeneity of the recognized groupings. Arbitrary
groupings have a very low predictive capacity. Consequently, it is indeed
important to have an "evolutionary" classification, that is, one based on
monophyletic groups that share much of their phenotype and genotype.

The question, Are all aspects of classification equally important? is
also sometimes asked. Surely classification at the species level is of first
importance because the results are of the most immediate concern to fellow
biologists in physiology, ecology, and behavior. The classification of the
classes and phyla, on the other hand, is of importance only for those who
ask phylogenetic questions, whether these concern macromolecules or
organ systems.

## 4.6  INFERENCES ON THE DELIMITATION OF TAXA

4.6.1    **Inferences on the Shared Genetic Program.** With few exceptions taxonomists agree that higher taxa (smaller or larger groupings of species) are the result of evolutionary divergence (descent with modification). The genetic program of the founder species of each evolutionary line gives a certain amount of genetic unity to its descendants. The genetic programs of the descendants will usually resemble each other more than the genetic programs of other phyletic lines not derived from this founder. If we knew the entire genotype of each organism, it would be possible to undertake a grouping of species that would accurately reflect their "natural affinity." Unfortunately, all attempts to determine the genotype of higher taxa directly have encountered insurmountable difficulties, among which three are preeminent:

1. Genetic analysis through crossbreeding is difficult, if not impossible, above the level of the species.
2. Correlation between observable phenotypic difference and genetic difference is rarely complete and often not even close. Intraspecific morphs are often far more different from each other than sibling species.
3. The characters of higher taxa are highly polygenic, but not even a guess is possible as to the number of genes involved.

Finally, the basic biochemical building stones of organisms are largely the same throughout the animal kingdom. Comparative biochemistry does of course permit an indirect genetic approach, because it can infer how many mutational steps separate the macromolecule of, let us say, a mammal from the homologous macromolecule of a lower invertebrate or microorganism (7.4.9). In the case of biochemical similarities, a particularly careful distinction must be made between similarities resulting from common descent and convergent similarities resulting from a response to similar selection pressures. Molecules are far simpler structures than anatomical structures or behavior patterns, and they may permit convergent evolution in nonrelated groups up to the point of essential identity. The independent evolution of a powerul nerve poison (Tetrodotoxin) in certain newts (e.g. *Taricha*) and fishes (particularly puffers) is an outstanding example. In principle, the most promising biochemical approach to the classification of the higher taxa is that of DNA matching (7.4.9), but the technical difficulties have not yet been overcome.

Recognition of the fact that characters are often the highly polygenic product of a complex genotype has resulted in a better understanding of two kinds of difficulties, the independent acquisition of a new character in parallel lines and the independent loss of characters in related lines

(10.2). Phenomena of parallelism always reveal a hidden genetic potential derived from a common ancestor.

**4.6.2    Inferences from the Phenotype.** Even though the direct genetic approach fails, there are a number of indirect methods which permit us to infer the genetic program of a taxon from a study of taxonomic characters. This process of inference is highly controversial, to a large extent owing to confused terminologies and unclear thinking. Simpson (1961, pp. 67–106) has clarified this complex problem in the most admirable manner, and we adopt his analysis almost without reservations.

A classification, as stated above (4.5.1), is a scientific theory. The members of a higher taxon are those species that show the greatest affinity with each other. Affinity, however, is not synonymous with superficial similarity, as is sometimes claimed. Nor is affinity, owing to the dual nature of phylogeny, synonymous with propinquity of descent, as believed by the cladists. How to determine affinity is the crucial problem in classification.

Phylogeny cannot be directly observed. It is something that happened in the past and must be reconstructed, it must be inferred from the available evidence. All scientific knowledge is in part based on inferences. As Simpson said (1961, p. 68), "This [dependence on inference] is not a feature peculiar to taxonomy. It is shared in greater or less degree by most of the inductive sciences." The Watson-Crick DNA model, the Krebs cycle, or indeed most findings of molecular biology cannot be observed directly. They are inferred from certain observations and are subsequently tested against further observations or experiments. It is therefore completely legitimate to define taxonomic categories in evolutionary (largely phylogenetic) terms, but to use evidence (comparative character analysis) that, as such, is almost entirely nonphylogenetic. Categories are defined in phylogenetic terms, but taxa are described by characteristics "that result from phylogeny and are evidence that the categorical definition is met" (Simpson, *loc. cit.*).

The evidence, used in the construction of classifications, consists of *taxonomic characters* (Chap. 7). Characters, however, differ in their phyletic information contents and must be evaluated accordingly. This procedure, called weighting, is treated in Chap. 10 (10.2).

## 4.7    THE POLYTHETIC CHARACTERIZATION OF TAXA

In classical taxonomy taxa were usually described by citing characters that were absolutely diagnostic for the given taxon. Linnaeus constantly revised his differentiae when the characters of newly discovered species showed that a character of a previously known taxon was no longer exclusive to that taxon. Higher taxa were characterized in terms of "characters in common." Sometimes a single character is fully diagnostic in this procedure, sometimes a set of characters. The combination of egg-laying, a flat bill, and certain anatomical characters is both sufficient and necessary for membership in the taxon *Platypus*. Beckner (1959, pp. 14–31) and Simpson (1961, pp. 42–43) have discussed such uniquely characterized taxa under the name monotypic, but since this term is used in taxonomy with a very different meaning, Sneath (1962, p. 291) has proposed the term *monothetic* for this concept.

After 1859 a new definition was adopted for what we now call a taxon. When the definition of the logicians—"individuals sharing common characters"—was replaced by "members of a group having descended from a common ancestor," a monothetic characterization of a taxon was no longer necessary. Actually, Adanson (in 1763) and other members of the empirical school had decided long before that a member of a taxon did not need to possess all the characters of the taxon and that such a deviant component of the taxon (e.g. species in a genus) did not need to be excluded and placed in a separate taxon. Taxa characterized by a set of characters of which each member has a majority are called *polythetic* taxa. For further details on this concept (called polytypic) see Beckner (*loc. cit.*) and Simpson (*loc. cit.*).

A higher taxon is polythetic if it satisfies the following three conditions:

1. Each species possesses a large (but unspecified) number of the total number of properties of the taxon.
2. Each property is possessed by a large number of the species.
3. No property is possessed by every species of the aggregate but is missing in the species of all other taxa.

Consequently, no single feature is essential for membership in a polythetically defined taxon nor is any feature sufficient for such membership.

Almost every monograph dealing with suprageneric taxa refers to characters that have some diagnostic value in conjunction with a couple of other characters but that are sufficiently irregular in distribution not to be reliably diagnostic by themselves. It is this incomplete correlation between characters and taxa which makes the construction of diagnostic keys so difficult. The fleas (Siphonaptera) supply abundant illustrations of suprageneric taxa that can be reliably defined only by combinations of characters, each one of which may also occur outside the given taxon or may occasionally be absent in a member of the taxon (Holland, 1964).

## 4.8  INFERRING RELATIONSHIP

Evolutionary classification demands the delimitation of taxa consisting of closest relatives. The evidence for relationship (10.1) consists primarily of weighted similarity. The interpretation of similarity is made difficult by a number of practical and theoretical problems which the taxonomist must clearly understand in order to avoid mistakes.

The polythetic diagnosis of taxa brings out the well-known fact that related taxa often overlap in some of their characteristics. This, together with convergence, is indeed the most formidable difficulty faced in routine taxonomy. It is one of the consequences of mosaic evolution that the character progressions of several character trends in a group are realized by the descendants independently and haphazardly. The living representatives of different phyletic lines derived from the same common ancestor may have a similar ratio of primitive and advanced characters, but a different assortment in each case (Table 4-1). The correct assortment of taxa under these circumstances depends on the finding of additional characters which will help in the weighting of the previously used ones. The methods of Maslin (1952), Wagner (1962), Hennig (1950, 1966), Wilson (1965), and Camin and Sokal (1965) are attempts to arrive at the construction of sound taxa in the face of widely overlapping character distributions (see Chap. 10).

**4.8.1    Homology.** The occurrence of convergence and various other kinds of similarity not resulting from common descent demonstrates that in the construction of classifications one must distinguish several kinds of similarities. Only similarities between homologous characters are of taxonomic importance. The term homology, like the terms species and classification, antedates evolutionary biology, but like these other terms it has acquired since 1859 a new, more precise, and biologically more significant meaning. Some workers have adopted different terminologies for structural and for nonstructural similarities, but with the ever greater utilization of nonmorphological characters in taxonomy, such a distinction would seem impractical and misleading [see discussions by Remane (1952, pp. 31–103),

TABLE 4-1.   MOSAIC EVOLUTION. Each of the six taxa shows a different assortment of primitive and advanced characters of features A–G.

| Taxon | Feature | | | | | | |
|---|---|---|---|---|---|---|---|
| | A | B | C | D | E | F | G |
| I | 1 | 5 | 3 | 2 | 4 | 1 | 6 |
| II | 3 | 6 | 1 | 4 | 1 | 3 | 3 |
| III | 6 | 4 | 1 | 5 | 2 | 2 | 5 |
| IV | 2 | 2 | 6 | 1 | 4 | 5 | 3 |
| V | 5 | 1 | 4 | 6 | 2 | 3 | 1 |
| VI | 1 | 2 | 5 | 6 | 3 | 5 | 1 |

1 = primitive, 6 = very advanced character.

TABLE 4-2.  DIFFERENCE BETWEEN HOMOLOGY AND ANALOGY

| Appearance | Derived from equivalent feature in common ancestor | |
| --- | --- | --- |
| | Yes | No |
| Similar | Homologous | Analogous |
| Not similar | Homologous | |

Simpson (1961, pp. 77–93), and Wickler (1961)]. With Bock (1963) we prefer to go back to Owen's two terms, but to define them as follows:

*Homologous* features (or states of the features) in two or more organisms are those that can be traced back to the same feature (or state) in the common ancestor of these organisms.

*Analogous* features (or states of the features) in two or more organisms are those that are similar but cannot be traced back to the same feature (or state) in the common ancestor of these organisms.

In the case of homology, similarity is not part of the definition because homologous structures are by no means necessarily similar (e.g. ear ossicles of mammals and the corresponding jaw bones in the lower vertebrates). Similarity must be referred to in the definition of analogy because non-homologous features that are not similar are not considered analogous (Table 4-2).

Having an unambiguous definition of homologous permits us to proceed in a similar way as in the application of the concepts biological species or biological classification. We must now seek the evidence that two features which we compare meet or do not meet our definition. Lists of criteria exist which help us in making the right decision; they will be discussed in Chap. 10.

**4.8.2  Phylogenetic Laws.** One of the reasons why the phyletic approach to classification came into discredit has been the reliance of some taxonomists on so-called phylogenetic laws and principles. Many of these are not only unreliable but totally false. Among so-called phylogenetic laws that should be rejected are the following:

1. Simple is always ancestral to complex.
2. Ontogeny (larval or embryonic stages) recapitulates phylogeny.
3. The "type" evolves harmoniously, and consequently all structures and organ systems evolve at equal rates.
4. There are goal-directed, teleological, evolutionary trends (orthogenesis).
5. New types of organisms originate by saltation.

Rensch (1947), Remane (1952, pp. 164–301), and Simpson (1953, 1961) have analyzed some of these so-called phylogenetic laws as well as additional ones not here enumerated. Hennig (1950, 1966) has attempted to develop new phyletic laws on the basis of cladism.

### SUMMARY

A classification based on phyletic weighting has numerous advantages. It is the only known system that has a sound theoretical basis, it has greater predictive value than other kinds of classifications, it stimulates a character-by-character comparison of organisms believed to be phylogenetically related, and it encourages the study of additional characters and character systems in order to improve the soundness of the classification and hence its information content and predictive value. Finally, it leads to the discovery of interesting evolutionary problems. Systems based on phyletic weighting thus not only have scientific advantages but are actually best able to answer the demands of the practice by having a greater total information content than artificial systems.

# Chapter 5 The Hierarchy of Categories and the Higher Taxa

*E*ach major group of animals can be subdivided into smaller and smaller subgroups. Within the vertebrates we can distinguish subgroups such as birds and mammals. Within the mammals, carnivores and rodents. Within the carnivores, those that are doglike, those that are catlike, and so forth. As Darwin said (1859, p. 411), "All organic beings are found to resemble each other in descending degrees, so that they can be classed in groups under groups," and if one wants to construct a classification of these species, "this classification is evidently not arbitrary like the grouping of the stars in constellations." The task of classification then is the delimitation of these groups and their arrangement in an orderly sequence. For this endeavor it is of the utmost importance that the student has a clear understanding of the meaning of the terms taxon and category (1.2).

## 5.1 HIGHER TAXA

The groups of species found in nature are higher taxa. Cats, carnivores, mammals, and vertebrates are higher taxa of different rank. The first step in classification, discussed in Chap. 4, is to determine which species show similarities indicating that they belong to one group, and furthermore to determine the delimitation of these groups, that is, the location of the discontinuities between neighboring groups.

*A higher taxon is an aggregate of related species separated from others by a discontinuity.* The ideal situation implied by this definition is not always met. For instance, in monotypic taxa the "aggregate of species" consists of a single species (because there are no close relatives), and the discontinuity, the gap separating one higher taxon from the next, varies greatly in extent. It is sometimes almost entirely bridged by intermediate species. See Chap. 10 and Simpson (1961, chap. 6). The definition fails to give any help in determining the rank of the taxon, because it is equally valid for higher taxa of all levels, from genus and family up to the phylum (see 5.4, 5.5, 5.6).

*The Meaning of Higher Taxa.* Most well-defined higher taxa, particularly at the genus and family level, occupy a well-defined niche or adaptive zone. They owe their origin to the invasion of this zone by a founder species and to the subsequent active and adaptive radiation which usually follows a successful adaptive shift (Simpson, 1953, 1959b, 1961; Mayr, 1960, 1963, chap. 19). It helps in the delimitation of higher taxa and in their ranking to be aware of their evolutionary origin. In recent years taxonomists have given much attention to the problem of the origin of higher taxa (for instance, Schaeffer and Hecht, 1965). It is advisable for two reasons to stress the ecological significance of higher taxa (5.3.2, 5.4.3, 10.4). Since the members of such a taxon are the descendants of the founder species that invaded the new adaptive zone, they clearly qualify as a monophyletic group. Secondly, since all the species occupy the same adaptive zone, in spite of some secondary radiation, they usually show a considerable amount of structural unity. Keeping this in mind may well prevent unnecessary "splitting" (10.3). Species or groups of species that are not separated by a well-defined discontinuity also fail, in most cases, to occupy a well-defined niche or adaptive zone.

As a heritage from the days when classification was considered synonymous with identification, there is an erroneous concept of the higher taxon, or rather the members of a higher taxon, as the carriers of an identifying character. A taxon is in fact a group of relatives, and whether or not they have the same "characters in common" is irrelevant. Many taxa are based on a combination of characters, and frequently not a single one of these characters is present in all members of the taxon, yet such a taxon may have a sound "polythetic" basis (4.7).

A group of related species consists of species descended from a common ancestor. Sound grouping of species is the indispensable foundation of a sound classification. In order to compensate for the possibility of the misleading effect of mosaic evolution, parallelism, and convergence, a careful phyletic weighting of numerous characters must be undertaken (10.4). The classifications of bees (Michener, 1944), Saturnid moths (Michener, 1952), and butterflies (Ehrlich, 1958) are exemplary analyses of this sort.

Where a fossil history is available, as in the case of mammals, its evaluation often leads to a better understanding of the higher taxa (Simpson, 1959b).

The practical aspects of the procedure of delimiting taxa and ranking them are treated in Chap. 10.

## 5.2  THE LINNAEAN HIERARCHY

The taxa of animals and plants, according to their comprehensiveness, are ranked in a hierarchy of categories. Within the animal kingdom the highest regularly used category is the phylum, and the lowest the species. Linnaeus, the first taxonomist to establish a definite hierarchy of taxonomic categories, recognized within the animal kingdom only five: *classis, ordo, genus, species,* and *varietas.* Two additional categories were soon generally adopted when the number of known animals grew—making finer divisions necessary: the family (between genus and order) and the phylum (between class and kingdom). The *varietas,* used by Linnaeus as an optional category for various types of infraspecific variants, was eventually discarded or replaced by the subspecies (3.3).

The remaining categories form the basic taxonomic hierarchy of animals. Any given species belongs thus to seven obligatory categories, as follows:

|  | WOLF | HONEY BEE |
|---|---|---|
| Kingdom | Animalia | Animalia |
| Phylum | Chordata | Arthropoda |
| Class | Mammalia | Insecta |
| Order | Carnivora | Hymenoptera |
| Family | Canidae | Apidae |
| Genus | *Canis* | *Apis* |
| Species | *lupus* | *mellifera* |

The basic five higher categories (genus, family, order, class, and phylum) permit the placing of a species of animals with a fair degree of accuracy. However, as the number of known species increased, and with it our knowledge of the degrees of relationship of these species, the need arose for a more precise indication of the taxonomic position of species. This was accomplished by splitting the original seven basic categories and inserting additional ones among them. Most of these are formed by combining the original category names with the prefixes super or sub. Thus there are superorders and suborders, superfamilies and subfamilies, etc. The most frequently used additional new category name is perhaps the term *tribe* for a category between genus and family. Vertebrate paleontologists also routinely use the category *cohort* between order and class. Some authors use terms for additional subdivisions, such as *cladus, legio,* and *sectio.* Some use infraclass below the subclass, and infraorder below the suborder.

The generally accepted categories are the following:


Kingdom
Phylum
Subphylum
Superclass
Class
Subclass
Cohort
Superorder
Order
Suborder
Superfamily (*-oidea*)
Family (*-idae*)
Subfamily (*-inae*)
Tribe (*-ini*)
Genus
Subgenus
Species
Subspecies


Indicated in parentheses are the standardized endings for the names of tribes, subfamilies, families, and superfamilies (13.33). Standardized endings for the categories above the family group have not yet been adopted in zoology.

The Linnaean hierarchy, with its need for arbitrary ranking, has often been attacked as an unscientific system of classification. Alternate methods, such as numerical schemes, have been proposed but have not found favor among taxonomists, primarily for two reasons. Assigning definite numerical values to taxa demands a far greater knowledge of the relationships of taxa than can be inferred from the available evidence. Secondly, an assignment of such values would freeze the system into a finality which would preclude any further improvements. It is the very subjectivity of the Linnaean hierarchy which gives it the flexibility required by the incompleteness of our knowledge of relationships. It permits the proposal of alternate models of relationship and gives different authors an opportunity to test which particular balance between splitting and lumping permits the presentation of a maximal amount of information. Like any other scientific theory it will forever be provisional (4.5). For the logical structure of the hierarchy, see Buck and Hull (1966).

## 5.3   THE HIGHER CATEGORIES

*Definition: A higher category is a class into which are placed all the taxa that rank at the same level in a hierarchic classification.* The category selected for a given taxon indicates its rank in the hierarchy. As explained in Chap. 1, taxa are based on zoological realities, categories are based on concepts. In that respect there is no difference between the category species and the higher categories from the genus up. In many other respects there is a great deal of difference between the concept of the species and the concepts of the higher categories.

The category species is "self-operationally" defined by the testing of isolating mechanisms in nature while nonarbitrary definitions for the supraspecific categories are not available. The species category signifies singularity, distinctness, and difference, while the higher categories have the function of grouping and ordering by not emphasizing differences between species, rather by emphasizing affinities among groups of species. They are collective concepts. Even though an operational definition for the higher categories does not exist, nor for the rank which they signify, they do have an objective basis because a taxon placed in a higher category (if correctly delimited) is "natural," consisting of descendants from a common ancestor. Higher taxa are often, if not usually, well delimited and separated from other taxa of the same rank by a pronounced gap.

Finally, comparative data furnish the evidence used for the delimitation of higher taxa and their ranking into categories, while interbreeding is the criterion used for ranking at the species level, because the species is a relational concept (Mayr, 1957b) and the higher categories are not.

Preevolutionary taxonomists, including Linnaeus, used higher categories, but they were unable to denote their significance in the framework of Aristotelianism, as was particularly evident for the categories above the genus level. They tried to explain the origin of the categories by deriving the lower ones from the higher through splitting, a remnant of Thomistic thinking. It was Darwin who supplied the scientific interpretation (1859, p. 422): "The natural system is genealogical in its arrangement, like a pedigree; but the degrees of modification which the different groups have undergone have to be expressed by ranking them under different so-called genera, sub-families, families, sections, orders, classes." The descendants of an aberrant species may evolve into a different genus, the genus in the course of geological history into a different family, and so forth. The origin of higher categories is thus exactly opposite from that envisioned by the scholastic philosophers. Higher categorical rank evolves through evolution, not lower rank through subdivision of higher categories.

Most taxa above the family level are sharply delimited. Mollusks, penguins, beetles, and indeed most higher taxa are separated from their nearest relatives by a decided gap, far more so than most genera and families. Nevertheless it remains true that the higher categories in which we place these taxa are ill-defined. Category means categorical rank, and no yardstick has yet been found for the nonarbitrary ranking of taxa. There is hardly a higher taxon that is not ranked higher by some and lower by other specialists. It is in the arbitrariness of definition that all higher categories differ from the species category. The criteria and operations used during the ranking procedure are discussed in detail in Chap. 10 (10.4).

## 5.4   THE GENUS

**5.4.1   Definition.** The genus is the lowest higher category and the lowest of all categories established strictly by comparative data (Cain, 1956). For the modern taxonomist the genus is no different in concept from family, order, or other higher categories. For Linnaeus, who based his theory of classification on the principles of Aristotelian logic, the genus occupied a very special place (Cain, 1958). This fact would have only historical interest for us if Linnaeus had not incorporated Aristotelian logic into the binominal system of nomenclature.

Since there is no operational definition available for any of the higher categories, one is forced to adopt a pragmatic definition: *A genus is a taxonomic category containing a single species, or a monophyletic group of species, which is separated from other taxa of the same rank [other genera] by a decided gap.* It is recommended for practical reasons that the size of the gap be in inverse ratio to the size of the taxon. In other words, the more species in a species group the smaller the gap needed to recognize it as a separate taxon, and the smaller the species group the larger the gap needed to recognize it. One of the functions of the genus, from Linnaeus' time on, is to relieve the memory (to facilitate information retrieval), and the "inverse ratio" recommendation prevents the recognition of a burdensome number of monotypic genera. To delimit as genera species groups of optimal size is an operation that requires experience, good judgment, and common sense. For a discussion of helpful criteria, see Chap. 10 (10.4).

An equivalent, nearly identical, pragmatic definition is applicable to the categories above the genus—family, order, class, etc.

In order to qualify for a given rank, a taxon must satisfy a number of conditions (see 10.5). It must be sufficiently different from other taxa of the same rank; it must be separated by a discontinuity; it should occupy a distinctive niche or adaptive zone; and in the absence of a marked discontinuity it should not display too great an internal diversity (heterogeneity). Finally it should, if possible, satisfy certain practical requirements, in consequence of which the recognition of a higher taxon is often a balanced compromise between the stated qualifications. Only the ideal genus is well separated by a gap, is of the proper size, is internally homogeneous, and fills a distinctive adaptive zone.

For the nomenclature of generic names see 13.38, see also 13.31 (gender) and 13.39 (collective groups).

**5.4.2   Generic Characters.** Taxonomic characters that prove generic distinctness do not exist (7.5). Taxonomic literature would have been spared innumerable generic synonyms if taxonomists had always remembered Linnaeus' (1737) dictum: "It is the genus that gives the characters, and not the characters that make the genus." This, in a sense, is still generally valid, even though we have abandoned the Aristotelian logic on which Linnaeus based his statement. The soundest genera are based on an overall appreciation and weighting of the various considerations previously cited (Michener, 1957).

The species included in a genus usually have many features in common, thus facilitating its delimitation. Recognition of a higher taxon is generally based on the occurrence of correlated character complexes (see 10.4). These may include some rather minute and inconspicuous characters, but, as Darwin said (1859, p. 417), "The importance, for classification, of trifling characters mainly depends on their being correlated with several other characters of more or less importance. The value indeed of an aggregate of characters is very evident in natural history." So important has this principle been considered by taxonomists that it led to much generic splitting when a taxonomist found a species which lacked one or another character of the correlated complex. Instead of revising his image of the genus, he named a new one.

Some genera are clearly natural groups yet cannot be diagnosed unequivocally by a single character. This occurs because every character, even though diagnostic for the majority of the species, is modified or absent in at least one or the other species of the genus (4.7). This is true, for example, of many genera and even families of birds.

**5.4.3   Meaning of the Genus.** When we assign generic rank to a group of species, we want to express a number of things that are characteristic of all the higher categories. A genus taxon is a phylogenetic unit, which means that the included species are descended from a common ancestor. Almost invariably it is also true that the genus is an ecological unit consisting of species adapted for a particular mode of life. The genus niche is obviously broader than the species niche, but both kinds of niches exist. Lack (1947) has convincingly shown the adaptive significance of genera for the Galapagos finches (Figs. 5–1, 5–2).

Like any operation in classification, recognition of a genus taxon cor-

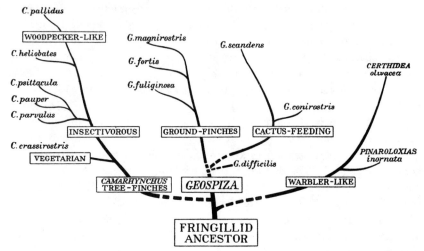

**Fig. 5-1.** Adaptive radiation of Darwin's finches (Geospizinae) on the Galapagos Islands into a number of different niches *(from Lack, 1947)*.

responds to the proposing of a scientific theory. Like all scientific theory it must have explanatory, heuristic, and predictive value (4.5). If there are several alternate ways of delimiting genera, we must be guided by the same principles as in the recognition of any scientific theory. "Where alternatives are available, we stand by the theory or concept that is most useful—the one that generalizes the most observations, and permits the most reliable predictions" (Inger, 1958, p. 383). For comments on the generic concept, see also Rosen and Bailey (1963).

## 5.5   THE FAMILY

As in the case of the genus and the other higher categories, it is not possible to give a nonarbitrary definition of the family category. What a lay person would often designate a "kind of animal" is often a family: The ladybird beetles (Coccinellidae), the long-horned beetles (Cerambycidae), the woodpeckers (Picidae), the swallows (Hirundinidae), and so forth. How distinctive a group of genera must be in order to be considered a family varies from one zoological group to another, for the various reasons previously indicated. If one wants to give a definition of the family category, it would be equivalent to that of the genus:

*A family is a taxonomic category containing a single genus or a monophyletic group of genera, which is separated from other families by a decided gap.* It is recommended, as in the case of the genus, that the size of the gap be in inverse ratio to the size of the family.

Like the genus, but perhaps to an even greater degree, the family is usually distinguished by certain adaptive characters which fit it for a particular niche or adaptive zone, e.g. the woodpeckers of the family Picidae, the leaf beetles of the family Chrysomelidae, etc. In most cases, families obviously are older than genera and have more often a worldwide distribution. An entomologist who knows the 422 families of British insects can go to Africa, or even Australia, and recognize nearly all the same families occupying similar niches.

Thus the family is a very useful category. The British entomologist would have to learn only 422 names to place a total of about 4,767 genera and 20,244 species. It is especially useful to the general zoologist because each family usually presents a general facies which is recognizable at a

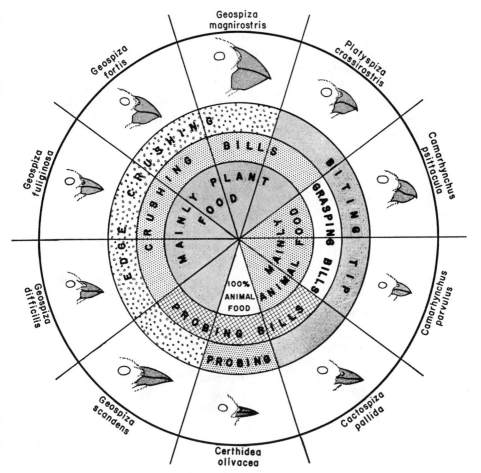

Fig. 5-2. Niche occupation, feeding habits, and bill structure in ten species of Geospizinae from Indefatigable Island (*from Bowman*, 1961).

glance, and all its species occupy a similar niche in their particular community, as, for instance, most of the thousands of species of Cerambycidae (long-horned beetles) in the world.

At any given locality the various families, like the various species, are generally distinct. Decided gaps between families are the rule rather than the exception, and little or no difficulty is encountered in "keying out" families in local faunal works. Unfortunately, the situation becomes much more complicated when a worldwide study is undertaken. Families are often found to break up into different distinctive groups on each continent, and types bridging the gap with other families are sometimes found. Relict groups may exist at the family level and defy efforts to attain a clear-cut classification. Thus, in many insect groups (scale insects, aphids, water striders, etc.) a choice has had to be made between enlarging the family concept beyond the limits of local convenience or recognizing connecting exotic types as separate families and using a superfamily category for the group as a whole. In entomology there appears to be a trend, not necessarily desirable, in the direction of the second of these choices. In ornithology a knowledge of the tropical relatives of the Temperate Zone forms has led to a reduction in the number of families. For instance, a study of tropical genera has induced many authors to consider the Old World flycatchers (Muscicapidae), warblers (Sylviidae), and thrushes (Turdidae) as only subfamilies of a more broadly conceived family, Muscicapidae.

Linnaeus did not recognize the family as a category, but it is significant that most of his genera have since been elevated to the rank of families. From this we may infer that his generic concept was not incompatible with our modern family concept, the difference between the genus and family being merely one of degree. With only 312 genera of animals in 1758, Linnaeus had no need for an intermediate category between genus and order. However, the number of newly discovered animal types increased so rapidly that the early nineteenth-century naturalists gradually evolved and universally applied the family concept to designate an intermediate level between genus and order.

The number of families continues to grow because of the advance in knowledge of existing animals and the discovery of new types. Thus by the end of the nineteenth century approximately 1,700 families of animals were recognized (Perrier, 1893–1932, *Traité de Zoologie*). That the trend is continuing is indicated by the fact that Brues, Melander, and Carpenter (1954) recognize 941 families of insects alone.

For matters relating to the nomenclature of families, see Chap. 12 and 13.30 (stem of family name), 13.33 (family names), 13.35–36 (coordinated and subordinate taxa), and 13.37 (homonymy).

## 5.6    ORDERS, CLASSES, AND PHYLA

The higher taxa above the family level are on the whole very well defined in the Recent fauna, and much less often connected by intermediates than families or genera. There are two exceptions to this broad statement. First, there is still doubt as to the significance of taxonomic characters in certain groups of lower invertebrates, for instance the sponges and the turbellarians. In some cases there has been a complete reclassification on the ordinal level, owing to the reevaluation of the weight of certain characters. The second reason is that even where there has been complete consensus as to the delimitation of the taxa, there has been strong disagreement as to the ranking. Instead of recognizing more suborders and superfamilies, certain authors have raised almost all taxa in rank, which has resulted in a great imbalance of the respective portions of the system (see 10.3, 10.4).

The taxa ranked in higher categories represent the main branches of the phylogenetic tree. They are characterized by a basic structural pattern laid down early in evolution, the special adaptive significance of which can now be perceived only dimly, if at all. Superimposed on it are seemingly endless adaptive modifications resulting from series of adaptive radiations that have taken place in the classes and phyla. In general, then, taxa in the higher categories are definable in terms of a basic structural pattern, but except for certain highly specialized groups, such as the order Siphonaptera (fleas), the order Chiroptera (bats), the order Impennes (penguins), etc., the higher taxa are not primarily or even predominantly distinguished by ad hoc adaptations. The taxa contained in the higher categories are in most cases widely distributed in space and time. For the names of higher taxa, see 13.34.

As in the case of genera and families, there has been a trend toward increase also in the number of recognized taxa above family rank. According to recent tabulations there are approximately 25 phyla, 80 classes, and 350 orders of Recent animals.

## 5.7    PRACTICAL CONSIDERATIONS IN THE CONSTRUCTION OF A CLASSIFICATION

Biological classification, as stressed in Chap. 4, is based on the fact that evolution produces groups of more nearly or more distantly related species. Evolution thus sets severe limits to the number of possible classifications. Nevertheless, even where there is complete consensus about the phylogeny of a group of organisms, it is nearly always possible to translate this knowledge into a number of alternate classifications. The reason for this is that three operations of the taxonomist cannot be carried out without an element of arbitrariness: (1) the delimitation of groups which we formally recognize as taxa (their "size"), (2) the rank in the hierarchy which we assign to a given taxon (e.g. tribe, subfamily, family), and (3) the position in the sequence of taxa. The necessity of having to translate a multidimensional phylogenetic tree into a linear sequence permits many alternate arrangements (Chap. 10.7). A taxonomist will be the more successful in his classification the more clearly he realizes that it is the major function of a classification to be useful. A classification is a communication system, and the best one is that which combines greatest information content with greatest ease of information retrieval. For instance, a classification which attempts to express every possible shade of relationship (by "splitting") makes information retrieval exceedingly difficult and hence defeats its own purpose. The taxonomist must remember that the key to a filing system does not contain all the information contained in the file.

The specific points that need to be taken into consideration in the making of a useful and stable classification are discussed in Chap. 10.

# Part II  Methods of Zoological Classification

# Chapter 6  Taxonomic Collections and the Process of Identification

$A$ ll classification is based on the comparison of specimens representing populations and species. We can determine the species-specific characteristics of a species only by comparing it with other similar species, preferably its nearest relatives. An adequate comparative collection is therefore as indispensable to the taxonomist as are electron microscopes, Warburg apparatuses, ultracentrifuges, and similar equipment for the cellular and molecular biologist.

Collections must either be borrowed from museums, which are repositories of systematic collections, or collected by the specialist himself. In most cases both sources must be tapped. Borrowed material is usually insufficient for certain crucial areas and does not give the biological information which is so vital in modern taxonomic research. On the other hand, it would require many years of effort for a single collector to achieve the broad geographic scope available in a museum collection accumulated during scores of years, possibly centuries.

## 6.1  SYSTEMATIC COLLECTIONS

**6.1.1  Value of Collections.** Museums serve an important role as centers of documentation. They supply a permanent record of faunas and floras, particularly where the biota has since been destroyed by natural catastrophes or the activities of man. This is particularly true of the localized biota of streams, lakes, and islands. Museums contain a sampling of many areas that are more or less inaccessible owing to their remoteness or for political reasons. Much of the material now preserved in various institutions can be replaced only at very high cost, or not at all. Other material is of unique value because it forms the basis of published research. It may be needed again at a later period for verification of the original data or for renewed study in the light of more recent knowledge or new techniques. For a survey of the scientific significance of taxonomic collections, see *Biological Material* (Mayr and Goodwin, 1955).

A taxonomic revision of a given group is possible only if adequate material of the majority of included species can be assembled. It is therefore legitimate, indeed necessary, that curators of museums build up their collections in groups which they or their associates plan to study in the future. The working taxonomist knows from experience how many revisions were started, only to be set aside for lack of adequate material. Currently unused material in a collection may be dormant, but this does not mean that it is dead. Collections are reference material, and in that respect they are just as necessary as books in a library, which likewise are not in continuous use yet must be available when needed.

6.1.2 **Purpose of a Scientific Collection.** For the old-style, typologically oriented taxonomist a collection was an identification collection. When he obtained additional specimens of species already represented in the collection, he considered them duplicates to be used for exchanges or to be given away. According to current thinking, biological classification is the ordering of populations (Chap. 4). Collecting, then, is the sampling of populations. Considering the great variability of most natural populations (Chap. 8), an adequate sample of every population should be collected and preserved (6.1.6).

But what is "adequate"? There are compulsive collectors, veritable pack rats, who insist on storing thousands upon thousands of specimens of each species. The ideal is somewhere between the typological extreme of preserving only a typical specimen of each species and the hoarding tendency of the compulsive collector. In making the right decision one must be guided by various considerations. More material is needed in a species with strong individual and geographical variation than in a uniform species. More material is needed for studies of specific and subspecific characters than of the characters of higher taxa. Birds, for instance, are on the whole more uniform in their anatomy than in their plumage characters. It would be illogical for this reason to preserve large series of spirit specimens of every species, such as are needed for the study of the geographic variation of plumage characters.

The size of the sample thus depends on the objectives of the research.

The presence or absence of an anatomical character diagnostic for a higher taxon is in most cases not subject to much variation, and one or two preserved specimens of a number of genera may be all the working material needed. However, when a population analysis is attempted, whether concerned with a study of size, proportions, coloration, or polymorphism, large samples from numerous localities are needed. In fishes, for instance, where each population of a species may have different statistics for various meristic characters (fin rays, vertebrae, etc.), a single specimen per locality would tell us very little. All phena of a species should, so far as possible, be represented in the collection.

**6.1.3   Collecting and Research.** On the whole, taxonomists spend only a small fraction of their time on the collecting of new material. Likewise, most specialists have caught up reasonably well with material they themselves collected. There are exceptions. Some big expeditions of the nineteenth century gathered material that is still not yet fully worked out. Some taxonomists cannot resist the temptation to use the summer to go in the field hoping to work out the material during the ensuing winter. However, those who are teaching at universities are often too busy during the academic year to continue their taxonomic research. A moratorium on further collecting is advisable in such cases to prevent too large an accumulation of unopened packing cases from previous field seasons. Large oceanographic expeditions also have a tendency to accumulate far more material than the extremely small number of specialists can work up at once. However, even such material is generally processed so as to make it readily available to specialists (see 6.3.1).

**6.1.4   Scope of Collections.** Only a few large national museums attempt worldwide coverage in all groups of animals. Most museums are largely restricted to a geographic area and to certain groups of animals. It is most important for the staff of a museum to have a clear-cut acquisition policy. Too broad a coverage inevitably leads to shallowness and a failure to obtain the depth required for monographic studies. There has been a steady and wholesome trend away from broad-purpose expeditions and faunal surveys in favor of intensive collecting of specific families or genera. The late Admiral H. Lynes, for instance, who was especially interested in *Cisticola,* a genus of African warblers with some 40 species, made a whole series of collecting trips to nearly every corner of Africa. He combined the collecting of specimens with a detailed study of the ecology, habits, songs, and nest construction of these birds. The result was that the genus *Cisticola,* formerly the despair of the bird taxonomists, is now reasonably well understood (Lynes, 1930). The work of J. Crane on the fiddler crabs (*Uca*) is another example.

A number of considerations are important if a faunal survey is attempted. It should be directed to a natural geographical area and not be hampered by a blind dependence on political boundaries. Faunal surveys have had the best results when concentrating on a particular taxonomic group. The most ambitious single undertaking along these lines in recent decades was probably the Whitney South Sea Expedition, operating under the auspices of the American Museum of Natural History in New York. This expedition visited practically every island in the South Pacific from the Tuamotus and Marquesas in the east to the Bismarck Archipelago, Palau, and the Marianas in the west, largely concentrating on bird collecting. It operated continuously from 1921 to 1934, and its work was continued by single collectors into 1940. The essentially complete collections made by this single expedition supplied the material not only for scores of detailed revisions but also for basic zoogeographic and evolutionary studies, such as *Systematics and the Origin of Species* (Mayr, 1942). The value of a single well-made collection is generally far greater than that of an equivalent number of specimens in casually made collections.

   **6.1.5   Where and How to Collect.** Any collecting trip must be carefully planned. All possible geographic information must be obtained beforehand, including the distribution of vegetation types, altitudes, seasons, means of public and private transportation, etc. In addition, previous collections must be carefully analyzed and existing type-localities mapped. A plotting of collecting stations and, in particular, a mapping of species distributions will reveal the location of crucial gaps. If the study of geographic variation is a major objective, the periphery of the range of each species should be given particular attention. This is where geographical isolates and incipient new species occur most frequently. If a species shows seasonal variation, the collections should be spaced seasonally. The season during which sexual maturity occurs is relatively short in many animals, so collecting should be done during that time. This is even more important if recordings of song and courtship, nest structure, egg proteins, embryos, or other materials are needed that can be obtained only during the breeding season. In the case of allopatric populations, the categorical status of which is uncertain (species or subspecies), a special effort should be made to collect in the intervening region to determine whether or not intergradation occurs. Owing to the rapid recent increase of human populations, resulting in more intensive agriculture and in drastic deforestation, there are many areas that are in desperate need of immediate collecting before the localized faunas become extinct. At the present time this task is far more urgent than collecting in remote uninhabited areas.

   Innumerable techniques for the collecting of different groups of animals are described in standard collecting manuals. New techniques are continually being developed, such as the use of mist nets for bird collecting and of "black light" (ultraviolet lamps) for insect collecting. Different kinds of traps, baits, poisons, and so on are well known to specialists, who generally share such information with beginners. Depending on the taxonomic group to be covered, one or the other of the following books (of which the full title can be found in the terminal bibliography) will be most useful:

Anderson, 1948 (vertebrates), Anthony, 1945 (mammals), Beer and Cook, 1958 (ectoparasites), Bianco, 1899 (marine animals), British Museum, 1936ff (various groups), Kirby, 1950 (protozoans), Kummel and Raup, 1960 (fossils), MacFadyen, 1955 (soil arthropods), Oldroyd, 1958 (insects) Peterson, 1934, 1937 (insects), Oman and Cushman, 1946 (insects), Russell, 1963 (marine), Van Tyne, 1952 (birds), Wagstaffe and Fidler, 1955 (invertebrates). Knudsen (1966) contains many references to special techniques.

**6.1.6   Contents of Collections.** The classical image of a systematic collection is that of preserved specimens, either dried or submersed in a preservative such as alcohol. Even though such specimens are the indispensable basis of all taxonomic research, the information which one can derive from dead, preserved specimens is limited. The modern systematist needs a great deal of additional information, much of which he has to collect himself while studying the living organism, either in its native environment or in the laboratory.

As far as possible, the collecting should provide unbiased population samples. No effort should be made to accumulate large numbers of "aberrations" or, in sexually dimorphic species, to concentrate entirely on the conspicuous sex. Not only adults should be collected, but adequate samples of all growth stages (including larvae) and associated parasites. Sampling should be done in such a way as to provide study material not only for the species describer but also for the evolutionist.

Collections of specimens may be augmented by collections of all sorts of recordings. These include films of courtship displays and other aspects of behavior; recordings of the vocalization of animals (sound libraries, tapes, sound spectrograms); collections or photographs or casts of the work of animals (nests, galls, spider webs, tracks, etc.). Collections of whole animals must be supplemented by collections that permit histological, cytological (chromosomal), and biochemical research (see 7.4).

To obtain such additional material requires, as a minimum, not only field work, but in many instances prolonged stays at a field station, particularly in the tropics, subtropics, or the Arctic. Museums increasingly include aquaria, terraria, insectaries, aviaries, and the like among their facilities in order to permit study and experimentation with living species. The museum of today is fully aware that the study of the diversity of nature requires a far broader approach than was envisioned by taxonomists of bygone generations.

**6.1.7   Preservation of Specimens.** This differs from one taxonomic group to the next. The basic rule is to preserve specimens in such a way as to make them least subject to any kind of deterioration, whether through the action of insect pests, mold, oxidation or bleaching by sunlight, drying out, protein decay, etc. Again we must refer to the appropriate manuals and handbooks for a description of specific methods (see above under 6.1.5). Properly preserved specimens of some groups are still in satisfactory condition 200 years after being collected. With more and more species becoming extinct, the problem of "permanent" preservation is raised increasingly often. Some recently suggested methods, such as embedding in plastics, are too new to permit predictions. Preservation in alcohol raises problems with sealing and appropriate containers, as discussed by Storey and Wilmowsky (1955) and Levi (1966).

    **6.1.8    Labeling.** A specimen that is not accurately labeled is worthless for most types of taxonomic research. So important, in fact, is the label that it is sometimes stated jocularly that the label is more important than the specimen. Many kinds of information are desirable, but by far the most important single piece of information is the exact locality of collection. In forms like certain land snails that may have racially distinct populations as little as 0.5 km apart, the locality must be stated with great precision. If the locality is a small community, farm, hill, creek, or other geographical feature which cannot easily be found on commercial or geodetic (e.g. United States topographic) maps, its position relative to a well-known place should be added on the label ("15 mi. NW of Ann Arbor, Mich."). The county or district name should be given with all less well known localities. If the specimen was collected in mountains, the altitude should always be given, and if in the ocean, the depth. Additional ecological information is valuable or even essential in forms like plant-feeding insects or host-specific parasites.

    Whenever possible, the label should be written in the field at the time the specimen is prepared. Any replacing of temporary labels with later permanent ones is a potential source of error. However, this cannot be avoided with insects when labels are printed for entire lots of specimens. All essential data should be recorded on the original labels. Data recorded in a field book are frequently overlooked and may be unavailable if the collection is divided. The original label should never be replaced by a museum label. A certain number of mistakes are always made in the transfer. If a museum label is desired, it should be added to the original label. Labels attached to specimens preserved in alcohol or formalin must be corrosion-proof. Any writing must be resistant to fading or washing out.

    What data, in addition to locality, are needed depends on the given group. Most good bird collectors, for instance, record on the label not only locality, date, and collector's name, but also the sex (based on autopsy), the actual size of the gonads, the degree of ossification of the skull (important for age determination), the weight (in grams), and the colors of the soft parts (Van Tyne, 1952).

The little extra time required to make these records is more than compensated for by the increased value of the specimens.

## 6.2  CURATING OF COLLECTIONS

Every taxonomist, sooner or later, is given the responsibility of curating collections. This requires a great deal of expert knowledge, but also, more importantly, a clear understanding of the function of collections, of the different kinds of collections needed in various areas of taxonomy, and of various policies concerning the use of collections. The journal *Curator* publishes much information on this subject.

6.2.1  **Preparation of Material for Study.** Bird and mammal skins are ready for study as sent from the field by the collector. Mammal skulls have to be cleaned. Some insects should never be placed in alcohol or other liquid preservatives; others are useless when dried. Invertebrates that are preserved in alcohol or formalin are usually ready for study as preserved. Microscopic slide mounts or slides of part of the organs may have to be prepared for the smaller forms. Instructions may be found in textbooks of microscopic technique [Clarck (1961), Francon (1961), Gray (1954, 1958), Jones (1950), Lee (1950), Needham (1958)]. Most insects are pinned, and the wings are spread if they are taxonomically important (or beautiful), as in butterflies and moths and some grasshoppers (for references, see 6.1.5). Species can be identified in many groups of insects only by a study of their genitalia. Microscopic slides or dry or liquid mounts of the genitalia may have to be prepared. Special techniques are needed for the preservation of protozoans (Corliss, 1963).

6.2.2  **Housing.** Research collections should be housed like research libraries in fireproof buildings that are reasonably dustproof. More and more museums keep their collections in air-conditioned buildings. Buildings originally designed for public exhibits are usually quite unsuitable for the housing of research collections. Rapid changes in temperature and humidity are detrimental to museum cases and to specimens. Photographs and films should be stored in air-conditioned rooms. Storage cases for specimens should be built sufficiently well to be insect-proof and, ideally, also dust-proof. Various firms now construct steel cases that satisfy all these requirements. It must be remembered that insect-proof museum cases reduce the labor of curating.

6.2.3  **Cataloging.** The method of cataloging depends on the group of animals. With the higher vertebrates, where collections consist of a limited number of specimens, it has been traditional to give each specimen a separate number and to catalog it separately. This very time-consuming procedure has been challenged, and it has been suggested that the methods of entomology and malacology be adopted for vertebrates (see below). All the specimens collected at a given locality or district or by one expedition are entered in the catalog together. This greatly facilitates the subsequent retrieval of distributional data and the preparation of faunisic analyses. Cataloging is usually done after the specimens have been identified, at least as far as the genus. This permits a permanent reference to the contents of the collection long after it has been broken up and distributed in the systematic collection or even dispersed to other institutions.

Catalog entries of vertebrates usually contain the following items:

1. Consecutive museum number
2. Original field number
3. Scientific name (or, at least, generic name)
4. Sex
5. Exact locality
6. Date of collecting
7. Name of collector
8. Remarks

In groups where the collections consist of large numbers of specimens, as, for example, insect collections, where additions of 100,000 specimens per year are not uncommon, it is customary to catalog accessions by lots, each lot consisting of a set of specimens from a given locality or region. Lot numbers, in turn, may refer to collector's diaries or to other sources of information on each collection. It is also customary to note whether a lot was received as a gift or by purchase or exchange; the names of the collector and donor are always given.

When museums and their collections were small, curators often had rather elaborate card-filing systems permitting easy retrieval of all sorts of information such as collecting station, name of the collector, etc. Maintenance of such files is so expensive in clerical as well as curatorial time that they have been abandoned in most museums where the emphasis is on research rather than providing information to the public. A properly organized and well-curated collection is a reference catalog in itself and permits rapid information retrieval. Experiments are now being made in some museums to place all information on each specimen on a separate IBM card. Study of pilot projects will reveal whether or not this plan is feasible (Squires, 1966). There is great danger that the preparation of the cards will cost more in staff time than it will eventually repay. Yet electronic data processing (EDP) is rapidly maturing, and taxonomists and curators must keep up to date with its advances.

The maintenance of catalogs and card files is very time-consuming and should never be carried to the point where it interferes with work on the collections. A list of the accessions is indispensable, however, since it often allows the recording of additional information on localities which cannot be entered in full on the specimen labels.

6.2.4    **Arrangement of the Collection.** As far as possible, the collection should be arranged in the same sequence as some generally adopted classification. The sequence of orders and families is reasonably standardized in many classes of animals. Unidentified material, whenever it is not to be worked out as a separate collection, is placed with the family or genus to which it belongs (see later in this chapter, under identification). The contents of trays and cases should be clearly indicated on the outside. The bird collection of the American Museum of Natural History lists on the tray labels the scientific names not only of all the available species and subspecies but also of those still lacking in the collection (indicated as such) (Fig. 6-1). The names on the collection cases and trays thus serve as a checklist of all the known species and subspecies of birds. Of course such a system is impossible in most groups, where the collections are far too incomplete. Where specimens are of highly unequal size (e.g. fish), storage in a strict taxonomic sequence is very wasteful of space. Extralarge specimens may have to be stored separately.

6.2.5    **Curating of Types.** The names of species are based on type-specimens (see type method, 13.48), which are thus official standards and, being virtually irreplaceable, must be curated with special care. Whenever

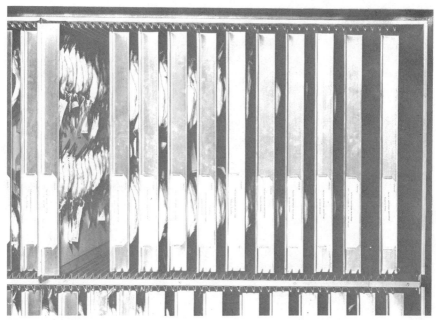

Fig. 6-1. Method of storing study specimens of birds in open trays (collection case in the American Museum of Natural History).

doubt arises as to the zoological identity of a nominal species, only reference to the type can resolve the doubt.

Many descriptions of classical authors are equally applicable to several related species. Early entomologists rarely, if ever, referred to the structural detail of the genitalia, now so indispensable for diagnostic purposes in most groups of insects. Only reference to the type can establish the basis for the classical name. The curator is responsible for making every effort to ensure the safety of these irreplaceable specimens. Types are customarily deposited in large collections of public or private institutions which have come to be recognized as standard type repositories.

When conducting an authoritative revision of a given genus, a specialist should be able to see all the existing types. If many of them are in a single institution, he should travel there for examination. He should obtain scattered types by mail loan. Modern curators are quite liberal in lending type specimens to qualified specialists. The number of recorded losses of shipments is very small, and if there is a real need for the replacement of a lost type, a neotype can be designated, as permitted by the Code under specified circumstances (13.55). Ideally types should be housed in a separate collection in order to facilitate rapid removal in case of emergency and to avoid the constant handling in a general-study collection. They should be clearly labeled in distinctive colors. If not previously cataloged, they should be numbered individually to facilitate referring to them in the literature and finding them in the collection. Since many types are those of synonyms, a card index by genera and another by species will save much time in locating the desired type. It is not economical to include in type collections all sorts of pseudotypes—that is, specimens that are not name bearers, in other words not holotypes, lectotypes, neotypes, or syntypes (13.48).

There is much to recommend the arrangement of type-collections alphabetically according to the given specific name. A type-collection is a reference collection (rather than a classification), and ease of reference should determine what system is adopted. When number systems are adopted, too many errors are usually committed for them to be practical.

Type-specimens assume such an important role in the taxonomy of lesser-known groups that many workers believe no one has the right, ethically, to retain a type in a private collection after study has been completed. Specialists sometimes donate their collections to a public institute but retain them on a permanent loan up to the end of their active research period. This includes the obligation to send out specimens on loan if such requests are made by cospecialists.

6.2.6    **Exchange of Material.** In the days when taxonomists considered most of their specimens "duplicates," exchanges were very popular. However, selecting material for exchanges and keeping records required by exchanges is very time-consuming, and except among private collectors exchanges are not as popular as they used to be. A specialist doing a monograph on a certain genus or family can always borrow material from other institutions and return it when he has completed his studies. Exchanges are most desirable in groups where series of unlimited size can be obtained and where the exchanges concern areas not readily accessible to the respective institutions (e.g. intercontinental exchanges). Exchanges are least defensible when they lead to the dispersal of biologically important population samples. Exchanges are sometimes necessary to build up complete identification collections. When an exchange is carried out, generosity is always the best policy. It is generally not advisable to insist upon exchanging specimen for specimen, except where institutional policies or other unusual factors demand it. Many specialists give away excess specimens of a large series as "open exchanges," not necessarily expecting any return.

**6.2.7   Expendable Material.** Improperly preserved or inadequately labeled specimens are usually valueless. Collections would be better off without them. Yet they are much less of a burden on a collection than would be the task of the curator to find and eliminate them. He must make sure with every specimen that it is not an unlabeled type or a specimen which, in one way or another, is unique or of historical value. The most efficient method for the elimination of useless material is to ask specialists to pull out such specimens when scrutinizing the material during a revision.

**6.2.8   Loans.** Modern curators are very generous in lending specimens to qualified experts. The axiom that systematic collections are the general property of science, not of a specific institution or curator, is being acknowledged more and more widely. Every loan, however, involves loss of time and effort, and the borrower should refund the lender for this outlay. Research grants now often include an item for such purposes, to cover not only the costs of postage but also of selecting the specimens, recording the loan, and getting the material packed for shipment. The modern curator, being essentially a research worker, must delegate these tasks to hired clerical help.

The borrower also has well-acknowledged obligations. Requests for loan of specimens should be as specific as possible, including a statement of the reason for the request and some indication as to the length of time for which the material is needed. The beginner may be unable to borrow certain material except through a loan to his institution, or a well-known colleague, or his major professor. In such cases any laxity in carrying out the conditions of the loan reflects not only on himself but also on the sponsoring individual or institution. If it should develop that the borrower is unable to complete his studies in the time designated, he should inform the party or institution which made the loan, without waiting and placing them in the embarrassing position of having to write and ask about the status of the study. The borrower should not ask for material that he does not actually need, nor should he ask for material that he could easily study by traveling to the museum.

If a specialist has agreed to identify a collection provided he receives certain specimens, he should make sure that the terms of the agreement are well understood and should return to the lender a list of the specimens which he has retained. All types and unique specimens must be returned to the lender in such cases. In the case of anatomical material, it is understood that dissection, the purpose for which the anatomical specimen was originally collected, will partially or entirely destroy the specimen. The borrower, in this case, is under obligation to preserve a pictorial record of the dissection. To cut down on loans, institutions sometimes make temporary or permanent transfers of collections. For instance, an inland museum, which has a small collection of marine invertebrates not used for exhibition or instruction may transfer it to a large museum that is active in marine research; but excessive consolidation of taxonomic collections would create not only monopolies but also an inherent danger. The concentration of so much irreplaceable material in a single institution makes it exceedingly vulnerable to destruction in the event of a catastrophe. Long-term loans to leading specialists are in most cases a better solution.

## 6.3    IDENTIFICATION

Under the word identification a number of very different activities are usually combined. They all involve the identification of previously unidentified material, but for very different purposes. They will be discussed separately here. [See also 1.2 and 4.3.1 for a definition of identification (in contrast to classification).]

**6.3.1    Sorting of Collections.** All material gathered on collecting trips and expeditions must be sorted and at least tentatively identified before it can be incorporated into a collection. The first rough sorting of freshly collected material is often done in the field. An entomologist may keep specimens of different species from different hosts in separate containers. Collections made by oceanographic expeditions are often at once roughly sorted, in part because different kinds of animals may require different methods of preservation. After the specimens are properly preserved and labeled, the usual practice is to segregate unstudied material down to orders and, whenever possible, to families or even genera. Such material is then available to the specialist, who can undertake the precise identification. Even such a tentative identification requires skill and experience. When large collections are involved, the establishment of a special sorting organization is advisable. The Smithsonian Institution in Washington maintains the Oceanographic Sorting Center (SOSC) charged with the preliminary sorting of the material gathered by the Indian Ocean Expedition and other oceanographic expeditions. When identified as to family or genus, the material is shipped to those specialists, all over the world, who have agreed to study it.

The ease of sorting newly received collections depends on the nature of the material (larvae or adults, microscopic in size or not) and on the maturity of the taxonomy of a given group and in a given region. In birds, for instance, identification at the species level is virtually never a problem, although subspecific identification may be difficult. In less well known groups for which no recent revisions, keys, or manuals are available, definite identification may be possible only down to the level of the family or at best to that of the genus.

Linnaeus thought that every zoologist should know every genus (only 312 genera of animals were known to him). In this spirit, at a time when the collections and the number of described species were still small, taxonomists attempted to identify every specimen, even when it belonged to groups of which they had no special knowledge. It is now realized that this is a very wasteful approach. When accessions are sorted in a modern zoological museum, they are identified only to the level (order or family) where they become available to the specialist. It is far more economical and important for a taxonomist to devote his time to the preparation of new monographs and keys than to attempt identification down to the species level in groups with which he is not familiar.

Unfortunately, identifications by nonspecialists are often erroneous. Most large collections contain numerous examples of misdeterminations. The original specimen in a series may have been quite authentic in such cases, but other specimens were subsequently added without critical analysis and without determination labels by experts. Such misdeterminations are more bothersome than leaving the material unidentified until it can be studied by an expert.

**6.3.2 Determination Labels.** Sooner or later all material in collections is seen by a qualified taxonomist or specialist who is able to identify it down to the species. Each specimen or each series should be labeled at the time such identification is made. The determination label should give the scientific (generic and specific) name and author, and in addition the name of the determiner and the year in which the identification was made. With this information on every specimen, the authenticity of the determination is established, and its dependability may be readily evaluated at any subsequent date on the basis of progress which may have been made in the study of the group during the intervening years. In bird and mammal collections these names are usually written in pencil so that they can be changed easily if there should be a change of nomenclature.

6.3.3   **Identification of Individual Specimens.** The taxonomist is frequently called upon to identify a particular specimen or species. If such identification is highly important for a special research project in applied biology or in experimental zoology, a taxonomist will gladly make every effort to identify the species on which this research is based. On the other hand, it is not the job of the taxonomist to undertake the routine identification of ecological collections or of archeological material. Such identification work is the responsibility of the ecologist or archeologist who wants his material identified.

There are, however, federal and state agencies charged with the responsibility of identifying economically important animals. These agencies employ specialists, to each of whom a particular group is assigned. The research taxonomist, on the other hand, does not have any responsibility whatsoever for routine identifications. Nothing is more detrimental to the productivity of a research museum than unreasonable identification demands by the public.

6.3.4   **Process of Identification.** Even a rank beginner trying to identify a specimen can usually tell that it is a bird, a spider, a grasshopper, or a butterfly. He can immediately go to keys and manuals specializing in these zoological groups. The real beginner, however, will have trouble with all but the most common kinds of animals. When in doubt about the order to which an animal belongs he should try the simple keys given in general textbooks and handbooks. Even the advanced student may encounter unusual species or immature or exotic forms which cannot be placed on sight in the proper family or order. However, modern works are generally available which provide family and subfamily keys and greatly assist in this stage of identification.

Driver's *Name That Animal* (1950) is a good elementary guide to the principal groups of animals. In addition to the general keys, a bibliography of the most important works on each group of animals is given. For example, Brues, Melander, and Carpenter (1954) is cited among the general works on insects; this book in turn gives keys to the orders, families, and subfamilies of insects and cites the more important monographs under each group. For North American freshwater animals, consult Pennak (1953) or Edmondson (1959); for marine animals, Light (1954) for the Pacific and Smith (1964) for the Atlantic. There are several series of very useful field guides now available, such as Putnam's *Nature Field Books,* the *Peterson Field Guide* series, the H. E. Jaques *Pictured Key Nature Series* (William C. Brown Company), the Doubleday *Nature Guides Series* and the *Golden Field Guides.* Other useful publications are those of Pratt (1935, invertebrates), Baker (1958, mites), Bishop (1952, salamanders), Stebbins (1954, reptiles and amphibians), Wright and Wright (1949, frogs), Smith (1946, lizards), Wright and Wright (1957, snakes).

There are many works on the animals of specific regions [e.g., Park, Allee, and Shelford (1939) for the Chicago region, or Eddy and Hodson (1955) for the North Central states]. There is an excellent survey of taxonomic works dealing with the British Isles (Smart, 1953). Other regional works for Europe are *Die Tierwelt Deutschlands* (Dahl, 1925 et seq.), *Faune de France* (1921–1950 and continuing), and *Die Tierwelt der Nord- und Ostsee* (Grimpe and Wagler, 1925 et seq.). Unfortunately there are no comprehensive bibliographies of regional taxonomic treatises available for the United States.

Identification is vastly more difficult when no convenient keys and manuals are available. The beginner is advised not to attempt it. If there is available a monograph or technical revision of recent date (see below how to find such literature), the specimen is run through the keys; the description of the appropriate species is checked, character by character; the specimen is compared with any illustration that may be given; and the recorded geographical distribution is checked. If all these points agree, the identification is considered as tentatively made, subject to comparison with authentic specimens and provided that no additional related species have subsequently been described. For further details on these steps, see the following sections. When there is not even a recent monograph or revision available, no one but a specialist should attempt a determination. Even he will not waste time trying to identify single specimens except under the exceptional conditions specified in 6.3.3.

### 6.4  MATERIAL FOR REVISIONARY OR MONOGRAPHIC WORK

The major preoccupation of the taxonomist, except perhaps of the specialist in the best-known groups of animals, is the preparation of revisions. The early part of the process, the determination of the species to be classified, resembles in some ways the process of identification, yet it is of a somewhat different nature. The details of this procedure are described in Chaps. 9 and 10.

However, before the actual work of revising can be started, the taxonomist must gather the needed specimens and the literature. Let us assume he wants to revise a certain tribe of beetles from South America. After he has completed a first examination of the material available in his own collection, he will write to the major museums whose scope includes the Coleoptera of South America and ask for the loan of their material. He should carefully observe the advice given in 6.2.8. During such correspondence he may discover that some other taxonomist has also started to revise this group; he will then have to negotiate with this specialist how to divide the task. In the case of large collections rich in types, he will have to make arrangements to visit these museums rather than ask for the loan of the material.

Far more difficult than tracing the available material is usually the tracing of the literature (Bottle and Wyatt, 1966). The zoologist may start with some general zoological bibliography such as Smith and Painter, *Guide to the Literature of the Zoological Sciences* (7th ed., 1967) and may consult various annual bibliographies and literature-abstracting services (see below).

When there is no recent monograph or revision, the most recent catalog for the group should be consulted. The catalog (11A.9) will give literature citations leading to the descriptions of all species known up to the time of its completion. Some catalogs furnish even more information, e.g., complete bibliographies under each genus and species, lists of synonyms, and geographical distribution. Taxonomic research is greatly facilitated by a good catalog, because it brings together the most significant published references to the group.

**6.4.1   Reference to Current Bibliographies.** Catalogs are inevitably out of date soon after they are published. This difficulty may be partially compensated for by the issuance of supplements. Nevertheless, it is not at all unusual to find that even the most recent catalog is twenty years old. In some of the major insect orders no general catalog has been prepared since 1900, and some groups have never been cataloged from a world standpoint.

Fortunately, there exists an unusual bibliography of the literature in systematic zoology, a great reference work entitled *The Zoological Record*. It is indispensable for taxonomic work. *The Zoological Record* has appeared every year from 1864 up to the present time. Each new scientific name is given, together with a reference to the place of publication and the type locality. The names are arranged alphabetically under families, but a systematic arrangement is followed for families and higher groups. Current numbers are available separately by purchase or subscription.

*The Zoological Record* is published by the Zoological Society of London in cooperation with the British Museum (Natural History) and the Commonwealth Institute of Entomology. The following 20 sections of *The Zoological Record* are published separately and may be obtained singly or as an entire volume each year: (1) Comprehensive Zoology, (2) Protozoa, (3) Porifera, (4) Coelenterata, (5) Echinodermata, (6) Vermes, (7) Brachiopoda, (8) Bryozoa, (9) Mollusca, (10) Crustacea, (11) Trilobita, (12) Arachnida[1] and Myriapoda, (12) Insecta, (14) Protochordata,[2] (15) Pisces, (16) Amphibia, (17) Reptilia, (18) Aves, (19)

---

[1] To Arachnida are added Merostomata, Pantopoda, Pentastomida, Tardigrada, Myriapoda, and Onychophora.

[2] Protochordata together with Pogonophora, Enteropneusta, Graptolithina, Pterobranchia, and Phoronidea.

Mammalia, (20) List of New Generic and Subgeneric Names. These are ordered from The Zoological Society of London, Regent's Park, London, N.W. 1, England (except for Insecta, to be ordered from the Commonwealth Institute of Entomology, 56 Queens Gate, London, S.W. 7).

The commonest method of using *The Zoological Record* is to start with the most recent volume and work back to the date of completion of the most recent catalog or revision. The particular genus or other group in question may be located in the table of contents of the section devoted to the particular class of animals involved. New names, synonymies, distributions, and in some instances even biological references are given. If the citation is not clear because of its abbreviated form, or if the exact title of the publication is important, reference may be made to the bibliography of papers arranged according to authors at the beginning of the section. For special needs, there is an elaborate subject index covering various phases of morphology, physiology, ecology, and biology.

Some groups of animals have never been cataloged or monographed. This is especially true of certain groups of insects. In such cases, it is necessary to work through the entire *Zoological Record* back to volume 1 (1864).

The best annual review of the taxonomic literature, prior to 1864, is found in the *Berichte über die wissenschaftlichen Leistungen* in different branches of zoology, including entomology and helminthology, published in Wiegmann's *Archiv für Naturgeschichte* (Berlin, 1835 et seq.). Additional important bibliographic aids covering this early period of zoology are Engelmann (1846), Agassiz and Strickland (1848), and the catalog of scientific papers published by the Royal Society (1800–1863). Sherborn's *Index Animalium* (1758–1800, 1801–1850) gives a complete list of generic and specific names proposed up to 1850 and Neave's *Nomenclator Zoologicus* (1939–1940, 1950, 1966) lists all generic names for the period 1758 to 1955.

The *Zoological Record* is always one or two years behind, so that other bibliographies must be consulted for the most recent literature. *Biological Abstracts* (1926 to date) is an important source of recent literature. Its section on systematic zoology contains abstracts of taxonomic papers, hence is a valuable source of information for papers which are not immediately available elsewhere. However, since *Biological Abstracts* covers the taxonomic field very incompletely, it is no substitute for the *Zoological Record*. There are important foreign abstracting journals which must be consulted. The *Berichte über die gesamte Biologie*, Abt. A, *Berichte über die wissenschaftliche Biologie* (1926 to date), and *Bulletin signalétique* 16, *Biologie et physiologie, Animales* (1962), are particularly important. In addition, Günther's reviews of the systematic literature (1956, 1962) are invaluable.

There are numerous zoological bibliographies dealing with zoology as a whole, or with vertebrate zoology (Wood, 1931), or with special groups, such as birds, fishes, or other taxonomic subdivisions. The student of taxonomy is advised to familiarize himself with the bibliographic aids that are available in his special field. Experienced specialists will be glad to give him advice.

**6.4.2   Nominal Species and Zoological Species.** Not all names found in the literature ("nominal species") represent actually different zoological species. In many cases individual variants of all sorts were erroneously described as separate species. Chapters 8 and 9 give instructions on how to determine what names refer to valid species and what others to intraspecific phena.

Equally troublesome are cases where similar (or not even very similar!) species masquerade under the same name. The correct determination of the actual zoological species in the group to be revised is the most important basic step in taxonomic research. This requires consultation of original descriptions (or improved redescriptions), a study of authentic specimens, and, in any authoritative revision or monograph, a study of the actual type specimens.

**6.4.3   Original Description.** Although the secondary literature is often a great help, reference should always be made to original or more recent authoritative descriptions. Unless this is done, there is a possibility that the species was misidentified in the secondary literature. Original descriptions are located by means of catalogs, monographs, the *Zoological Record,* or other bibliographical sources, as described above.

Copies of original descriptions may be difficult to find. Even the largest libraries are not complete, and the average university library will often be found wanting. This is not so much a reflection on the caliber of libraries as it is evidence of the extent and diversity of scientific publications throughout the world. Although largely confined to a half-dozen languages, taxonomic papers are published in practically every country in the world. This poses a very real problem for libraries with limited budgets. The situation is further complicated because priority places a premium on the earlier works. No taxonomic work published since 1758 becomes "out of date" if it contains new names, and as a result of limited editions, losses through the years, and other factors, there are not enough copies available to supply all biological libraries. To an increasing extent, specialists make photocopies of those pages of periodicals which they have to consult most frequently. As long as this is not done commercially, it is not likely to infringe on copyright (but see *Science,* vol. 152, pp. 291–292).

The search for original descriptions is often involved. Helpful, if not essential, to the searcher are: full use of the familiarity with all locally available scientific libraries; reference to the *Union List of Serials* to locate publications in other libraries for interlibrary loan; extensive use of microfilm services and other copying devices; and accumulation of reprints, by purchase or by exchange with other workers, and of photostats (Xerox copies).

Descriptions are the foundation of taxonomy, since only the printed word is indestructible. Types may be lost, and the original author is available for only a brief span of years to pass on "his" species.

The original description should be read several times, first to obtain a general impression or mental picture of the actual specimen which the original author had before him. Then, particular characters which the original or subsequent authors considered important should be extracted and checked against the specimens in question. Finally, any comparative notes given by the original author should be checked. In many cases such comparative characters are the most useful clues to identification.

Original descriptions are normally the court of last appeal for purposes of general identification. However, many of them are totally inadequate, particularly those published prior to 1800. The value of a description is in direct proportion to the judgment of the author and his ability to select significant characters and describe them in words, also to the extent and nature of the material available to him at the time of description. For these reasons descriptions given in a thorough and authoritative monograph of recent date are usually more usable than original descriptions.

Illustrations are often as valuable as, or more valuable than, original descriptions, particularly when there are language difficulties. In popular groups, such as birds or butterflies, there are many works with colored plates. Such works are often a great help in the rapid identification of specimens. Colored plates are not always well reproduced, however, and there are many opportunities for error if too much dependence is placed upon them.

If the original description is accompanied by an illustration, the difficulty occasionally arises that characters of illustration and description are in conflict. It can sometimes be proved in such cases that the artist did not have access to the type specimen and used another in the belief that it agreed with the type. Such discrepancies are not infrequent in the works of early authors.

**6.4.4 Comparison with Authentically Determined Specimens.** It is sometimes impossible to make a satisfactory determination from the literature alone. Such a situation exists if the group has been neglected and the descriptions are poor. Even under ideal conditions identification is greatly facilitated if "authentic material" is available for comparison—material named by a specialist and preferably compared by him with the holotype. Paratypes do not always necessarily qualify, because series of syntypes are sometimes composites of several species.

Comparison with specimens is a highly technical task requiring a considerable background of knowledge of and preparation in the particular group in question. For this reason preliminary identification made by direct comparisons with authentic collections, without first studying the literature and the significant characters of the group, is often valueless.

Reference collections are often accumulated for the express purpose of facilitating identification. In such cases comparison is made with whatever series of specimens is available, and one must judge whether the specimen in hand falls within the possible range of variation of a given species.

Care should be taken not to rely exclusively on comparison with supposedly authentic specimens. Even "authoritative" collections may contain wrong identifications, or may be incomplete. In such cases hasty comparison made without following the other steps in identification may lead to erroneous conclusions.

6.4.5    **Comparison with the Type.** Type-specimens are the most authentic of all but are much too valuable to be used for routine identifications. Ideally, in the course of a monographic study of a group all type-specimens should be reexamined. At this time the significant characters are usually known and can be checked using the same technique and the same interpretation of the characters as are applied to the rest of the material.

In work with subspecies it is not always necessary to have type-specimens for comparison (if there is no question as to the identity of the species). On the other hand, a series of specimens from the type-locality ("topotypical specimens") is desirable to provide information on the characters and variability of the subspecies.

# Chapter 7  Taxonomic Characters

**B**iological classification consists of the assembling of organisms into groups that are similar as a result of their common descent. One of the main operations of taxonomy is thus the determination of this special kind of similarity. This, as the history of taxonomy shows, is a difficult and highly controversial procedure (10.1). The experienced taxonomist compares two kinds of organisms and, merely by inspection, integrates a large number of attributes into a "similarity value." His judgment is based on years of analysis, comparison, and weighting of so-called taxonomic characters.

## 7.1  NATURE OF TAXONOMIC CHARACTERS

*Definition. A taxonomic character is any attribute of a member of a taxon by which it differs or may differ from a member of a different taxon.*

This definition implies a number of important points. A characteristic by which members of two taxa agree but differ from members of a third taxon is a taxonomic character (a statement true for the characters of all higher taxa). To define taxonomic character as "any attribute of an organism" would not be correct. Features by which individuals of the same population differ from each other, such as differences between the sexes and age classes, are not taxonomic characters. However, when populations (taxa) differ from each other by the presence or absence of sexual dimorphism or in larval characteristics or in any other manifestation of sex or age, then such a difference becomes a taxonomic character. Taxonomic characters are population characteristics. The comparison of populations and taxa is the standard method for the study of taxonomic characters, and any attribute qualifies if it is established as different through such comparison.

"Taxonomic character" means actual or potential taxonomic difference, such as "eyes red" versus "eyes white." The definition of taxonomic character as that by which taxon A differs from taxon B has been universally accepted in the taxonomic literature of the last 200 years. The computer taxonomists, perhaps not fully aware of the previous definition, have suggested changing this terminology. They would call "eyes" the character, and "red" or "white" character states. However, "red eyes," "spiny legs," etc. are what the literature has universally called taxonomic characters, and it is the *feature* (eyes, legs, etc.) which displays this taxonomic variation for which a new term needs to be found. To transfer the term character from its traditional usage to a new one is bound to prove confusing.

Taxonomic characters have a double function:

1. They have a diagnostic aspect uniquely specifying a given taxon; an emphasis on the differentiating properties of taxa is particularly strong at the level of the lower categories.
2. They function as indicators of relationship; this property makes them especially useful in the study of the higher taxa.

Not every publication in zoology deals with taxonomic characters. Work published on species that are expecially suited for experimental research, such as *Canis lupus, Macaca rhesus, Rattus norvegicus, Mus musculus, Drosophila melanogaster,* or *Tribolium confusum,* deal with general biological phenomena, such as the behavior of genes, the chemistry of enzymes, the physiology of muscle action, or the nature of learning. Most of the results of such studies are of wide applicability. They are valid for all mammals, all terrestrial warm-blooded vertebrates, all metazoans, or all life. Literature of this type of research is of taxonomic interest only to the extent that comparison between different taxa reveals differences. But even eliminating such literature, there are at least 20,000, but more likely about 50,000 papers published annually which present evidence on taxonomic characters and in part discuss them quite fully. It is impossible in this chapter to present more than the barest outline of this important topic. Chapters 3 to 5 of *Animal Species and Evolution* (Mayr, 1963, pp. 31–109) deal with certain kinds of taxonomic characters, particularly characters such as isolating mechanisms and differences in niche utilization, which permit the sympatry of closely related species.


## 7.2  CHARACTERS AND CLASSIFICATION

Taxonomic practice reveals that not all taxonomic characters are equally useful. Some are powerful indicators of relationship, others are not. The usefulness of a character depends on its *information content,* that is, on its correlation with the natural grouping of taxa produced by evolution. The taxonomist, either consciously or unconsciously, assigns a certain *weight* to each character according to its "goodness," that is, according to its ability to reveal relationship. The problem of weighting will be dealt with more fully in Chap. 10 (10.4).

Some discussion has taken place in recent years whether to define taxonomic characters in terms of the phenotype or of the underlying genetic program. Some authors felt that the latter would be "more scientific." There is considerable doubt as to the validity of this claim. The pathway from the gene to the phenotype is a long one in higher organisms and the "one gene, one character" hypothesis is now thoroughly discredited. What the taxonomist deals with, his taxonomic characters, are components of the phenotype and have almost invariably a highly complex genetic basis. The genotype of most taxonomic characters is now, and probably will remain so for a long time, refractory to precise genetic analysis. On the other hand, a number of recent studies on gene-controlled enzymatic intrapopulation polymorphism indicate that much of the variation of the genotype is a poor source for taxonomic characters. It is thus evident that the direct study of the genetic basis of taxonomic characters is either technically impossible or else not very helpful.

Unfortunately, it is likewise impossible to take the needed information directly from the phenotype. Size illustrates the danger of an unevaluated use of the phenotype particularly well. A large-sized species may differ from a small-sized species in every measurable character, hence in literally thousands of phenotypic characters. There is, however, almost total redundancy, since the size difference might be caused by a single genotypic factor. The literature abounds in cases where such redundancy was not eliminated in the phenetic character listing, so that an absurdly distorted classification resulted. A single adaptation (wing diving or seed cracking in birds; see below) may lead to an impressive number of changes in the overall phenotype, most of which are redundant as far as information on relationship is concerned (see also "redundant characters" in 10.4.3).

Until he has a proper understanding of the value of various potential characters, a taxonomist must consider a large number of characters and compromise between using too many and using too few. Considering too many characters is uneconomical because it requires too much time: it introduces redundancy because different sets of characters often lead to essentially the same answer. Giving equal weight to many characters is sometimes actually misleading because it may conceal the effects of special adaptations, parallelism, and convergence. For advice on the taxonomic weight of single characters see 10.2.3.

*Unreliable Characters.* Every specialist knows characters which in his group are "unreliable," that is, poor indicators of relationship; such characters are said to have low weight (10.4.3). All highly variable characters are in this category. The pattern of branching in arteries of vertebrates may be different not only in different individuals of the same population but even in the left and the right side of the body. Differences in arterial pattern are not nearly as helpful for classification as some authors thought they were. Wing venation provides important characters for the classification of insects. However, Sotavalta (1964) showed that there was far more variation in this character in the tiger moths (Arctiinae) than known and that the traditional generic arrangement of the family, based on wing venation, is thoroughly in need of revision. For a further discussion of variability of characters see 10.4.3.

Any kind of *regressive character* is usually of low taxonomic weight. Taxa based on the loss of eyes, wings, toes or other appendages, or certain teeth, etc., are often unnatural (10.4.3).

In order to get away from unreliable single characters taxonomists in recent years have been searching for an "overall character" which would represent a single denominator for the taxonomic position of a taxon. Phenetic distance was believed by the pheneticists to be such an indicator (10.2). The serologist thought that protein interaction was such an overall measure, while DNA matching is the most recent candidate for such a measure (7.4.9). All these methods are helpful, but since all have failed on occasion (sometimes frequently) they cannot be considered to be the hoped-for panacea.

*Dual Function of Characters in Classification and Identification.* Characters serve different functions in different operations of taxonomy, as discussed in Chap. 4. They are used during the analytical phase to determine the units of classification (species and other lower taxa) and during the synthetic phase to help in the correct delimitation and ranking of higher taxa. The same characters, but usually only a selected small number of them, are used in a very different operation, namely that of *identification. Key characters* are characters that are easily perceived, of very low variability, usually present in preserved material, and useful as convenient labels for the taxa distinguished by the process of classification. Many taxonomic characters, such as chemical, chromosomal, physiological, and behavioral ones, may have high value for purposes of classification but are poor or useless for identification because they are inaccessible in preserved material or because their determination requires difficult techniques. Chapter 11 contains a discussion of characters most useful in *keys*.

## 7.3  TAXONOMIC CHARACTERS AND ADAPTATION

Formerly taxonomists tended to consider characters as if they were specially created for their convenience. Now we realize that as phenotypic manifestations of the genotype taxonomic characters are the product of a long history of natural selection.

*Nature of Adaptation.* The taxonomic weight of a character depends to a large extent on the nature of its adaptive significance. The following comments might be helpful to those undertaking such evaluation. For a more detailed discussion, see Chap. 10.

1. *Adaptation to the General Environment.* A given type of adaptation may be very widespread, yet the specific mode in which it is expressed may be of high taxonomic significance. For instance, the pelagic larvae of most marine invertebrates have ciliary bands, yet the particular configuration of the bands is characteristic for classes and sometimes even phyla. All pelagic organisms have floating devices, yet the particular manner in which floating is achieved differs from taxon to taxon. Pigments, poisons, and other chemical constituents of the body may by convergence have become similar in appearance or effect, yet a proper analysis almost invariably shows highly specific correlation with taxonomic groups.

2. *Specific Adaptations.* Most adaptations for special niches are far less revealing taxonomically than they are conspicuous. Typical examples of this are substrate adaptations, such as white coloration in arctic mammals and birds, sandy coloration of desert animals, the bright coloration of tropical treetop birds, warning colorations, and various types of Müllerian mimicry.

Occupation of a special food niche and the correlated adaptations have a particularly low taxonomic value. The bill in birds and the teeth in cichlid fish may be very different in closely related species, in response to a shift in niche occupation. See 10.4.2 and 10.4.3.

3. *Isolating Mechanisms.* Any species-specific character that serves as an isolating mechanism may be reinforced by natural selection and become exceedingly conspicuous. In birds this is true particularly for species without pair formation (Mayr, 1942, Sibley, 1957). However, such characters are of low weight for classification on the generic and higher levels.

4. *Competitive Character Divergence.* Some differences, particularly in niche utilization, between closely related sympatric species are a consequence of competitive exclusion (Mayr, 1963, chap. 4). A process which Brown and Wilson (1956) called *character displacement* may lead to a stronger morphological divergence between such species than would be expected on the basis of relationship.

There is always a residue of taxonomic characters of which the functional and hence selective significance is not evident. To determine whether such characters have an unknown significance or are incidental phenotypic by-products of a selected genotype must remain for later research. One can summarize the relation between adaptation and the value of taxonomic characters by saying that narrowly adaptive characters generally do not give as much information on relationship as do characters that owe their conformation to the constitution of the overall genotype, which selection has shaped for thousands and millions of generations.

## 7.4  KINDS OF CHARACTER

Almost any attribute of an organism might be useful as a taxonomic character, if it differs from the equivalent feature in members of another taxon. However, proper classifying work is possible only when adequate material of many species is simultaneously available for comparison. Museums provide this opportunity and this is the reason why the taxonomist prefers characters that can be easily observed in preserved specimens (e.g. morphological characters).

In each group of organisms, whether they are birds, butterflies, sea urchins or snails, different taxonomic characters exist. It is part of the training of the taxonomist to become familiar with the characters that are most useful in the particular taxon in which he plans to specialize. Monographs and handbooks usually give detailed descriptions of the characters used. In his revision of the North African scorpions, for instance, Vachon (1952) devotes 27 pages to a detailed description and illustration of the taxonomic characters of that group.

*Morphological Characters.* Features of the external morphology vary according to kind of animals. They range from such superficial features as plumage and pelage characters of birds and mammals through scale counts of fishes and reptiles, to the highly conservative and phylogenetically significant sutures and sclerites of the arthropod body. The internal anatomy provides an abundant source of taxonomic characters in practically all groups of higher animals. The extent to which such characters are used routinely varies from group to group, generally in inverse ratio to the abundance and usefulness of easily observed external characters. In mammals the skull (with teeth) is routinely preserved and used in classification, while reptiles, amphibians, and fish are normally preserved in alcohol and are always available for dissection. On the whole, aspects of the internal anatomy more often supply characters for the classification of the higher taxa than for discrimination at the species level. Fossils consist almost en-

TABLE 7-1. KINDS OF TAXONOMIC CHARACTERS

1. Morphological characters
   *a.* General external morphology
   *b.* Special structures (e.g., genitalia)
   *c.* Internal morphology ( = anatomy)
   *d.* Embryology
   *e.* Karyology (and other cytological differences)
2. Physiological characters
   *a.* Metabolic factors
   *b.* Serological, protein, and other biochemical differences
   *c.* Body secretions
   *d.* Genic sterility factors
3. Ecological characters
   *a.* Habitats and hosts
   *b.* Food
   *c.* Seasonal variations
   *d.* Parasites
   *e.* Host reactions
4. Ethological characters
   *a.* Courtship and other ethological isolating mechanisms
   *b.* Other behavior patterns
5. Geographical characters
   *a.* General biogeographical distribution patterns
   *b.* Sympatric-allopatric relationship of populations

tirely of preserved hard parts—in the case of Mesozoic mammals, for instance, largely of teeth.

Even in this traditional area great advances have been made in recent decades. Descriptions have become more detailed and better standardized. The taking of measurements is being automated (Garn and Helmrich, 1967). In the case of lower invertebrates careful microscopic analysis has revealed an abundance of characters even in such seemingly nondescript forms as nematodes. The development of new silver impregnation techniques has revealed a wealth of characters even among protozoans, particularly ciliates.

New organs and structures are steadily added to those that show taxonomically important differences. The *spermatozoa* of many taxa, for instance, have a highly peculiar and specific morphology and may serve as useful indicators of relationship.

**7.4.1 Hard Parts** (shells, external skeletons, etc.) **and the Work of Animals** (mines, tracks, etc.). It would be senseless to worry whether to consider hard parts, etc., as morphological, physiological, or behavioral characters. Much of the classification of the invertebrates is based on characters of exoskeletons and shells. Likewise, among the protozoans tests, shells, thecal plates, cysts, and other hard parts are vital in the classification of foraminiferans, radiolarians, testaceous rhizopods, flagellates, and others. The orientation types of calcite crystals in the skeleton of echinoderms agrees well with their classification in families and orders (Raup, 1962).

Many taxa of dinosaurs have been based on fossil tracks. In the classification of gall insects the gall sometimes yields as good a clue to relationship as do the insects themselves. The form of the mines is an important taxonomic character in mining insects, and it even sheds light on their history since these mines are sometimes well preserved in fossil leaves. However, since 1930 it has no longer been permissible to base the name of new species exclusively "on the work of an animal" [Arts. 13a and 24b (iii)].

Fig. 7-1. Diagnostic mine patterns caused by six species of leaf miners of the genus *Phytomyza* on the leaves of *Angelica* (50, 51) and *Aquilegia* (54). The letters *a, b, c* refer to different species of *Phytomyza* on the same host plant (*from Hering*, 1957).

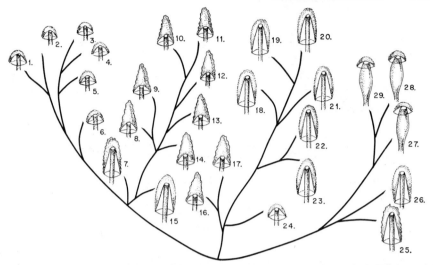

**Fig. 7-2.** Types of spermathecae in the *Drosophila repleta* group placed on a phylogenetic tree constructed on the basis of chromosomal evolution (*after Throckmorton*, 1962).

**7.4.2   Coloration.** Color pattern and other aspects of coloration are among the most easily recognized and thus the most convenient characters in certain groups of animals. Every species of birds can be recognized by its coloration except for a few genera with sibling species (e.g., *Collocalia, Empidonax*). The same is true for certain reef fishes, butterflies, and other animals. Even where coloration is not completely diagnostic, it often helps to narrow down drastically the number of species to be considered. In groups in which subspecies are recognized routinely, such as mammals, birds, butterflies, or certain wasps, color again plays an important role; many subspecies are entirely based on coloration.

The quality of color is not easy to describe in words, so that a description by one author is often misunderstood by his readers. In original revisionary work, therefore, it is preferable not to rely on descriptions but to base one's judgment on the comparison of specimens if at all possible. There are, however, various ways by which greater precision can be given to color determination and description, as described in Chap. 10.

**7.4.3   Genitalic Structures.** For reasons that are not yet fully understood (Mayr, 1963, p. 103), the genitalia of many animals, particularly arthropods, not only show a great deal of structural detail but are also highly species-specific. Being three dimensional structures they have to be carefully prepared to be strictly comparable (Dreisbach, 1952). In many groups of insects and spiders genitalic structures are more important for species diagnosis than is any other character. Yet even here it has been found either that a single species may have a good deal of variation or else that two related species have indistinguishable structures. Genitalic

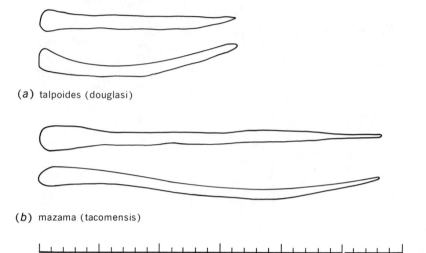

(a) talpoides (douglasi)

(b) mazama (tacomensis)

| | | | | | | |
|---|---|---|---|---|---|---|
| 0 | 5 | 10 | 15 | 20 | 25 | 30 mm. |

**Fig. 7-3.** Shape of baculum as taxonomic character. (a) *Thomomys talpoides;* (b) *Th. mazama.*

structures are most useful among the arthropods. In most vertebrates they are soft, but the gonopodium of some fishes, the hemipenis of snakes, and the baculum of mammals supply good taxonomic information.

**7.4.4   Other Characters.** Morphological characters of adult specimens are still used more frequently than any others but they are supplemented to an ever-increasing extent by other characters, as listed and discussed below (e.g., Blair, 1962; Munroe, 1960). This is particularly true for "difficult" species, genera, and families where the evidence from morphology has been equivocal or contradictory. There has been an increasing utilization of *new characters,* because (1) morphology reflects only part of the genotype and may not reflect genetic relationship accurately; (2) morphology in certain taxa does not supply sufficient characters; and (3) external morphology is sometimes misleading owing to special adaptations. The introduction of new kinds of taxonomic characters has been one of the aspects of the so-called new systematics. The new characters (proteins, chromosomes, behavior, etc.) supplement but do not displace the use of morphological characters.

A particularly important reason for the utilization of new characters is that they serve as check on the conventional morphological characters. When discrepancy between a morphology-based and a non-morphology-based classification occurs, still other sets of taxonomic characters must be used. Fortunately, it turns out that the newer characters usually confirm the classifications based on the more classical approaches. It seems that the general morphology is a reflection of a large part of the genotype and thus permits, on the whole, reliable conclusions on relationships.

**7.4.5 Larval Stages and Embryology.** Various immature or larval stages, the embryology, and sometimes even the eggs may provide taxonomic information. The various sibling species of the *Anopheles maculipennis* complex (malaria mosquitoes) were discovered owing to differences in egg structure. The classification of the Aleyrodidae (whiteflies) is based primarily on the pupae. Discovery of the leptocephalus-like larval stage of the fish order Heteromi confirmed the previously suspected relationship with the eels.

In groups with complete *metamorphosis,* entirely different sets of characters evolve in larvae and adults, and conclusions drawn from the characters of one stage form a very useful check on conclusions from those of another. This was shown for frogs (Anura) by Orton (1957) and Inger (1967). A careful comparison of larval and adult characters of the digger wasps (Sphecidae) permitted Evans (1964) to show that some features of the adult structure (emarginate eyes, loss of wing veins) had been overvalued. A proper evaluation of larval characters led to an improved arrangement. Similar findings have been made in other groups of insects (Van Emden, 1957). Larval and adult characters are visible manifestations of the same genotype. There can be different identification schemes for larvae and adults, but there must be only one classification for a given group of organisms. This must be based on the proper weighting of both adult and larval characters. It is sometimes the adults and sometimes the larvae which have acquired specialized adaptations, and only a biological analysis can lead to the proper evaluation. In groups like the sponges (Porifera) with extremely few morphological characters, a due consideration of the embryology has been a great help in classification (Lévi, 1956).

**7.4.6 Genetic Characters and Sterility.** There has been much confusion as to the meaning of the term genetic characters. Broadly speaking, all characters, except nongenetic modifications of the phenotype, are genetic characters. Isolating mechanisms and monogenic characters have sometimes been designated as genetic characters, but this is an unjustified restriction. The term genetic characters is actually meaningless.

On the whole, only closely related species can hybridize successfully. In groups of related forms the presence or absence of cross-sterility is important information, but this information must be employed carefully. Ducks, for instance, show much cross-fertility not only of species but even of related genera. Yet two such closely related species as the Wood Duck (*Aix sponsa*) and the Mandarin Duck (*A. galericulata*) are completely sterile owing to a chromosomal rearrangement. In birds, on the whole, there is much fertility among closely related species, while in some groups of insects there is already much sterility when geographically remote populations of the same species are crossed. In other groups of insects congeneric species may be fully fertile with each other. A careful analysis of the genetic compatibility of various species of toads has led to a regrouping of species in the genus *Bufo* (Blair, 1963). Indeed, a few exceptions notwithstanding, the relative degree of fertility in species crosses is, on the whole, a very sensitive measure of relationship.

**7.4.7   Chromosomes.** Botanists have made use of the abundant information provided by the chromosome pattern far longer than have zoologists. Improvements in cytological technique during the past twenty years now permit chromosomal studies even in so-called difficult groups, such as the mammals. Birds and lepidopterans with their small chromosomes and high chromosome numbers create the greatest difficulties. Diptera, particularly those with giant chromosomes, and Orthoptera are among the groups that are most suitable for chromosomal studies. White (1954, 1957) has provided useful summaries, already made partially obsolete by the recent rapid advances. The current abstract and review literature provides convenient access to this field. Matthey for mammals and reptiles, S. G. Smith for beetles, Hughes-Schrader for mantids, White for grasshoppers, Stone, Carson, and Wasserman for Drosophila, Keyl for chironomids, and Halkka for homopterans are among the numerous authors who have advanced the classification of animals, particularly that of insects, through studies of chromosomal patterns. The study of primate chromosomes is another very active field, yielding much information on relationships (Chiarelli, 1966).

Chromosomes are particularly useful on two different levels. On the one hand, they aid in the comparison of closely related species, including sibling species. These are often far more different chromosomally than in their external morphology (Mayr, 1963, chap. 3). On the other hand, chromosomal patterns are of extreme importance in establishing phyletic lines. Most chromosomal changes are unique events which are then characteristic for all descendants of the ancestral population in which the new pattern first became established. Changes in sex determination, in all sorts of rearrangements of the chromosomes and centromeres, in fusions, fissions or translocations, in the acquisition of supernumeraries, etc., often give unequivocal clues to relationship. For instance, the similarity in the spermatogenesis of mallophaga and anoplura (true lice) strongly supports the belief in a close relationship of these taxa. The neotenic salamanders *Necturus* and *Proteus* were believed to have become similar by convergence, but both genera have a haploid chromosome number of 19, otherwise unknown among urodeles, and the homologous chromosomes are very similar (Kezer et al., 1965).

Polyploidy is rare in animals, as compared with plants, but there are numerous other ways by which increases or decreases in chromosome number may occur. The most frequent evolutionary trend, curiously, in many groups is from high to low chromosome number through chromosomal fusion. Even though the chromosomes represent the genetic material, it is not true that the amount of chromosomal change reflects the amount of genetic change. Close relatives may show considerable rearrangement; indeed many species are polymorphic for various types of chromosomal rearrangements. On the other hand, cases are known in which a considerable degree of genic change is not or only lightly reflected in the chromosomal pattern, as in the Hawaiian Drosophila (Carson, 1967).

**7.4.8    Physiological Characters.** This group of characters is hard to define. All structures are the product of growth processes, that is of physiological processes, and are thus ultimately physiological characters. Also all physiology is regulated by enzymes and other macromolecules and thus is not separable from biochemical characters. By "physiological characters" one generally means growth constants, temperature tolerances, and the various processes studied by the comparative physiologist. Species differences are abundant in these characters, but since they are not present in preserved material and usually require special apparatus for their study, they are rarely used by the taxonomist. Some representative cases are discussed by Mayr (1963, pp. 60–65), and fuller treatments are given in textbooks of comparative physiology (e.g. Prosser and Brown, 1961).

**7.4.9    Biochemical Characters.** The major molecular inventions were made by the earliest living organisms. Even the most primitive procaryotes have on the whole the same kinds of macromolecules and metabolic processes that we find in the highest animals and plants. Still, there is enormous specificity at every taxonomic level, and this is being increasingly exploited by the taxonomist.

Serology provided the earliest widely used method of comparing proteins. This method is based on the principle that the proteins of one orga-

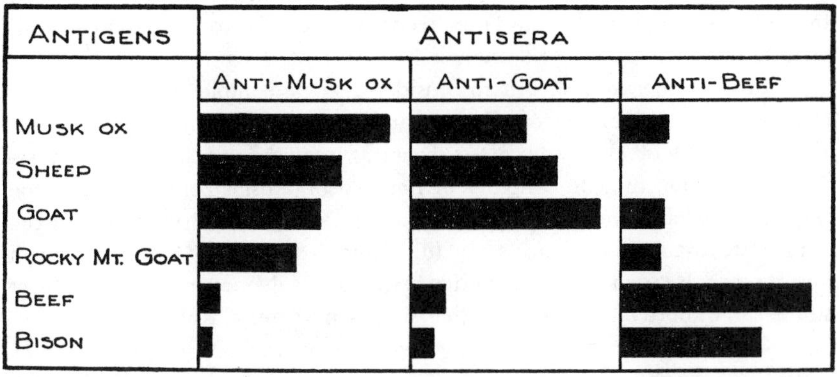

Fig. 7-4. Reaction between the blood sera of six species of bovids with the antisera against three species (*from Moody, 1958*).

nism will react more strongly with antibodies to the proteins of a closely related organism than to those of one more distantly related. Unfortunately, the method encounters various technical difficulties, and though used for more than forty years it has not contributed as much to a clarification of otherwise ambiguous cases as had been hoped. A summary of some of the achievements of this method is found in the *Kansas Symposium* (Leone, 1964). As a result of improvements in the techniques there has recently been a revival of interest in the quantitative study of antigenic reactions (e.g. Williams, 1964). The study of blood-group genes ("immuno-genetics") has shed light on relationships between species of pigeons (Irwin, 1947, and later papers of his school) and is now used extensively in the study of primates.

*Chemical Taxonomy.* Much recent work has been devoted to a tax-onomy of specific chemical components and macromolecules. Paper chroma-tography has been used by numerous authors to compare the chemical composition of closely related species, with particular attention to amino acids and peptides revealed by ninhydrin treatment, and the purines, pyrimi-dines, or other substances that either fluoresce or absorb ultraviolet light. Various methods of electrophoresis reveal the molecular composition of complex proteins. Sibley (1960) analyzed the egg-white proteins of more than 100 species of birds and was able not only to add clear-cut indications of relationship in previously equivocal cases, but also to raise doubts con-cerning previous arrangements and to make suggestions as to relationship of taxa previously considered highly isolated. Throckmorton (1962) did simi-lar studies for the species of Drosophila. Paper electrophoresis, which was used in these earlier studies, has been largely replaced by newer techniques that permit a much finer resolution, but still newer techniques are con-tinually being introduced. These techniques have been used also in the study of reptiles, amphibians, insects, mollusks, and other animals, and there is no doubt as to their increasing importance. It is necessary to consult the current literature for the latest techniques and instrumentation in pro-tein, peptide, and amino acid analysis (e.g. Wright, 1966).

Still another approach is to consider a single complex molecule for in-stance the hemoglobin, of one species, and compare its amino acid composi-tion with that of closely or more distantly related species. Patterns of re-placement often indicate whether or not two organisms belong to the same phyletic line. Two recent symposia (Handler, 1964, and Bryson and Vogel, 1965) present such evidence from this rapidly developing field. Any given enzyme can be used for these studies, but the results are, of course, subject to the same shortcomings as any other single-character classification.

The newest and most exciting development is the attempt to go back to the very basis of relationship, the genetic program itself. Techniques of DNA matching are now being developed (Hoyer et al., 1964) which hold much promise. It is not yet certain that these methods can be made sensitive enough to add to the existing information, but the study of evolving molecules superimposed on the background of classical taxonomy is bound to reveal discrepancies and inaccuracies which will lead to improvements in classification.

The study of structural characters has revealed that each organ or organ system may have its own specific rate of evolutionary change (mosaic evolution, 10.4.5). Much evidence indicates that this principle is equally valid for chemical characters. A comparison of man (*Homo*) with the African apes (*Pan*) shows, for instance, that there has been little evolutionary divergence in their hemoglobins and serum proteins since they branched from each other, even though the hominid line has since entered an entirely new adaptive zone (see Fig. 4-5). When using taxonomic characters to draw inferences on classification one must always balance the potentially conflicting information derived from different characters. One must also understand the subtle differences between evolutionary phenomena at the molecular and at the organismic level (Mayr, 1964*b*; Simpson, 1964).

**7.4.10   Behavior.** Behavior is undoubtedly one of the most important sources of taxonomic characters. Indeed, behavioral characters are often clearly superior to morphological characters in the study of closely related species, particularly sibling species (Mayr, 1963). Yet there are two major technical drawbacks. Behavior cannot be studied in preserved material, and it is intermittent even in the living animal. Certain types of behavior

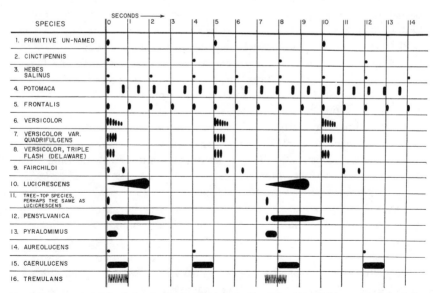

**Fig. 7-5.** Pattern of light flashes in North American fireflies (*Photuris*). Height and length of the marks indicate intensity and pattern of the flashes (*from Barber, 1951*).

occur only during the breeding season or during part of the 24-hour period. The comparative study of behavior of related species has become an autonomous discipline, comparative ethology. It has already made major contributions to the improvement of classifications of birds, bees, wasps, orthopterans, frogs, fishes, and other groups.

The reason for the importance of behavior is obvious. Behavior characteristics are the most important isolating mechanisms in animals, and new adaptations are often initiated by changes in behavior. The rapidly expanding literature on behavioral systematics has, in part, been summarized in a number of reviews: Mayr (1958), Baerends (1959), Cullen (1959), Alexander (1962), and Wickler (1961, 1967). Exemplary studies are those of Evans (1957, 1966) on the digger wasps (Sphecidae), of Sakagami and Michener (1962) on sweat bees, of Spieth (1952) on the genus *Drosophila,* of Johnsgard (1965) on ducks (Anatidae), of Tinbergen (1959) and Moynihan (1959) on gulls (Laridae), and of Smith (1966) on *Tyrannus.*

A great technical advance in the study of behavior has been the development of accurate sound-recording devices and of further devices, such as sonagraphs, for the translation of sounds into graphic patterns (Fig. 7-6). More than 40 species of North American crickets were either discovered or rediscovered as a result of the careful analysis of their songs by B. B. Fulton and his followers, (Walker, 1964, Alexander, 1967). The classification of species in several avian genera (for instance, *Myiarchus, Empidonax, Tyrannus*) has been greatly helped by an analysis of sound recordings. A comparison of the calls of frogs and toads has not only led to the discovery of previously unrecognized sibling species but has also shed light on the relationship between previously established species. Important studies of comparative sound analysis in anurans were made by Barrio, Blair, Bogert, Littlejohn, Main, and Mecham. The acoustics of behavior of animals has been summarized in a number of recent volumes (Tavolga and Lanyon, 1960, and Busnel, 1963).

In addition to courtship behavior and acoustic behavior, various other kinds of behavior elements have taxonomic value. For example, the pattern of the webbing constructed by various spiders, mites, and caterpillars may be used at various levels in the classification. The two bee genera *Anthidium* and *Dianthidium* were slow to be recognized on morphological grounds, yet all known species of the former construct their nests of cottony plant fibers and those of the latter construct theirs from resinous plant exudations and sand or small pebbles.

The use of extraneous materials in the construction of nests or of larval or pupal cases provides characters at various levels in the classification of caddisworms and bagworms. The manner in which such materials are attached to the shell is a useful taxonomic character in distinguishing species of the Molluscan genus *Xenophora*. The egg cases of praying mantids have a species-specific form.

Behavior patterns that are characteristic of higher taxa are far more rare. Examples are the use of mud in nest building by barn swallows and crag martins (*Hirundo*), certain "comfort movements" (scratching, stretching, bathing, etc.) in birds (McKinney, 1965), and grooming movements in insects. Attention up to now has been directed so strongly at the compari-

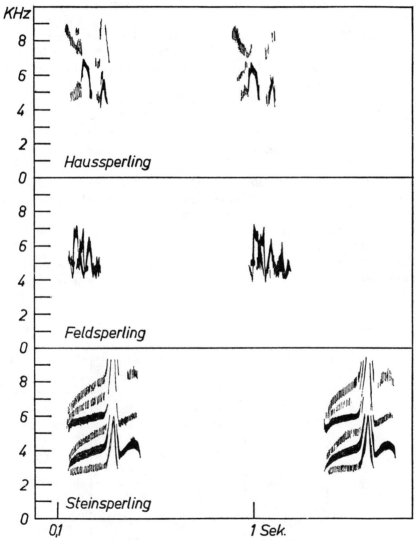

Fig. 7-6. Graphic representation of the songs of three species of sparrows (*Passer domesticus, Passer montanus, Petronia petronia*) (*from Thielcke*, 1964).

son of closely related species that the diagnostic value of behavior patterns at the level of higher taxa has been relatively neglected.

**7.4.11   Ecological Characters.** It is now well established that every species has its own niche in nature, differing from its nearest relatives in food preference, breeding season, tolerance to various physical factors, resistance to predators, competitors and pathogens, and in other ecological factors. Where two closely related species coexist in the same general habitat, they avoid fatal competition by these species-specific niche characteristics (principle of competitive exclusion, Mayr, 1963, chap. 4). A number of sibling species were discovered as a result of discrepancies in food preference ("host specificity")—for example, the apple and blueberry maggot—or habitat preference (*op. cit.,* chap. 3). Many aspects of the life cycle, such as life span, fecundity, and length or time of breeding season, may be different in closely related species (*op. cit.*).

Niche specificity is quite pronounced even in species such as birds, mammals, or mollusks that are not particularly substrate-specific. Kohn (1959) found that every species of the genus *Conus* in the Hawaiian Islands differs ecologically from related species. Two sibling species (*ebraeus* and *chaldaeus*) feed on nereid polychaetes, but of 199 *ebraeus* 136 contained nereid species *a* and none species *b*, while of 106 *chaldaeus* 5 contained species *a* and 98 species *b*. The larvae of *Drosophila mulleri* and *aldrichi* live simultaneously in the decaying pulp of the fruits of the cactus *Opuntia lindheimeri*. The two species are, however, markedly specialized in their preference for certain yeasts and bacteria (Wagner, 1944).

Niche specialization is even more important for animals that are substrate-specific. This includes host-specific plant feeders among insects and mites as well as host-specific parasites. Unfortunately there is no up-to-date

TABLE 7-2.   ECOLOGICAL DIFFERENCES BETWEEN SIBLING SPECIES IN THE
*Anopheles maculipennis* COMPLEX

| Species | Habitat | Water type | Hibernation | Malaria carrier |
|---------|---------|------------|-------------|-----------------|
| *melanoon* | Rice fields | Fresh water | No | No |
| *messeae* | Cool standing water | Fresh water | Yes | Almost never |
| *maculipennis* | Cool running water | Fresh water | Yes | No |
| *atroparvus* | Cool waters | Brackish | No | Slightly |
| *labranchiae* | Mostly warm waters | Brackish | No | Very dangerous |
| *sacharovi* | Shallow standing water | Often brackish | No | Very dangerous |

summary of the field except the excellent treatment of Dethier (1947), the emphasis of which is on the sensory aspects of host selection. Many new species of insects were discovered by comparing populations of the "same" species occurring on different plant hosts. Some enthusiasts carried this principle too far and invariably made the occurrence on a different host an excuse for the description of a new species. In the North American Cryphalini (bark beetles) alone, 53 of the "species" described by Hopkins turned out to be synonyms. Downey (1962) and Kohn and Orians (1962) have given useful summaries of some of the relevant literature. Host specificity of external parasites was discussed by Clay (1949), Hopkins (1949), and Holland (1964).

Ecological differences between subspecies are frequent. Indeed most widespread species exhibit differences in the ecology of local populations, particularly of peripherally isolated populations (Mayr, 1963, pp. 312ff, 355).

Again, as with behavioral characters, differences on the species level have been studied far more extensively than have ecological differences between higher taxa. Many of these are simply taken for granted. That whales occupy a different adaptive zone from that of bats is too obvious to be mentioned. Yet a close study shows that even most genera, when well-founded, occupy definably different niches or adaptive zones. Lack (1947) showed for instance that each genus of Galapagos finches is characterized by its utilization of the environment. *Geospiza* is a ground finch (chief food, seeds), *Camarhynchus* a tree finch (chief food, insects), and *Certhidea* a warbler finch (chief food, small insects). For more on this topic, see 5.3.2 and 10.5.3*b*, also Bowman (1961).

**7.4.12 Parasites and Symbionts.** In several instances sibling species were discovered because their parasites were different. Two species of *Octopus* in California are hosts to two different species of mesozoan parasites. A new species of termites was discovered because its nests contain a different set of termitophile staphylinid beetles than those of a previously known species. Parasites are also important in contributing to our knowledge of the relationship of higher taxa. Parasites evolve together with their hosts and are in some cases more conservative than their hosts. Unfortunately, they shift to new hosts more frequently than is sometimes admitted, and the evidence on parasites must be evaluated very carefully (Baer, 1957). For instance, the flamingos (Phoenicopteri) exhibit characteristics which they share with both storks and geese. Their bird lice (Mallophaga) belong to the same genera that occur on the geese. At first sight this might suggest close relationship, but it actually indicates merely a comparatively recent transfer of the lice from the geese to the flamingos. If the bird lice had been derived from a common ancestor, we would expect related, but slightly different, parasites in the two orders of birds. The anatomical evidence indicates intermediacy, but on the whole, closer affinity with the storks than with the geese. Man (*Homo*) and the African apes (*Pan*) share more external and internal parasites with each other than does *Pan* with the orang (*Pongo*). This strengthens the case for a close relationship between *Homo* and *Pan* (established on other grounds). However, it might also be due to cross-infection between sympatric host species, although man also occurs in the habitat of *Pongo*.

The fact that intracellular symbionts supply important taxonomic characters is the discovery of P. Buchner (1966*a*) and his school. For instance, the most primitive tribes of the coccids (Steingeliini, etc.) have no symbionts, but once a coccid taxon has acquired them this symbiont (with all its highly specific adaptations) will be found in the derived phyletic lines of coccids. Repeatedly, unnatural taxa of coccids could be unmasked because they had heterogeneous complements of symbionts (see also Buchner, 1966*b*). The same is true for the symbionts of other groups of insects. The protozoan faunas in the intestines of termites evolved together with their hosts and are potentially useful indicators of relationship in cases of ambiguity in the termite classification (Kirby, 1950*b*).

**7.4.13 Geographical Characters.** Geographical characters are among the most useful tools for clarifying a confused taxonomic picture and for testing taxonomic hypotheses. Most sound classifications show some correlation with geographic or associated ecologic features. The taxonomist is primarily interested in two kinds of geographical characters, (1) general biogeographic patterns, which are especially useful in the arrangement and interpretation of higher taxa, and (2) the allopatric-sympatric relationship, which is most helpful in determining whether or not two populations are conspecific (see Chap. 9). For a detailed treatment of various aspects of the relation between taxonomy and geography, see also *Systematics Association Publication* no. 4 (D. Nichols, 1962), and Simpson (1965).

Studies of the distribution of large numbers of groups of plants and animals have revealed well-defined geographic patterns. Biogeographers have divided the world into various realms, regions, provinces, subprovinces, etc. based upon generalized comparisons of faunas and floras. These are not rigidly defined, but in general they represent distributional centers which exist today or have existed in the past. The units of distribution are not static but may be expanding or retreating, and we thus find it more useful to refer to them as faunas, floras, or biotas rather than as zones or areas. A taxonomist should have an understanding of the geological history of the regions in which such biotas center, as well as a knowledge of the past relationships of the faunas and floras concerned (Mayr, 1964*a*). Possession of this information will permit a much sounder interpretation of various higher taxa.

The reason for the great taxonomic value of geographical distribution

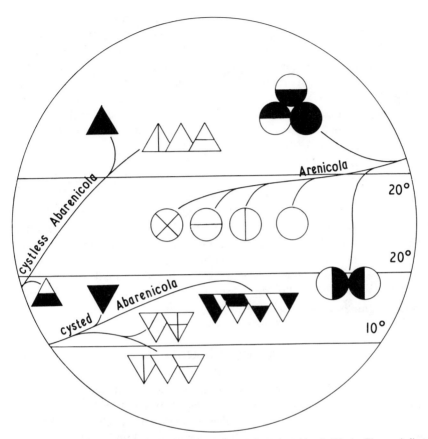

Fig. 7-7. Distribution pattern of groups of species of lugworms (Arenicolidae). Figures indicate isotherms of ocean surface waters (*from Wells, 1963*).

is evolution and monophyly. Just as every taxon is descended from a common ancestor, so is every colonization across a barrier effected by a founder species. In other words, there is high probability that related species in an area are descendants of a common ancestor and that in the majority of the cases no other species of this group occur disjunctly at a far distant place. For instance, the mammals of South America are either not related to those of Africa or, if of common ancestry, have presumably reached South America by way of North America. The hystricomorph rodents, seemingly close to the African porcupines, appeared to be an exception, owing to the absence of porcupines in the early Tertiary period of North America. A reexamination of these rodents, prompted by this zoogeographical puzzle, revealed that the porcupines of South America and of Africa are of independent origin (Wood, 1950). Distributional difficulties have shed light on taxonomic relationships in many other instances. They suggest, for example, that the New Zealand "thrushes" (*Turnagra*) are not thrushes but presumably Pachycephalinae, and the New Zealand "tits" not Paridae but Malurinae. Both reassignments lead not only to an improved classification but also to a considerable zoogeographical simplification. Distribution is thus shown to be an important tool in taxonomic analysis. Its most meaningful application consists in always asking whether our inferences on the relationship of taxa are consistent with their pattern of distribution.

**7.4.14    The Relative Value of Different Kinds of Characters.** Each kind of character adds to the total information. Very often the taxonomist is limited to preserved material, in other words essentially to morphological characters and to such geographical and ecological information as came with the specimens. Whenever the morphological analysis permits ambiguity, and this is often the case, every effort should be made to supplement the basic information with biochemical, behavioral, or other additional information. Different kinds of characters are used in the discrimination of species than in the determination of relationships among higher taxa. It is only within the most recent past that an effort has been made to determine the relative weight of different kinds of characters at different levels of the taxonomic hierarchy.

## 7.5    CHARACTERS AND CATEGORICAL RANK

Characters that unequivocally designate categorical rank do not exist. We cannot say that species $A$ must be placed in a separate genus because it has a generic character (5.4.2). The degree of morphological difference between species and genera is much greater in some families than in others. There are stronger behavioral differences among closely related species in some groups than in others. The same character that is associated with generic difference in one family may vary individually in species of another family. Nor does the total level of difference tell us much. Morphs within a single population may differ more from each other than do good sibling species, at least in visible morphological characters. Again and again we must remember that it is the taxon which gives us the characters, not the reverse.

Empirically, the experienced taxonomist knows what kind of characters have what kind of value in the particular group in which he is a specialist. Any difference that functions as an isolating mechanism or facilitates competitive exclusion is primarily useful for the discrimination of species. What particular system of communication is used in isolating mechanisms (light signals in fireflies, calls in frogs, etc.) may serve as character of a higher taxon, while a specific isolating mechanism may characterize a species. On the other hand, characters that indicate occupation of a particular adaptive zone or that are by-products of a well-defined, well-integrated genotype generally help in the discrimination and delimitation of higher taxa. Recent studies of biochemical characters have clearly shown that some of them are most useful at the species level, for instance some immunological methods, while others are not sensitive at that level but are full of information concerning the higher taxa; still others are subject to local polymorphism and have no taxonomic value. It is up to the taxonomist to find out what character will be most useful at a given categorical level.

Even though broad generalizations cannot yet be made, a differential significance of characters is gradually becoming apparent within each higher taxon. Different kinds of characters appear to be most useful at the following categorical levels:

1. The recognition of subspecies (geographic variation)
2. Distinguishing closely related species, particularly sibling species
3. The grouping of related species into genera
4. Determining the relationship of higher taxa, from families to phyla

## CONCLUSIONS

A taxonomic character is any attribute by which a member of a population may differ from members of other populations of organisms. This attribute may relate to any feature of the dead or living organism that is amenable to comparison.

1. Taxonomic characters that are conservative (i.e., that evolve slowly) are most useful in the recognition of higher taxa; those that change most rapidly or concern isolating mechanisms are most useful in the lower taxa.
2. Taxonomic characters that are subject to parallelism, especially those involving loss or reduction, should be used only with great caution.
3. Taxonomic characters are expressions of the biology of their carriers. An understanding of this biology is a prerequisite for the proper evaluation of these characters.
4. The same phenotypic character may vary in value and constancy from taxon to taxon and even within a single phyletic series. The weight to be given to such a character depends largely on its constancy in the given group.

The entire zoological classification is based on the proper evaluation (weighting) of taxonomic characters. This operation, then, is the most important as well as one of the most difficult tasks of the taxonomist (10.4).

# Chapter 8 The Qualitative and Quantitative Analysis of Variation

*I*n his work the taxonomist does not encounter species *as such;* he works with specimens and samples, and his first task is to assemble these into groups of similar specimens, *phena* (Chap. 1). These in turn he assigns to species. The history of taxonomy, particularly the evidence supplied by the long lists of synonyms in many groups, proves that this is not an easy task, owing to the great variability of most species.

A student of inanimate objects establishes classes merely on the basis of similarity. He would not hesitate to place things as different as the caterpillar, chrysalis, and imago of a butterfly in three different classes, while he might place the adults of two similar species of butterflies in the same class. The biologist, however, knows that he must take into consideration other factors than mere similarity. The continuity of the genotype from the fertilized egg through all larval stages to the adult is one such factor. Sexual reproduction—which results in a genetic cohesion among all the individuals belonging to the same local population—is another such factor. All members of a local population are, so to speak, products of the same gene pool and are thus a single taxonomic entity.

It is the knowledge of these biological phenomena which permits the taxonomist to assign phena correctly to species. In a species with sexual dimorphism, for instance, males and females will belong to two different phena. The fact that one phenon consists entirely of males and the other entirely of females indicates that these phena are not species. Additional information will now be brought to bear on the situation. For instance, if the two phena are the only ones in the collection made at a given locality representing a certain genus and were collected simultaneously in the same habitat, the probability is high that they represent males and females of the same species. Breeding tests, the raising of larvae, and the study of courtship and copulation in nature furnish additional sets of biological information permitting the correct assignment of phena. The number of possible inferences from the available information is usually large. A knowledge of the nature and amount of sexual dimorphism found in living (Recent) species often permits the correct assignment of fossil phena to species, as is the case, for instance, among fossil ostracods.

The greater the amount of information available about the phena in question, the easier their classification. All the information must be considered that is specified in Chap. 6, such as correct locality, habitat (and other relevant ecological information), and season of capture. The reasons for precision in this information are twofold:

1. Many aspects of ecology and life history are species-specific (see Chap. 7).
2. The phenotype of animal populations of the same species often varies according to locality, season, or habitat (see below).

Differences between phena may thus reflect either a species difference or intraspecific variation. A complete understanding of intraspecific variation is therefore necessary before we can make the probabilistic statement that phenon $B$ belongs to a different species from phenon $A$. This is the reason for the immense importance of a thorough understanding of individual and geographic variation.

The taxonomic literature still contains numerous named phena that have not yet been correctly combined into biological species. This includes males and females in sexually dimorphic groups of insects (8B.1), workers and sexual castes in social insects, stages in the life cycle of parasites, and juvenile stages as well as morphs. It is one of the continuing preoccupations of the taxonomist to unmask nominal species that are not genuine biological species.

*Basic Questions Concerning Phena.* When a taxonomist has before him two or more phena which he must correctly assign to the species to which they belong, he must ask himself these questions:

1. *Do the phena or samples come from the same locality?* If the answer is yes, only two alternatives are possible. Either the phena are individual variants of a single species or else they are several good species (Chap. 9A.1). If the phena are from different localities, there is a third possibility, namely that they represent different subspecies of a geographically variable species (Chap. 9B).

*2. Is there any evidence for reproductive isolation between the populations from which the phena were sampled?* This is a much more difficult question to answer. Collected material may provide only a few clues helpful in this problem.

If the investigator has the answer to these questions, he can read from the discrimination grid (Table 8-1) what the taxonomic disposition of the phena ought to be.

Presence or absence of morphological difference as well as geographic relationship can nearly always be determined in properly labeled material. It is the lack of information on reproductive isolation which causes most of the difficulties. By looking at the discrimination grid we can see that without information on reproductive isolation we shall have difficulty in choosing between alternatives (1) or (5), (2) or (6), (3) or (7), (4) or (8). A high percentage of taxonomic errors are the result of a wrong choice between one of these pairs. What can we do, in the absence of direct information on reproductive isolation, to avoid error? The experienced taxonomist knows that a full understanding of both individual and geographic variation is more helpful in the avoidance of error than is anything else. Most of the present chapter will therefore be devoted to a thorough discussion of variability. Even though every kind of animal has its own pattern of variability, it is apt to fall somewhere into the framework presented in this chapter.

Phena collected at the same locality (or geological horizon) are either different biological species or manifestations of individual variation (alternatives 3 and 7). In order to be able to discriminate between the two alternatives, in the absence of information on reproductive isolation, it is necessary to review the total potential of a species for individual variation. Such variation is far greater than the beginner realizes, and it sometimes deceives even experienced taxonomists. It is estimated that more than half of all synonyms owe their origin to an underestimation of individual variation.

TABLE 8-1. DISCRIMINATION GRID

|  | Not reproductively isolated | Reproductively isolated |
|---|---|---|
| Morphologically identical: | | |
|   Sympatric | (1) Same population | (5) Sibling species |
|   Allopatric | (2) Same subspecies | (6) Sibling species |
| Morphologically different: | | |
|   Sympatric | (3) Phena of the same population | (7) Different species |
|   Allopatric | (4) Different subspecies | (8) Different species |

TABLE 8-2.  MAJOR TYPES OF VARIATION WITHIN A SINGLE POPULATION

I. Nongenetic variation
    A. Individual variation in time
        1. Age variation
        2. Seasonal variation of an individual
        3. Seasonal variation of generations
    B. Social variation (insect castes)
    C. Ecological variation
        1. Habitat variation (ecophenotypic)
        2. Variation induced by temporary climatic conditions
        3. Host-determined variation
        4. Density-dependent variation
        5. Allometric variation
        6. Neurogenic color variation
    D. Traumatic variation
        1. Parasite-induced variation
        2. Accidental and teratological variation
II. Genetic variation
    A. Sex-associated variation
        1. Primary sex differences
        2. Secondary sex differences
        3. Alternating generations
        4. Gynandromorphs and intersexes
    B. Non-sex-associated variation
        1. Continuous variation
        2. Discontinuous variation (genetic polymorphism)

Various forms of individual variation are listed in Table 8-2. The two most important classes are nongenetic and genetic variation.

## 8A. NONGENETIC VARIATION

It is of course impossible in a preserved museum specimen to determine directly whether a given variant has a genetic basis. Nevertheless, it is important for the taxonomist to understand that various types of variation exist and that in the better-known groups it is possible in most cases to make a valid inference as to the status of a given variant on the basis of field observations and available experimental evidence. For a discussion of the evolutionary aspects of individual variation see Mayr (1963, pp. 138–158). As a broad generalization, one can state that nongenetic variation adapts the individual, while genetic variation adapts the population and the species.

Animals, as a whole, are developmentally far better canalized than plants. In addition, through their power of locomotion and their sensory abilities they have the capacity for habitat selection. As a result, some well-known exceptions notwithstanding, nongenetic changes of the phenotype are far less of a problem for animal taxonomists than they are for botanists. However, every zoologist must familiarize himself with the types of nongenetic variability which he might encounter in his group.

## 8A.1   INDIVIDUAL VARIATION IN TIME

**8A.1.1   Age Variation.** Whether they are born more or less developed or whether they hatch from an egg, animals in general pass through a series of juvenile or larval stages in which they may be quite different from adults. The catalogs of any group of animals list numerous synonyms that resulted from the failure of taxonomists to recognize the relationship between various age classes of the same species.

In reptiles, birds, and mammals there are no larval stages, but immature individuals may be rather different from the adults, particularly in birds. For example, Linnaeus described the striped immature goshawk (*Accipiter gentilis* Linnaeus) as a different species (*gentilis*) from the adult (*palumbarius*) with its crossbarred underparts. Several hundred bird synonyms are based on juvenal plumages. By finding specimens that molt from the immature into the adult plumage it is usually easy to clear up this difficulty.

In many small mammals (e.g., rodents and insectivores), samples collected in different seasons differ in the proportion of adults and juvenals. Adults and immatures must be separated before samples from different localities are compared.

In many fishes the immature forms are so different that they have been described in different genera or even different families. The immature stages of the eel (*Anguilla*) were originally described as *Leptocephalus brevirostris* Kaup. The unmasking may be especially difficult in neotenic animals, that is, animals that become sexually mature in a larval stage.

The difficulties for the taxonomist are even greater in groups with larval stages which are so different that they have not even the faintest resemblance to the adult (e.g., caterpillar and butterfly). The floating or free-swimming larvae of sessile coelenterates, echinoderms, mollusks, and crustaceans are often extremely different from the adults. The taxonomic status of such larval stages can be settled either by assembling a complete sequence of intermediate stages or by rearing them through a complete life cycle.

The taxonomic identification of larval stages of parasites is made particularly difficult in groups where the different stages occur on different hosts. It is customary in helminthology to assign formal taxonomic names

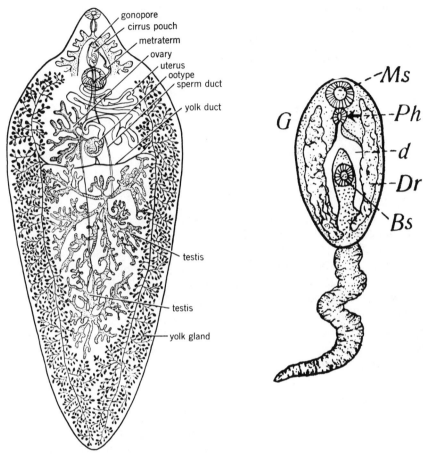

gonopore
cirrus pouch
metraterm
ovary
uterus
ootype
sperm duct
yolk duct

Ms
G
Ph
d
Dr
Bs

testis

testis

yolk gland

Fig. 8-1. Difference between adult liver fluke (*Fasciola hepatica*) and its larval stage (*Cercaria*) (*after Chandler and Read*).

to the larval (*Cercaria*) stage of flukes (trematodes), in order to facilitate their ready identification. Such dual nomenclature (Art. 42c) is dropped as soon as it becomes known to what trematode species a given cercaria belongs. This can usually be established only through rearing.

Age variation is not restricted to differences between larval stages and adults but occurs also between "young" and "old" adults. For example, in various species of deer (*Cervus,* etc.) it is known that older stags often have antlers with more points than younger ones. The shape of the antlers may also change. This age variation must be considered when the antlers of different species or subspecies are compared. There is probably no more addition of points (or only an irregular one) after a certain age has been reached. It would be as futile to try to determine the exact age of a stag by the number of points of its antlers as to determine the age of a rattlesnake (*Crotalus*) by the number of rings in the rattle, or the age of a hornbill (*Aceros plicatus* Forster) by the number of folds in the casque on the bill.

The taxonomist aims to work with samples that are as homogeneous as possible. It is much easier to achieve this in animals that have a definite adult stage (after the larval one) than it is in those that show continuous growth, such as snakes or fishes, which may reach maturity after having attained only half their potential size, or less. In such forms it is advisable. to work with regressions rather than with absolute measurements. Meristic characters (e.g., number of scales or fin rays) are in many species not increased after they are formed, in spite of the enormous subsequent growth. These characters are therefore especially important in herpetology and ichthyology.

In birds it is generally assumed that final size is reached with the first completely adult plumage. There is now good evidence that this is not always so. In the hornbill of the Papuan region, *Aceros plicatus,* it is well established that those adults which have only two or three folds on the bill are younger than those with five or more folds. Birds in adult plumage with five to eight folds have a bill length of 198 to 227 mm, while birds in equally adult plumage with two or three folds have a bill of 185–199 mm. In some passerine birds it has been possible through banding to show that the size of known adult individuals increases slightly over the years.

**8A.1.2    Seasonal Variation of the Same Individual.** In animals that live as adults through several breeding seasons, it sometimes happens that the same individual has a very different appearance in different parts of the year. Many birds have a bright nuptial dress which they exchange for a dull ("eclipse") plumage at the end of the breeding season. Among North American birds this is true, for example, of many ducks, shore birds, warblers, tanagers, and others. In many species a change of plumage is restricted to the males.

In arctic and subarctic birds and mammals, such as ptarmigans (*Lagopus*) and weasels (*Mustela*), there may be a change from a cryptic white winter dress to a "normally" colored summer dress. In other birds the colors of the soft parts change with the seasons. In the common egret (*Egretta alba* Linnaeus) and in the European starling (*Sturnus vulgaris* Linnaeus), the bill may change from yellow to black. Plumage changes in birds are usually effected by molting, but wear alone may produce striking changes. In the European starling (*Sturnus vulgaris*), for example, the freshly molted bird of October is covered with white spots, and all the feathers show whitish or buffy margins. During the winter the edges of the feathers wear off, and in the spring, at the beginning of the breeding season, the whole bird is a beautiful glossy black without the molt of a single feather. A similar process of wear brings out the full colors of the nuptial plumage in the males of many other birds. In arid regions, particularly in real deserts, the sun bleaches the pigments. A bird before the molt will look much paler than one in freshly molted plumage.

In all these cases it is the same individual which in different parts of the year looks very different. Such seasonal variation is particularly common among vertebrates, with their elaborate endocrine systems. Many such seasonal variants were described as distinct species before their true nature was realized.

**8A.1.3   Seasonal Variation of Consecutive Generations.** Many species of short-lived invertebrates, particularly insects, produce several generations in the course of a single year. In such species it is not uncommon that the individuals which hatch in the cool spring are quite different from those produced in the summer, or that the dry-season individuals are different (e.g. paler) from the wet-season population.

Such seasonal forms can usually be recognized not only by the occurrence of intermediates in the intervening season, but also through identity of wing venation, genitalia, etc.

*Cyclomorphosis.* A special kind of seasonal variation is found in certain freshwater organisms, particularly rotifers and cladocerans. The populations of a species undergo quite regular morphological changes through the seasons, in connection with changes in the temperature, turbulence, and other properties of the water. Many "species" have been named, particularly in the genus *Daphnia,* that are nothing but seasonal variants. There are several recent analyses of this cyclomorphosis by Brooks (cladocerans), by Buchner (rotifers), and by their associates. In rotifers different morphs may result from different kinds of food.

## 8A.2   SOCIAL VARIATION (INSECT CASTES)

In the social insects, such as some bees and wasps, but particularly among ants and termites, castes have developed. These are definite groups of individuals within a colony in addition to the reproductive castes (queens, and males or drones): workers (sometimes of different types), and soldiers (also sometimes of different types). In the Hymenoptera, these castes are most commonly modified females and genetically identical (except for the workers in some social bees), but in the Isoptera (termites) both sexes may be involved. The structural types observed may result from different larval food or may be due to hormonal or other controls. Obviously, taxonomic names should not be applied to these intracolonial variants; but invalid species have sometimes been described because it was not realized that there were different types of soldiers or workers in the same colony.

## 8A.3   ECOLOGICAL VARIATION

8A.3.1   **Habit Variation (Ecophenotypic).** Populations of a single species that occur in different habitats in the same region are often visibly different. The taxonomic treatment of such local variants has fluctuated between two extremes: some authors have described them as different species while others consider them all as nongenetic variants. Actually they may be (1) microsubspecies (or ecological races) or (2) nongenetic ecophenotypes. The latter are particularly common in plastic species, such as some mollusks.

Dall (1898) gave a very instructive account of all the variations he observed in a study of the oyster (*Crassostrea virginica* Gmelin):

> The characteristics due to situs may be partially summarized: When a specimen grows in still water, it tends to assume a more rounded or broader form, like a solitary tree compared with its relatives in a crowded grove. When it grows in a tideway or strong current the valves become narrow and elongated, usually also quite straight. Specimens which have been removed from one situs to another will immediately alter their mode of growth, so that these facts may be taken as established. When specimens are crowded together on a reef, the elongated form is necessitated by the struggle for existence, but, instead of the shells being straight they will be irregular, and more or less compressed laterally. When the reef is dry at low stages of the tide, the lower shell tends to become deeper, probably from the need of retaining more water during the dry period. . . . When an oyster grows in clean water on a pebble or shell, which raises it slightly above the bottom level, the lower valve is usually deep and more or less sharply radially ribbed, acquiring thus a strength which is not needed when the attachment is to a perfectly flat surface which acts as a shield on that side of the shell. Perhaps for the same reason oysters which lie on a muddy bottom with only part of the valves above the surface of the ooze are less commonly ribbed. When the oyster grows to a twig, vertical mangrove root, or stem of a gorgonian, it manifests a tendency to spread laterally near the hinge, to turn in such a way as to bring the distal margin of the valves uppermost, and the attached valve is usually rather deep, the cavity often extending under and beyond the hinge margin; while the same species on a flattish surface will spread out in oval form with little depth and no cavity under the hinge.

In freshwater snails and mussels, such habitat forms are particularly common. The upper parts of rivers, with cooler temperatures and a more rapid flow of water, have different forms from the lower reaches, with warmer and more stagnant waters. In limestone districts the shells are heavy and of a different shape from those which grow in waters poor in lime. This dependence of certain taxonomic characters on environmental factors was, curiously enough, entirely overlooked by some earlier workers, a fact which resulted in absurd systematics. Schnitter (1922), who largely cleared up the situation, described these absurdities as follows:

> The last step in the splitting of the freshwater mussels of Europe was done by the malacozoologists Bourguignat and Locard. According to the shape and the outline of the shell, they split up the few well-known species into countless new ones. Locard lists from France alone no less than 251 species of *Anodonta*. On the other hand, two mussels were given the same name, if they had the same outline of the shell, even though one may have come from Spain and the other from Brittany. It seems incredible to us that it never occurred to these authors to collect a large series at one locality, to examine the specimens, to compare all the individuals and to record the intermediates between all these forms. It is equally incomprehensible that these people did not see the correlation between environment and shape of shell, even though they spent their entire lives in collecting mussels.

All these "species" of *Anodonta* are now considered to be habitat forms of two species, and the other names have been sunk into the synonymy of the two valid species.

Whether a given habitat form is an ecophenotype or a microgeographic race is not always evident. It is sometimes necessary to transplant it or to raise it in the laboratory in order to solve this question. Much work of this sort still remains to be done.

8A.3.2 **Variation Induced by Temporary Climatic Conditions.** Some animals with a highly plastic phenotype may produce year classes that differ visibly from the norm owing to unusual conditions (drought, cold, food supply, etc.) in a given year. Fish of a given year class may be stunted or, on the contrary, have proportions indicative of particularly rapid growth. Samples of susceptible species must be collected in such a way as to compensate for distortions caused by this factor (Harrison, 1959; Mayr, 1963, p. 145).

8A.3.3 **Host-determined Variation.** Host-determined variations in parasites of plants and animals provide a potential source of taxonomic error and permit confusion with microgeographical races or with sympatric species. This phenomenon is most commonly expressed in size differences but may involve other morphological or physiological characters.

Gerould (1921) has reported that the braconid wasp, *Apanteles flaviconchae* Riley, spins white cocoons when reared from blue-green caterpillars of *Colias philodice* Godart, but golden cocoons when reared from yellow-green caterpillars from the same species. In another parasitic wasp, *Trichogramma semblidis* Aurivillius, Salt (1941) found that males tend to be wingless and otherwise modified when they develop in the eggs of the alderfly, *Sialis lutaria* (Fabricius) (Megaloptera), but not when reared from lepidopterous hosts.

Another kind of difficulty is cited by H. S. Smith (1942). He states that the encyrtid wasp, *Habrolepsis rouxi* Compère, readily parasitizes red scale on citrus but is unable to do so when the red scale is reared on *Cycas*. This apparent immunization by the plant host might well confuse interpretations which utilize parasites as a taxonomic character.

Host-induced variation is not as common as it was formerly believed to be. Many so-called host races of the literature have been found in recent decades to be valid sibling species (Mayr, 1963, chap. 3).

**8A.3.4   Density-dependent Variation.** The effects of crowding are sometimes reflected in morphological variation. This is not uncommon where crowding produces a shortage of food materials. However, density-dependent variation need not be related to food supply. Uvarov, Kennedy, and others have shown that gregarious species of locusts exist in various unstable biological phases. (Kennedy, 1956, 1961; Albrecht, 1962). These phases differ in anatomy (Fig. 8-2), color, and behavior characteristics and have often been described as distinct species. When newly hatched nymphs are reared under crowded conditions, they mostly develop into the gregarious phase; under less crowded conditions into the transitional phase; and when isolated and reared separately, into the solitary phase. Similar phases have been reported by Faure (1943*a*, 1943*b*) in two species of armyworms (Lepidoptera), *Laphygma exigua* (Hübner) and *L. exempta* (Walker).

**8A.3.5   Allometric Variation.** Allometric growth may result in the disproportionate size of some structure in relation to that of the rest of the body. If individuals of a population show allometric growth, animals of different size will show allometric (heterogonic) variability. This is par-ticularly well marked among insects. It involves such features as the heads of ants (Fig. 8-3), the mandibles of stag beetles (Lucanidae), the frontal

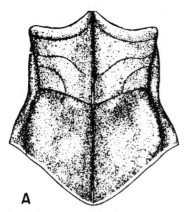
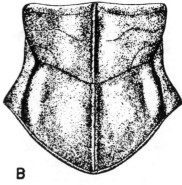

**A**                              **B**

Fig. 8-2. Morphological differences between sedentary and migratory phena of *Locusta migratoria* Linnaeus. Pronotum of female in dorsal view, *A* solitary, *B* migratory phase (*from Uvarov,* 1921).

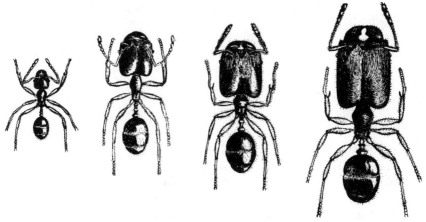

Fig. 8-3. Allometric variability. Neuters of *Pheidole instabilis,* showing increase in the relative size of the head with absolute size of the body (*after Wheeler,* 1910).

horns and thoraces in scarabs, antennal segments of thrips, etc. Failure to recognize the nature of such variations has resulted in much synonymy.

The exact causes of much of this variation are unknown. In species with continuous growth, it is actually a form of age variation (see 8A.1). Some of it has a genetic basis and is more properly classified under 8B.2. In holometabolic insects, however, where this phenomenon is particularly common, it is closely correlated with size, and this in turn is thought to be the result of variation in food supply, which causes the larva or nymph to metamorphose at different growth stages. For a recent discussion of allometry see Gould (1966).

**8A.3.6   Neurogenic or Neurohumoral Variation.** Neurogenic or neurohumoral variation is color change in individual animals in response to the environment. Such changes are accomplished through the concentration or dispersal of color-bearing bodies known as *chromatophores.* This type of variation was first thoroughly studied in the chameleon. It occurs sporadically in the lower animals but is best developed among the crustaceans, cephalopods, and cold-blooded vertebrates (cyclostomes, elasmobranchs, teleost fishes, amphibians, reptiles). Space will not permit a discussion of this specialized type of variation. For details the reader is referred to Fingerman (1963), Gersch (1964), and Waring (1963).

## 8A.4   TRAUMATIC VARIATION

Traumatic variation occurs with varying frequency in different groups of animals. The abnormal nature of this type of variation is usually obvious, but in some cases it is subtle and may be misleading.

8A.4.1   **Parasite-induced Variation.** Aside from such familiar effects of parasitism as swelling, distortion, and mechanical injury, parasites may produce conspicuous structural modifications. In the bee genus *Andrena,* for instance, parasitism by *Stylops* frequently results in reduction in the size of the head, enlargement of the abdomen, and changes in puncturation, pubescence, and wing venation. It commonly results, also, in intersexes. Since *Andrena* is markedly sexually dimorphic, these intersexes have been a source of taxonomic confusion and synonymy. However, in one case (Linsley, 1937) a stylopized intersex proved of value in associating the sexes of a bee which had been described as two different species.

Salt (1927) has made the most comprehensive study of the morphological effects of stylopization in *Andrena.* In females he found reduction of the pollen-collecting organs, loss of anal fimbriae, changes in relative length of antennal segments, reduction of facial foveae, reduction of the sting and accessory organs, paling of ventral abdominal pubescence, acquisition of angular cheeks, and yellow on the normally dark clypeus. In males he reports the development of long hairs resembling the female flocculi, broadening of the posterior basitarsus, changes in proportions of antennal segments, loss of cheek angles and some yellow from the clypeus, indications of facial foveae, and reduction in size of genitalia.

Some strikingly different termite soldiers from the Orient were assigned to a new genus and species, *Gnathotermes aurivillii.* Later it was shown that these modified soldiers are nothing but parasitized individuals from colonies of *Macrotermes malaccensis* (Haviland).

8A.4.2   **Accidental and Teratological Variation.** Accidental variation is usually externally induced, although it may work internally through some developmental or hormonal system. The external stimuli may be mechanical, physical, or chemical. Such variation is extremely diverse; in most animals it may be readily identified because the individuals involved either deviate so markedly from type as to be recognized as freaks, or because the injuries or abnormalities involved are asymmetrical. However, in those forms which undergo metamorphosis, injuries to an earlier stage may produce later abnormalities which are not so easily recognized as such. This is especially true when the anomalies involve characters which are normally of taxonomic value in the group concerned. For instance, certain types of pupal injury in beetles may produce symmetrical abnormalities in punctation, surface sculpturing, or segmentation of appendages; in butterflies, symmetrical modification of wing patterns. In most cases, however, even with such subtle difference, the abnormal nature of the variation may be detected by the specialist without much difficulty.

Teratological variation has been elaborately studied and classified by Cappe de Baillon (1927) and Balazuc (1948). The student interested in pursuing this subject further is referred to these works for details and for further references.

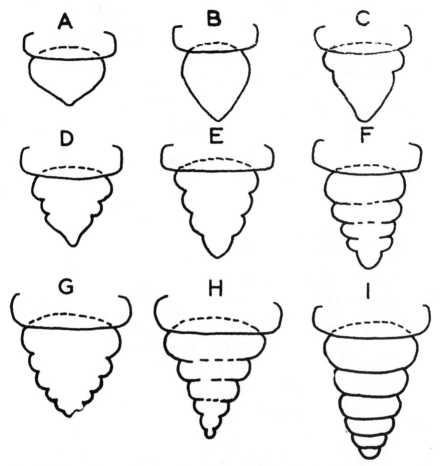

Fig. 8-4. Different degrees of segmentation (A–I) in the pleon of males of the parasitic copepod *Pleurocrypta porcellanae*. Stage I was originally described as a new species in a different genus (*from Bourdon,* 1965).

## 8A.5  POST-MORTEM CHANGES

The taxonomist must guard against one further type of individual variation. In many groups of animals it is impossible to prevent post-mortem changes of preserved specimens. Some extreme cases are known in birds. The deep orange-yellow plumes of the twelve-wire birds of paradise (*Seleucidis ignotus* Forster) fade in collections to white. Skins of the Chinese jay (*Kitta chinensis* Boddaert), whose plumage is green in life, turn blue in collections, owing to the loss of the volatile yellow component in the pigment. Many birds that are clear gray or olive-gray when freshly collected become more and more rufous through oxidation of the black pigment ("foxing"). Many synonyms have been created in ornithology owing to the comparison of freshly collected material with old museum specimens.

Other post-mortem changes result from the chemical action of preservatives or killing agents. A common color change of this nature takes place when certain yellow insects, especially wasps, are overexposed to cyanide. The specimens turn bright red, and so far no method has been found for reversing this reaction without injury to the specimens.

When preserving specimens with evanescent colors (corals, marine slugs, etc.), it is essential to take full notes and preferably color photographs or water-color sketches. This will make possible an accurate description of the living animal.

## 8B. GENETIC VARIATION

In the cases of variation discussed in the preceding section, the same individual is actually or potentially subject to a change in appearance. In addition to this noninherited variation, there is much intrapopulation variation that is primarily due to differences in genetic constitution. This genetically induced individual variation can—somewhat arbitrarily—be divided into two classes.

### 8B.1   SEX-ASSOCIATED VARIATION

Among genetically determined variants within a population, there are many which are sex-associated. They may be sex-limited (express themselves in one sex only) or be otherwise associated with one or the other sex, or they may involve sex characters or modes of reproduction. Some of these are as follows:

8B.1.1   **Primary Sex Differences.** These are differences involving the primary sex organs utilized in reproduction (gonads, genitalia, etc.). Where the two sexes are otherwise quite similar, primary sex differences will rarely be a source of taxonomic confusion.

8B.1.2   **Secondary Sex Differences.** There is more or less pronounced sexual dimorphism in most groups of animals. The differences between male and female are often very striking, as for instance in the birds of paradise, hummingbirds, and ducks. In many cases the different sexes were originally described as different species and retained this status until painstaking work by naturalists established their true relationship. A celebrated case is that of the king parrot *Eclectus roratus* (Müller) of the Papuan region, in which the male is green with an orange bill, the female red and blue with a black bill. The two sexes were considered different species for nearly one hundred years (1776–1873), until naturalists proved conclusively that they belonged together.

Striking sexual dimorphism is particularly frequent in the Hymenoptera. The males of the African ant *Dorylus* are so unlike other ants that they were not recognized as such and were for a long time considered to belong to a different family. In the mutillid wasps (Mutillidae) the small wingless female and the large winged male are so different that some taxonomists use a different nomenclature for the two sexes. Whole "genera" consist entirely of males, others of females. The best way of determining with which female of "genus" *B* a given male in "genus" *A* belongs is to find a pair *in copula* or to watch a female in the field and catch the males as they are attracted to her. Once it has been established that *B* is the female of *A*, it is sometimes possible to associate several other "species pairs" in the same genus by utilizing additional information on distribution, frequency, color characters, etc.

**8B.1.3    Alternating Generations.** In many insects there is an alternation of generations that is very confusing to the taxonomist. In the genus *Cynips* (gall wasps), the agamic generation is so different from the bisexual one that it has been quite customary to apply different scientific names to the two (Kinsey, 1930). In the aphids (plant lice) the parthenogenetic wingless females are usually different from the winged females of the sexual generations (Fig. 8-5).

**8B.1.4    Gynandromorphs and Intersexes.** Gynandromorphs are individuals that show male characters in one part of the body and female characters in another part. Thus the two halves of the body may be of opposite sexes, or the division may be transverse, or the sex characters may be scattered in a mosaic. In the latter case symmetrical variants may be produced. Usually gynandromorphs are easily recognized as such and rarely provide a source of taxonomic confusion. Gynandromorphism is produced by an unequal somatic distribution of chromosomes, particularly the sex chromosomes.

Unlike gynandromorphs, intersexes are likely to exhibit a blending of male and female characters. Intersexes are generally thought to result from an upset in the balance between male-tendency and female-tendency genes. This upset may result from irregularities in fertilization or mitosis, or from physiological disturbances associated with parasitism. Intersexes are particularly apt to appear in populations of interspecific or inter-subspecific hybrids. They have been studied in greatest detail in *Lymantria* (Goldschmidt, 1933) but are well known in many other animals.

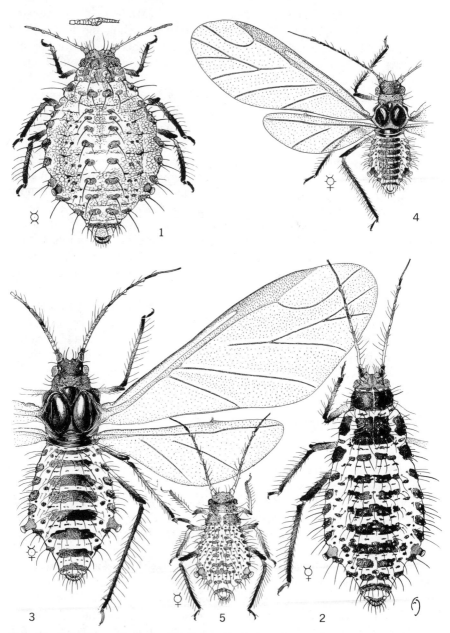

**Fig. 8-5.** *Periphyllus californiensis* (Shinji). 1, Fundatrix or stemmother; 2, normal apterous parthenogenetic viviparous female; 3, alate of same; 4, smallest spring alate viviparous female (*Essig and Abernathy*, 1952).

## 8B.2  NON-SEX-ASSOCIATED INDIVIDUAL VARIATION

This term is simply one of convenience applied to intrapopulation variation which is not sex-limited or does not primarily involve sex characters.

**8B.2.1  Continuous Variation.** The most common type of individual variation is that which is due to the slight genetic differences which exist between individuals. No two individuals (except monozygotic twins) in a population of sexually reproducing animals are exactly alike, genetically or morphologically. One of the outstanding contributions of population genetics has been the establishment of this fact. The differences are in general slight and are often not discovered unless special techniques are employed.

The study of this variation is one of the foremost tasks of the taxonomist. It is now evident that no one individual is "typical" of the characters of a population. Only the statistics of the whole population can give a true picture of the population (Chap. 8C). A model study of variability, based on 2,877 skins of the house sparrow (*Passer domesticus*) is presented by Selander and Johnston (1967).

Each character may show a different degree of variability within a single population. Likewise there are different degrees of variability among related species. Just why one species should be highly variable and another one not is not always clear. A taxonomist who has adequate material concerning one species should not hastily assume that this would permit him to be certain of the variability of related species.

Some early taxonomists vastly underrated individual variation in many genera of animals. The species of the snail genus *Melania* (fresh and brackish water) have been described largely on the basis of shell characters, such as the presence or absence of spines and of diagonal and spiral ribs. However, spined and spineless specimens occur in the species in *M. scabra,* *M. rudis,* and *M. costata,* sculptured and smooth specimens in *M. granifera,* and so forth. In a revision of this genus, no less than 114 "species" were found to be nothing but individual variants and had to be added to the synonymy of other species (Riech, 1937).

**8B.2.2  Discontinuous Variation (Polymorphism).** The differences between individuals of a population are, in general, slight and intergrading. In certain species, however, the members of a population can be grouped into very definite classes, determined by the presence of certain conspicuous characters. Such discontinuous individual variation is called *polymorphism.* Frequently such polymorphism is controlled by a single gene, subject to simple Mendelian inheritance.

Polymorphism is more pronounced in some groups of animals than

TABLE 8-3. MIMETIC POLYMORPHISM IN WEST AFRICAN *Papilio dardanus* BROWN (*after Goldschmidt,* 1945)

| Male | Nonmimetic females | Mimetic females | Models |
|---|---|---|---|
| Typical *dardanus* | Basic type ♀ similar to ♂<br><br>*dionysus* Doubleday and Hewitson | *hippocoon* Fabricius<br><br>*trophonissa* Aurivillius<br><br>*niobe* Aurivillius | *Amauris niavius* Linnaeus<br><br>*Danaus chrysippus* Linnaeus<br><br>*Bematistes tellus* Aurivillius |

in others. The spotting in lady beetles (Coccinellidae) is a well-known example of genetic polymorphism, as is industrial melanism in moths. Polymorphism has great biological importance, since it proves the existence of selective differences between apparently neutral characters. For a more detailed discussion of the topic, see Ford (1945, 1966) and Mayr (1963). Its practical importance to the taxonomist is that it has led to the description of many so-called "species" that are nothing but polymorphic variants (morphs). In ornithology alone, about 100 species names were given to polymorphs. The establishment of their true nature has led to a considerable simplification of taxonomy.

Perhaps the most spectacular cases of polymorphism are to be seen in the Lepidoptera and more particularly in certain species of butterflies. The common alfalfa butterfly, *Colias eurytheme* (Boisduval), for example, has two strikingly different female forms, one resembling the orange-colored male while the other is largely white. The most complicated cases of sex-limited polymorphism which have been studied genetically are the examples of mimetic polymorphism in African swallow-tail butterflies of the genus *Papilio*. Quite apart from the fact that allopatric populations throughout Africa show distinct subspecific differences which are correlated with differences in the species of butterflies which they mimic, we find that several distinct female forms exist within a single population. Thus in West Africa one finds, in the same population of *Papilio dardanus* Brown, one male form and five female forms, three of the latter mimicking different models which belong to the families Danaidae and Nymphalidae (see Table 8-3 and Fig. 8-6). The most remarkable feature of this polymorphism is that although the various forms are so distinct as to resemble representatives of three different families of Lepidoptera, breeding experiments have shown that the differences are caused by a few Mendelian genes. Another celebrated case is that of *Pseudacraea eurytus* Linnaeus (Carpenter, 1949).

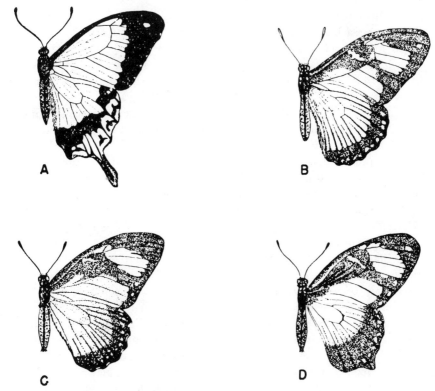

Fig. 8-6. Mimetic polymorphism in the *Papilio dardanus* complex. *A*, male of *cenea*—also basic type of nonmimetic female, ground color yellow; *B*, *dionysus*, nonmimetic female, group color of forewings white, hind wings yellowish; *C*, *trophonissa*, mimetic female, ground color for forewings white, hind wings brownish; *D*, *hippocoon*, mimetic female, ground color white (*redrawn from Eltringham*, 1910, *by Goldschmidt*, 1945).

## 8C. STATISTICAL ANALYSIS OF INDIVIDUAL VARIATION

### 8C.1  TAXONOMY AND ELEMENTARY STATISTICS

It may be useful to a beginner to single out for discussion a few terms and elementary methods in statistical analysis that are routinely used in taxonomy. These discussions are not meant as a substitute for the technical literature. In order to find out how to calculate standard deviations or regression coefficients, one should consult Simpson, Roe, and Lewontin (referred to as SRL in the ensuing discussions) or a more advanced text.

The replacement of typological thinking by population thinking has had a far-reaching impact not only on the concepts of taxonomy but also on its methods. To consider taxa as populations and aggregates of populations led automatically to a statistical approach. The taxonomist deals with samples from natural populations and can make estimates on the characteristics of these populations only with the help of statistics.

A quantitative analysis adds greatly to the precision of a description. Actual measurements of a series of specimens are far more useful than the meaningless statement "of medium size." "Sex comb with seven teeth" is more precise than merely "sex comb present." Such precision is all the more important whenever related species differ not by the presence or absence of a character, but rather by its size, proportions, or number. Whenever we compare two populations or two other taxa, we cannot make a concrete statement on the degree of difference until we have determined the amount of variation within the compared entities.

Statistical methods should be used whenever they contribute to the taxonomic analysis. They are not to be used as window dressing. Taxonomic papers are occasionally published with highly elaborate statistics that make no contribution whatsoever to the taxonomic analysis. To undertake the calculations is pure waste of time in such a case. Statistics cannot improve heterogeneous original data or unreliable measurements. The simplest method yielding the desired information is always the best.

A detailed presentation of the principles of statistics and of the application of the various statistical methods is beyond the scope of this manual of taxonomy. Many excellent texts are available. The best for the purposes of the animal taxonomist is *Quantitative Zoology* by Simpson, Roe, and Lewontin (1960), referred to in this text as SRL. The splendid biometric analysis of Nigerian ostracods by Reyment (1960, 1963) is a good introduction to an application of modern methods, particularly of multivariate analysis. Seal (1964) gives an exposition of some of the more complex multivariate methods. Special applications to paleontology are described in Miller and Kahn (1962). The coming of the computer has greatly facilitated the use of many of the more elaborate methods. Computer programs are now available for many methods for which the calculations used to be altogether forbidding. See also Sokal (1965).

Statistical methods never produce an automatic answer to questions of taxonomic evaluation. In this respect the situation is quite different from that in experiments where the worker is often satisfied with a simple "yes" or "no" answer to the question whether a change in experimental conditions has resulted in a "significant" difference from a set of controls.

However, statistical analysis may give us important information on the nature of our data. Experience has shown that morphological measurements usually show a normal distribution. A strong deviation from normality requires that we examine the material for "bias" and the possibility of "heterogeneity" (see statistics texts). Statistical analysis may also give us important information as to the weight which we should assign to certain characters. Highly variable characters, as well as some that are closely correlated with other characters, are given low weight in classification.

*What Statistics Cannot Do.* Statistics is no cure-all. It cannot make taxonomic decisions for us. Statistical methods do not tell us whether two phena belong to the same population or not, whether two allopatric populations belong to the same species or represent two species, or whether two sympatric phena are individual variants or sympatric species. Reproductive isolation (species criterion) and degree of morphological difference are not always closely correlated. Sibling species may be almost identical morphologically, while intrapopulation variants and subspecies are often strikingly different. The taxonomist must also keep in mind that what may be "significant" for the statistician may not be at all significant biologically or at least taxonomically.

**8C.1.1    Samples and Sampling.** It is rarely, if ever, possible to study an entire natural population. The taxonomist must be satisfied with a sample from which he attempts to reconstruct the properties of the population from which the sample was drawn. In order to permit reliable conclusions, the sample should be homogeneous, adequate, and unbiased (Cochran, 1953).

A heterogeneous sample can often be segregated into smaller homogeneous samples by separating the specimens according to age, sex, locality, or other factors that had introduced heterogeneity. Great care must be taken when segregating a sample to avoid bias (see below). Homogeneity is particularly important in comparative studies because samples which differ in their components owing to heterogeneity cannot be legitimately compared.

The sample should not be *biased;* that is, the method of getting the sample should be such that the variations of the pertinent characters in the sample occur at the same frequency as in the population. Fossils, for instance, are sometimes deposited according to size, and a sample drawn from one of these size classes is not an unbiased representation of the population from which it is drawn. Collectors sometimes concentrate on unusual specimens and thus introduce bias. Instead they should adopt the same safeguards of randomization as experimentalists and pay due respect to the variation introduced by locality, season, and time of day. One should never discard part of a collection and keep only those specimens that seem either typical or particularly interesting for being atypical. In polymorphic populations special efforts should be made to collect specimens in the true population frequency. In order to reduce collecting bias, it is often advisable to employ several different collecting techniques at the same locality.

What constitutes an "adequate" sample depends on the nature of the studied taxon and on the objective of the investigation. In the case of the living coelacanth, *Latimeria,* a single specimen was adequate to prove that this was a new species and genus of a class of vertebrates believed to have been extinct for more than 70 million years. In some groups of animals diagnostic species characters are well defined in only one sex, and the description of a new species may not be feasible if the collection contains only representatives of the other. The study of polytypic species requires a comparison of samples from different populations which may differ only in quantitative characters. Large samples are a necessity in such studies. This is also true for the study of variation in polymorphic species. As a general rule one can say that an adequate sample is a sample which allows a reasonable estimate of the total variability of a species. Whenever knowledge of a group has reached the degree of maturity where taxonomic analysis concentrates on a study of individual and geographic variation of species, the availability of large samples for study is a necessity. A knowledge of the variability of species is valuable not only for the taxonomist but for anyone dealing with the biology of species. The evolutionist in particular, but also the ecologist and population biologist, are interested in the nature and the extent of variation within and between populations of species.

8C.1.2    **Measurements and Counts.** Only quantitative data can be subjected to a statistical analysis. In this fact lies the importance of characters that can be counted or measured. Meristic ( = countable) characters, such as number of spines, scales, or fin rays, permit greater accuracy than do measurements. They are therefore favored wherever possible—by students of echinoderms, fishes, and reptiles, for example. SRL (pp. 20–30) gives various requirements of good measurements. Most important is that measurements be standardized (applying to a specified distance) and accurate. For instance, the length of the bill in birds may be measured in several ways: (1) from the nostril to the tip, (2) from the beginning of the feathering to the tip, or (3) from the beginning of the bony forehead to the tip. Observations have shown that the first can be measured very accurately but does not give the full length of the bill; the third can be measured fairly accurately in all birds with a steep forehead; and the second can rarely be measured with any accuracy. Consequently, in some genera of birds the third is the preferred measurement, in others the first. In this case as well as in all similar cases, the record should show which of several possible measurements was actually taken.

Having to take many measurements of a large series of specimens is very time consuming. Automation is beginning to come to the rescue of the investigator, particularly through a computer-connected recording of the measurements (Garn and Helmrich, 1967).

It is seldom possible to predict which of a set of possible measurements will be most important in the comparison of several samples, and so it is advisable to measure all variates that may possibly be of importance. Subsequent analysis will show that many of these measurements either fail to show significant differences or are merely duplications of other data. In view of the high costs of printing, such superfluous data should not be published. They may be placed in the archives of a public institution (museum or library) where they are available to other students.

When measuring an important lot of specimens, or when measuring specimens before the method of doing so has been completely standardized, one should measure each variate repeatedly. The duplicate sets of measurements should be taken on different days and on new record sheets. When completed, the various sets of measurements should be compared and averaged. Particularly deviating measurements should be checked for possible errors in the measuring technique or in recording.

**8C.1.3 Measurable Characteristics.** Total length is usually a very important measurement, particularly when it is used as the yardstick for ratios and proportional measurements. In each case what is meant by *total length* should be specified. Is it taken before or after preservation? Does it include or exclude appendages on the head and the tail? Total length is most satisfactory in beetles and other rather rigid, hard-shelled animals. In birds, the wing length ( = actually the length of the longest primary) is a much less variable quantity than is total length measured in the flesh. The cube root of the weight may, under certain conditions, replace the total length in calculations of allometric ratios. Body length, i.e. total length minus tail length, is usually a more accurate measure of size than total length.

Different measurements are used for nearly every category of animal. In mammals, for instance, body and tail length are measured, as well as length of the hind foot and ear and the various dimensions of the skull. In birds, wing, tail, bill, and tarsus are the most commonly measured variates. In most groups of insects not only length but also width and antennal and tarsal formulas should be given. These data should be recorded as a routine matter regardless of their immediate diagnostic value. Special measurements are traditionally given in particular taxonomic groups, such as the length of the rostrum in Hemiptera, length of the wings in some Diptera, etc. It is important for comparative purposes to give measurements that conform with the system which is customary in the group under study and to present them in a standardized sequence.

**8C.1.4 Technical Aspects of Measuring.** Zoological measurements are now universally given in terms of the metric system. However, many descriptions written in the nineteenth century use inches and lines (1 line = $\frac{1}{12}$ in.). (See Table 8-4.)

TABLE 8-4.   CONVERSION OF LINES INTO
              MILLIMETERS

| | |
|---|---|
| 1 line  = 2.11$\frac{2}{3}$ mm | 7 lines = 14.81$\frac{2}{3}$ mm |
| 2 lines = 4.23$\frac{1}{3}$ mm | 8 lines = 16.93$\frac{1}{3}$ mm |
| 3 lines = 6.35 mm | 9 lines = 19.05 mm |
| 4 lines = 8.46$\frac{2}{3}$ mm | 10 lines = 21.16$\frac{2}{3}$ mm |
| 5 lines = 10.58$\frac{1}{3}$ mm | 11 lines = 23.28$\frac{1}{3}$ mm |
| 6 lines = 12.7 mm | 12 lines = 25.4 mm |

Various measuring tools are used for different groups of animals. A millimeter rule (often with a "zero stop") and dividers (calipers) are used for most larger animals. The eyepiece micrometer is used to measure microscopic objects. It may be divided into small or large units and may be arranged as a linear scale or in squares. The individual units must be translated into the metric system by calibration with a stage micrometer.

Projection devices, such as microscopic projectors, are sometimes useful. By means of such devices, the specimens can be drawn from a projected outline, and the various parts can then be measured in the enlargement. This method is particularly useful for the measurement of relative sizes and angles. Care should be taken to avoid the danger of distortion in the two-dimensional projection of the three-dimensional object.

As far as refinement is concerned, it is important to carry measurements to whatever decimal point may be necessary, but not to waste effort by an unreasonable desire for accuracy. It would be useless to give the height of a person as 176.583 cm.

How, then, shall the proper degree of refinement be decided? The recommended unit of measurement is one-twentieth of the difference between the largest and the smallest specimen, if an adequate series is available. Thus if the measurements range from 10 to 12 mm, one should measure to 0.1 mm; if they range from 40 to 50 mm, to 0.5 mm. If they range from 70 to 90 mm, no decimal places need be recorded. If fractions are rounded off, they should consistently be rounded to the nearest full number, halves to the nearest even number. When fractions are measured, a bias in favor of integral numbers should be avoided.

8C.1.5  **Recording of Measurements.** Whenever large numbers of measurements are taken, it is advisable to enter them on special data sheets. If adequate samples are available, each sample should be recorded on a separate sheet. Each specimen should be entered separately, its field or museum number (if any), age, and sex recorded, and then the various measurements recorded in separate columns. If there is room, the calculated ratios between measurements can be entered on the same sheets.

Color differences are important characters in the taxonomy of virtually all animals. Various instrumentations are used to determine the color values objectively and to translate them into precise measurements of reflected light of selected wavelengths (Chap. 10). This permits a statistical analysis.

## 8C.2  STATISTICS OF A POPULATION SAMPLE

The beginner often asks, "What statistics should I use?" When one deals with a single character (univariate analysis), the two essential statistics are the arithmetic mean, M, and the total variance, V (or the square root of the variance, designated as the standard deviation, SD). With these two values, one can calculate most other statistics.

If the values of a sufficiently large sample are plotted, it is usually found that the resulting frequency curve corresponds to the so-called *normal curve* (see statistics texts). The reason why most biological characters seem to show the pattern of variation of the normal curve is probably that they depend on a great number of genetic factors each of which makes either a positive or a negative contribution to the phenotype of the character. For example, literally hundreds of genes tend to increase body size and a similar number to decrease it. The probability is very low that an individual will have all plus genes or all minus genes; most individuals will have a balance between the opposing tendencies. This is why the majority of individuals of a population are relatively close to the mean value.

8C.2.1  **Standard Deviation (SD).** The standard deviation measures the variability of a sample. The broader the scattering of values around the mean, the "flatter" the curve, the greater the standard deviation. Knowledge of the SD of a population permits predictions as to the observed range of variation, because in a normal curve

M $\pm$ 1 SD includes 68.27% of the population
M $\pm$ 2 SD includes 95.45% of the population
M $\pm$ 3 SD includes 99.73% of the population

With desk calculators universally available, the standard deviation is now calculated by a machine method, as described in SRL, p. 87.

8C.2.2  **Coefficient of Variability.** The numerical value of the SD is meaningful only in relation to the mean value of the same sample. An SD of 2 indicates extremely low variability if the mean is 120, but very high variability if the mean is 8. In order to permit a comparison of degrees of variability in different species, it is advisable to calculate the coefficient of variability (CV). This expresses the standard deviation as percentage of the mean:

$$CV = \frac{SD \times 100}{M}$$

The numerical value of the CV depends on the measured character and on the particular taxonomic group. There are different CVs for meristic quantities, linear measurements, and ratios. The number of eyes (a meristic quantity) in the human species has a CV that is virtually 0; height in man (even in a homogeneous sample) has a CV exceeding 4. The CV is often a sensitive indicator of the homogeneity of samples. If, for instance, the CV of a certain statistic fluctuates around 4.5 in a series of samples, but is 9.2 in one sample, such a sample should be reinvestigated. It may include an additional sibling species, wrongly sexed specimens, or some other alien component. Zones of secondary intergradation between subspecies are often characterized by a greatly increased CV. The calculation of CV is particularly useful when comparable samples of the same species from different localities are investigated or when the variability of different variates of the same sample is compared. Even though widely used by taxonomists, the coefficient of variability has the great disadvantage that there are no statistical tests for comparing different CVs. Lewontin (1966), following earlier authors, has therefore proposed that the variance (or the standard deviation) of the logarithms of measurements be used instead. This measure of intrinsic variability is invariant under a multiplicative change. This means also that it does not matter what units of measurements (metric or nonmetric) are used. A simplified approach is possible for CVs of less than 30 (which includes most that are used in taxonomy). Here the squared CV equals the variance of the natural logarithms (to the base of $e$) and can be used in statistical tests.

8C.2.3    **Linear Measurements.** Absolute size of the whole body or of selected measurements of parts is remarkably invariant in adult mammals and even more so in adult birds.

A series of 49 adult males and 29 adult females of the kingfisher *Halcyon chloris pealei* Finsch and Hartlaub from Tutuila Island, Samoa, had the measurements and coefficients of variability shown in Table 8-5. In carefully measured homogeneous samples of adult birds, the CV of wing length is usually between 1 and 2.5, rarely above 3. In mammals the CV for linear dimensions is usually between 4 and 10, occasionally between 3 and 4.

In insects which reach the imago stage through molt or metamorphosis, a small CV of linear measurement might be expected, since there is no further growth after the sclerotic exoskeleton has hardened. However, the final size of the imago depends to some extent on the feeding conditions of the larvae or nymphs, and as a result there is usually considerable variability in the size of the imago. Linear measurements are extremely variable in all animals that continue to grow throughout life, such as fishes and snakes, not to mention such forms as corals and bryozoans. Employment of mean values and coefficients of variability are rather meaningless in

TABLE 8-5.  MEASUREMENTS OF A SAMPLE OF *Halcyon chloris pealei* FROM TUTUILA, SAMOA

|  | N | Range | Mean | SD | CV |
|---|---|---|---|---|---|
| **Adult males:** | | | | | |
| Wing | 49 | 94.0–101.0 | 97.48 | 1.71 | 1.75 |
| Tail | 49 | 63.5– 69.5 | 66.44 | 1.33 | 2.00 |
| Bill | 49 | 31.0– 39.0 | 34.46 | 1.56 | 4.54 |
| **Adult females:** | | | | | |
| Wing | 29 | 95.5–102.5 | 98.86 | 0.88 | 0.90 |
| Tail | 29 | 64.0– 72.0 | 67.62 | 1.56 | 2.29 |
| Bill | 28 | 33.5– 37.5 | 35.20 | 0.95 | 2.69 |

such cases. Instead, bi- and multivariate analysis, particularly multiple-regression analysis, yield statistics that are useful in taxonomic comparisons.

**8C.2.4  Meristic Quantities.** When countable characters, such as the number of segments, scales, spines, or chaetae vary, we speak of meristic variation. Some meristic characters are exceedingly constant, such as the number of eyes in mammals; others may have a greater or lesser variability which is usually characteristic for a given species. Examples are the number of scales in snakes or fin rays in fishes. Meristic characters are not always strictly discontinuous, particularly at the beginning or end of a series, so that the total number may have to be decided somewhat arbitrarily.

The CV of most meristic characters is smaller than that of linear measurements, and it is not permissible to compare the CV of the two kinds of variates. One should only compare relatively equivalent sets of data, such as the coefficients of linear measurements with those of other linear measurements, ratios with ratios, etc.

**8C.2.5  Ratios.** The statistics so far discussed deal with the variation of absolute size of single characters. Shape and proportions, however, are often more important in taxonomy than absolute size. Ratios between two dimensions or two parts of the body are therefore often more informative than linear measurements. The ratio is most meaningful if it is given in a formula which expresses the size of the smaller value as a percentage of the larger:

$$R \text{ (ratio)} = \frac{s \times 100}{l}$$

where $s$ equals the smaller of the two values and $l$ equals the larger. Commonly used ratios are the length of the head against length of the body (without head), length of the tail against length of the body, width of the skull against length, etc. Such ratios are quickly calculated with the help of a slide rule. If $R$ is near 100, it may happen that $s$ is larger than $l$ in some samples. It is obvious that the positions of $s$ and $l$ cannot be reversed in such cases even when $R$ becomes larger than 100.

Ratios are best demonstrated visually in the form of scatter diagrams in which one value is plotted on the abscissa, the other on the ordinate (Fig. 8-7).

It is important that the proper standard of comparison is chosen when one wants to determine the relative size of an organ or appendage. For instance, relative head width in insects is calculated against head length (without rostrum). Relative tail length in birds is usually calculated against wing length (as standard of general size). However, the wing is not an accurate yardstick for size in migratory and high-altitude birds, nor in some birds in which the wing is used in courtship. The cube root $(\sqrt[3]{\phantom{x}})$ of body weight might be a better measure in such species. If an appendage is calculated against the whole, as tail against body, the appendage should not be included in the whole; the trunk without the tail should be used as standard of the "whole."

A ratio has a typological connotation, as does the mean. It may therefore be a misleading way of expressing dimensional relationships when our

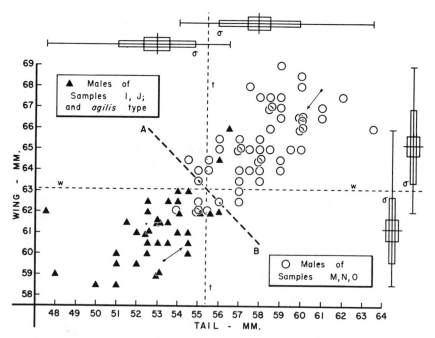

Fig. 8-7. Separation of two subspecies of *Parus carolinensis* on basis of wing and tail length; triangles = *agilis*, circles = *atricapilloides*, and AB = line of best separation (*from Lunk,* 1952).

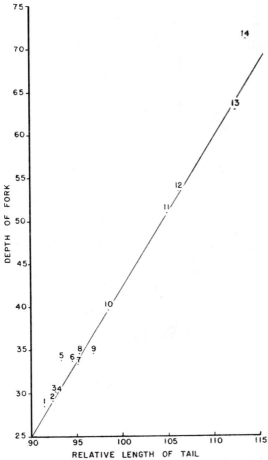

**Fig. 8-8.** Increase in the depth of the tail fork (in millimeters) correlated with increase in relative length of tail (in percentage of wing length) in *Dicrurus leucophaeus*. Each point indicates the mean for the available sample of adult males of one of 14 subspecies (*from Mayr and Vaurie, 1948*).

sample is highly heterogeneous with respect to age and size or when the various compared body parts display allometric growth. In such cases it is better to undertake a *regression analysis* (SRL, p. 213) and in the case of allometric growth to plot the variation on logarithmic paper. Good straight-line relations are sometimes obtained by plotting absolute size against the relative size of a body part (Fig. 8-8).

In the case of ratios and regressions we are comparing two variates. Whenever we want to deal simultaneously with more than two variables, we must employ some method of multivariate analysis.

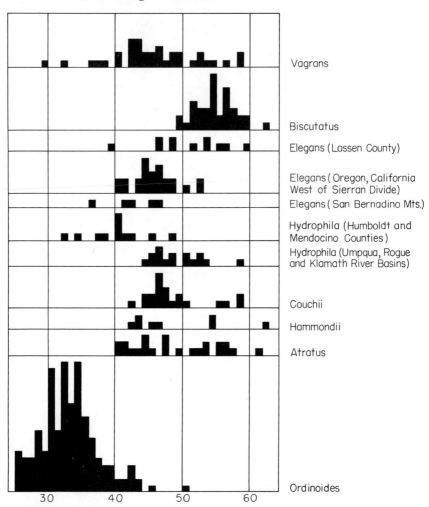

**Fig. 8-9.** Histograms showing head and body length in centimeters of adult males of *Thamnophis ordinoides*. Each square represents a specimen (*Fitch*, 1940).

**8C.2.6 The *t* Test.** Statistically this is the most satisfactory method of determining whether or not the difference between two samples is "significant" (see SRL, p. 176). In cases where a multivariate analysis is employed, the corresponding test is called the $T^2$ test.

**8C.2.7 Comparison of Frequencies.** The taxonomist is often confronted with the problem of having to determine whether two morphs or other variants occur at the same frequency in two or more populations. If he had infinitely large samples, he would simply express frequencies in percentage. With samples of limited size, a test is employed that tells us what the probability is that the populations from which the samples are taken are really different or, respectively, what the probability is that the observed differences are merely due to accidents of sampling. This test is called the $\chi^2$ (chi-square) test (SRL, pp. 306–338).

The degree of significance of a given $\chi^2$ value is given in $P$ tables, which can be found in all standard statistical texts. $\chi^2$ tests are highly sensitive to sample size.

## 8C.3   GRAPHIC PRESENTATION OF QUANTITATIVE DATA

It is often desirable to present numerical data visually. Such a presentation by graphical methods not only permits a rapid visual survey of all the data, but actually often brings out fine points that are not apparent in the raw data. SRL, chap. 14, gives an excellent survey of such graphic

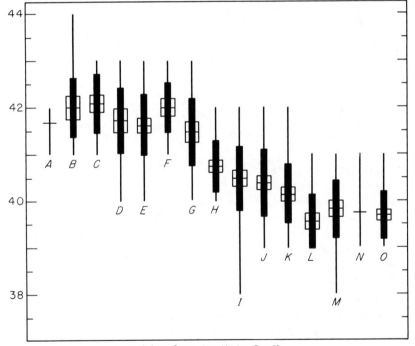

Localities from North to South

**Fig. 8-10.** Population-range diagram. Variation in the number of vertebrae of the anchovy, *Anchoviella mitchilli.* The letters A to O refer to 15 population samples, arranged from north (A) to south (O). In each sample the vertical line indicates the total variation of the sample; the broad portion of the line, 1 SD on each side of the mean; the hollow rectangle, twice the standard error on each side of the mean; and the crossbar, the mean (*Hubbs and Perlmutter,* 1942).

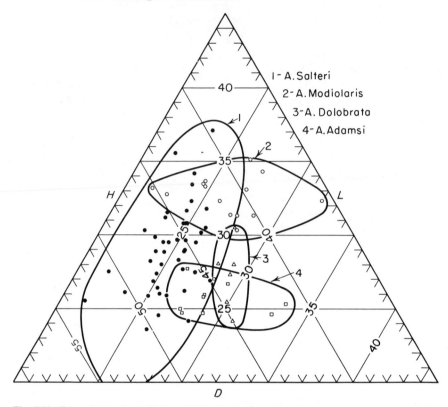

**Fig. 8-11.** Triangular graph of the length (L), height (H), and distance to maximum down-bulge (D) of four species of *Anthracomys* (*Burma*, 1948).

methods. In the present text only those few simple methods will be mentioned that are most frequently used in taxonomic publications.

**8C.3.1   Histograms.** Unreduced samples are best shown as histograms. A histogram consists of a set of rectangles in which the midpoints of class intervals are plotted on the abscissa and the frequencies (usually number of specimens) on the ordinate. This presentation has several advantages, the principal one being that it presents the original data in minimum space. Whatever form of statistical analysis a subsequent author may want to apply, he will find the actual number of specimens given for each size class. A quick comparison of different populations is made possible by arranging a series of histograms above one another (Fig. 8-9).

**8C.3.2   Population Statistics Diagrams.** Even more data can be compressed into minimum space by giving sample range, mean, one or more SDs, and two standard errors. This is the method of Dice and Leraas (1936) (see Fig. 8-10). Several modifications of this method have been proposed. For instance, one can give the size of the sample ($N$) with each bar and replace standard error by the 95% confidence limits of the mean. Nonoverlap of 1.5 SD (of each compared sample) indicates a degree of difference usually considered sufficient for subspecific separation, as discussed later.

8C.3.3   **Scatter Diagrams.** The difference between two or more populations in respect to two characters is best illustrated by a scatter diagram. Each individual is indicated by a spot or other symbol which is placed where the value for one character (read off the ordinate) intersects the value for the other character (read off the abscissa); each population is indicated by a different symbol (circles, squares, triangles, solid or empty, etc.—see Fig. 8-7). Scatter diagrams have many advantages. They help the student to visualize allometric relationships and facilitate the plotting of regression lines. They also sometimes disclose errors of measurement or sexing that might otherwise go undiscovered.

If three characters are involved, triangular charts can be employed.

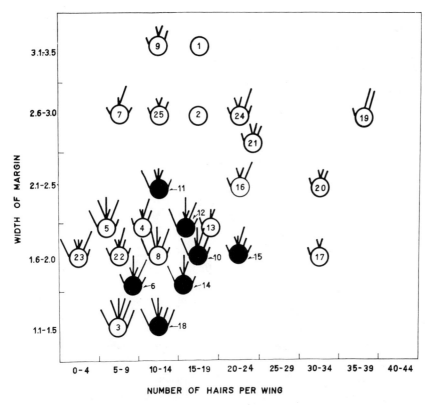

Fig. 8-12. Pictorialized scatter diagram (pictogram) of 25 individuals from a population of stemless white violets. Two of the characters are indicated along the margins, five by position and length of rays; filled circles indicate individuals with a heavy blotch on the spur petal (*from Hatheway, 1962, after Anderson, 1954*).

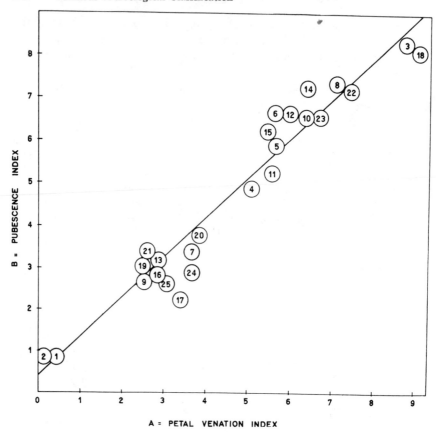

Fig. 8-13. Weighted scatter diagram, ordering the same 25 individuals shown in Fig. 8-12. The horizontal scale is a weighted index composed of four aspects of petal venation. The vertical scale combines five measures of hairiness (*from Hatheway,* 1962).

In this case the actual values are not plotted, but rather their percentage contribution to the sum of the characters. For example, if character $a = 80$ mm, $b = 32$ mm, and $c = 48$ mm,

$$a + b + c = 160 \text{ mm} = 100\%$$

Then $a = 50\%$, $b = 20\%$, and $c = 30\%$ of the whole. These percentages are plotted on the graph, which thus shows proportions rather than absolute sizes. In each individual case the triangular graph is scaled in such a way as to produce a maximum spread of the points. As an illustration a triangular chart from a paper by Burma (1948) is reproduced in Fig. 8-11.

More than three characters can be presented in a pictorial scattergram (Anderson, 1949, chap. 6) (Figs. 8-12, 8-13). It is advantageous to select those two characters for plotting on ordinate and abscissa which result in the clearest separation of the clusters.

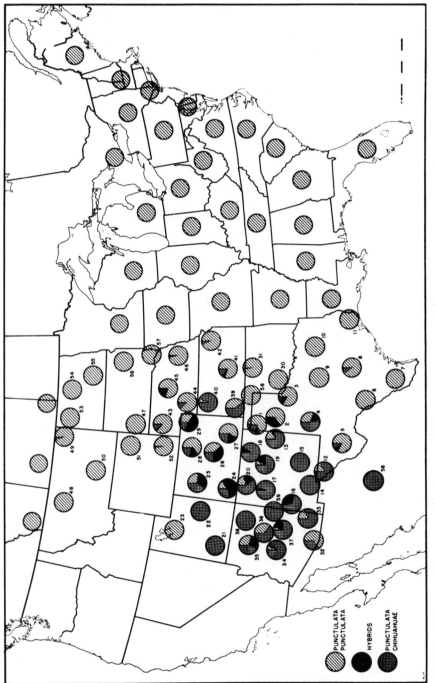

Fig. 8-14. Map of the distribution of the tiger beetle *Cicindela punctulata*, illustrating the pie-graph method. The size of the sectors of each circle indicates the relative frequency in the collected samples of three variants: *punctulata*, *chihuahuae*, and intermediates. Many populations are pure *punctulata*, others are pure *chihuahuae* (*unpublished, courtesy of Dr. M. A. Cazier*).

179

8C.3.4  **Mapping of Quantitative Data.** It is often desirable to illustrate the geographical relationships of populations which differ in quantitative characters. In the case of continuous characters (size, etc.), the simplest method is to record the means of the various populations on a base map, and if there is regularity to draw in the isophenes ( = lines connecting points of equal expression of a character). For instance, if the means of a series of populations in a species vary from 142 to 187, it is helpful to draw in the isophenes of 140, 150, 160, 170, 180, and 190 (see Chap. 3, fig. 3-4). Subspecies borders frequently fall where the gradients are particularly steep. A method for determining such contour lines is described by Lidicker 1962.

If qualitative or semiqualitative characters are to be plotted, it is sometimes helpful to choose a different symbol for each class of characters. The relative size of the symbol can be used to indicate sample size.

To present frequencies of polymorph characters on a map, the "pie graph" is the most convenient method. The percentage occurrence within the population is indicated by the size of the sectors (Fig. 8-14).

# Chapter 9  Taxonomic Decisions on the Species Level

*I*n his sorting of phena and populations into subspecies and species the taxonomist is forever forced to make decisions. In this chapter we shall discuss the various alternatives and suggest appropriate methods (some statistical) that may help in reaching the right decision. A study of the discrimination grid (Table 8-1) will facilitate asking the right questions.

Doubts arise in the analysis of sympatric samples and in the comparison of allopatric (or allochronic) samples. The two parts of this chapter, 9A and 9B, are devoted to these topics.

## 9A. ANALYSIS OF SYMPATRIC SAMPLES

The taxonomist is frequently in doubt whether certain phena sampled in the same general area belong to a single species or to several. On the whole there are three very different kinds of situations that are responsible for these difficulties: (1) extreme difference of phena belonging to a single species, (2) extreme similarity of good biological species (sibling species), and (3) wide variability and phenotypic overlap of two species. The first hint that—in the case of extreme difference—only one species is involved, or—in the case of extreme similarity—more than one species is involved, is usually provided by behavioral, ecological, or distributional data. The subsequent analysis serves to confirm (or refute) the earlier supposition.

### 9A.1 PHENA (INDIVIDUAL VARIANTS) OR DIFFERENT SPECIES?

There is hardly a species that does not contain several if not dozens of phena. To add to the complexity of the situation several other species with a similar assortment of phena may be sympatric. Often a phenon of one species resembles a corresponding phenon of another species more than other phena of the same species. For instance, females in certain species of many genera of birds and insects are more similar to the females of closely related species than to the males of their own species. There is nothing in the phenotype of caterpillars that would permit correct association with imagos. Only breeding can do this, or the careful evaluation of other biological information.

The correct assignment of many phena is possible on the basis of a correct interpretation of morphological information. If a large sample of a population is available, intermediate forms between the more extreme variants are usually found. Also, certain characters in every group are less subject to individual variation than others. The genitalic armature in most insects, the palpus in spiders, the radula in snails, and the structure of the hinge in bivalves are such characters. If several sympatric phena agree in their genitalic armatures (or one of the other mentioned characters), it is very probable that they are conspecific. However, even here one has to apply a balanced judgment. Although in most genera of Lepidoptera there are characteristic differences between the genitalia of related species, there are cases known in which forms have identical genitalia, even though they are different species by every other criterion. Parasitic animals present some rather special problems (Manwell, 1957).

The establishment of correlations is often very helpful. If two forms which differ in character $a$ can be shown to differ also in the less conspicuous and functionally unrelated characters $b$, $c$, and $d$, it becomes very probable that they are different species. Some years ago Mayr (1940) found that among birds identified as the southeast Asiatic minivet (*Pericrocotus brevirostris* Vigors), some had the innermost secondaries all black while others had a narrow red margin on these feathers. A detailed study revealed that those birds with red on the innermost secondaries had seven additional minor characters: a more yellowish red of the underparts, a different distribution of black and red on the second innermost tail feather, a narrow whitish margin along the outer web of the first primary, and four other minor characters. Slight though they were, these characters were well correlated with each other and with geographical and vertical distribution. The conclusion that two full species were involved has since been confirmed by several authors.

As a general rule, one finds that the decisions of a superior taxonomist, when based on a careful evaluation of the morphological evidence, are confirmed when a species recognized by him is subjected to genetic tests or to an evaluation of nonmorphological characters. A purely phenetic approach, on the other hand, is unable to discriminate between phena and species.

## 9A.2 SIBLING SPECIES

Biological species are reproductively isolated gene pools (Chap. 2). When two populations (gene pools) become geographically isolated, they diverge in their isolation genetically and may eventually acquire isolating mechanisms. As a by-product of the genetic divergence during this process, species normally also acquire morphological differences suitable for diagnostic purposes. A few species fail to acquire conspicuous morphological differences during the process of speciation. Such very similar, cryptic species are called *sibling species*. All the available evidence indicates that minuteness of morphological difference is the only aspect in which they differ from ordinary species. They are merely those species that are near the invisible end of the spectrum of morphological species differences. They grade imperceptibly into species that are morphologically more and more distinct from one another. Once discovered, and thoroughly studied, sibling species are usually found to have previously overlooked morphological differences.

Mayr (1963, pp. 33–58) has shown how widespread in the animal kingdom sibling species are. Most of them were discovered not during routine taxonomic analysis but during the study of species that are medically (e.g. *Anopheles*), genetically (e.g. *Drosophila, Paramecium*), cytologically, agriculturally, or otherwise of special importance. It is therefore impossible to indicate what percentage of species are sibling species. In the case of North American crickets, about 50 percent of the species were discovered through differences in their sounds (Walker, 1964) and in certain genera of protozoans (e.g. *Paramecium*) the percentage of cryptic species seems to be even higher.

The discovery of sibling species is possible because they may differ in various attributes even when they are extremely similar in the morphological characteristics normally employed in taxonomic analysis. Mayr (1963, p. 50) has listed a number of characteristics that facilitate the recognition of sibling species. Precise measurements sometimes display bi-modal characteristics, and the two modes can be correlated with additional characters. Very often there are differences in the number or structure of the chromosomes, a fact which has led to the recognition of sibling species in *Drosophila, Sciara, Chironomus, Prosimulium,* and other dip-terans, as well as in orthopterans, beetles, and other insects. Various aspects of behavior—such as differences in visual and vocal displays, nest construc-tion, breeding season, migratory behavior, prey selection, and host prefer-ence—have perhaps led to the discovery of more sibling species than any other characteristic. Sibling species may differ in their pathogenicity (e.g. *Anopheles*), or in their susceptibility to parasites and suitability to serve as hosts. Various biochemical methods, particularly those testing protein specificity, are suitable for checking on the probability of a real difference between "stocks" discovered by one of the other methods.

Kohn and Orians (1962) point out that the ecological situation makes the occurrence of sibling species more probable in some groups than in others. Sibling species will be common whenever species are able to occupy different niches in the same community without appreciable differentiation in those morphological characters that are used by taxonomists in their classification. This means, in particular, protozoans, small crustaceans, and insects. Every case of sympatry among sibling species is a case where no sympatric character divergence, in the sense of Darwin, has occurred in taxonomically useful morphological characters.

Sibling species are obviously inconvenient to the museum taxonomist. Specimens of such species cannot in some cases be recognized in preserved material. However, since species are not the creation of museum taxonomists but phenomena of nature, it is impossible to ignore their existence. The museum worker will be unable in many cases to do better than label mu-seum specimens from a group of sibling species by the group name, e.g. *Anopheles maculipennis* group. Once a sibling species is discovered, morpho-logical differences are usually found subsequently which will permit the identification of preserved material. It is generally possible in a group of sibling species to identify old type-specimens as to the particular species to which they belong. Sibling species are fortunately rare among vertebrates and other groups of special interest to the paleontologist. Whenever the recognition of a sibling species depends entirely on the behavior of living animals, it is of course impossible to recognize them in fossil specimens. The paleontologist is forced to proceed under the assumption that the genera he studies do not include sibling species.

Sibling species among protozoans are not formally described as species and not designated with binominals. The approximately 16 species of the *Paramecium aurelia* complex, for instance, are designated as varieties.

One particular class of sibling species, polyploids, raises special difficulties. Polyploids are unable to engage in normal gene exchange with individuals of different ploidy, and because of this reproductive isolation they have the biological characteristics of good species. They may, however, be morphologically indistinguishable, particularly if they are autopolyploids. Fortunately polyploidy is rare in the animal kingdom and essentially restricted to genera with parthenogenesis. Occasionally there may be a secondary return to sexual reproduction—in the oligochaetes, for instance. It is up to the specialist to decide whether or not to give taxonomic recognition to what is, biologically, surely a species. In instances of morphological identity such recognition would be unwise.

Until recently, polyploidy was believed to be restricted to invertebrates. A number of parthenogenetic (or gynogenetic) populations have been found, however, among fishes (several genera), salamanders (*Ambystoma jeffersonianum* group), and lizards (*Lacerta, Cnemidophorus*), and at least some of these populations are triploid. On the basis of their morphological and ecological differences (and absence of gene exchange), several of these populations have been described as separate species.

## 9A.3   VARIATIONAL OVERLAP

Closely related species are sometimes so variable and their variation so overlapping that no single character seems to have absolute diagnostic value. A combination of characters usually permits the correct assignment of all seemingly intermediate specimens. As Anderson (1954) perceptively pointed out, mechanical reliance on a biometric analysis or on straight diagnostic characters is often less effective than a purely intuitive approach based on the totality of characters as revealed by inspection.

A combination of two characters is often sufficient for diagnosis. In such cases it is simplest to plot the two characters on a "scatter diagram." This method usually yields several clusters of points when heterogeneous material is involved (see Chap. 8C).

A number of multivariate methods are available when the bivariate scattergram does not yield satisfactory results. Anderson's pictogram (8C.3.3) is one type of multivariate scattergram (Fig. 8-12).

In most cases there is no doubt as to the validity of species recognized by these methods, but there may be uncertainty about certain specimens which appear to be intermediate. Calculation of a *character index* is the simplest quick method for placing such specimens (see Table 9-1). Since this index was first proposed by Anderson (1936) and Meise (1936) to evaluate the hybrid nature of intermediate specimens, it was originally proposed as "hybrid index." I prefer the broader term character index, since the method is equally suitable for a quantitative treatment of non-

TABLE 9-1.  WEIGHTED DECIMAL CHARACTER INDEX
(after Meise, 1936)

| Character | Passer domesticus | Passer hispaniolensis |
|---|---|---|
| Crown | gray, 0 | rust brown, 30 |
| Ear coverts | gray, 0 | white, 10 |
| Sides of nape | gray, 0 | black, 15 |
| Flanks | plain, 0 | striped, 10 |
| Other | 0 | 35 |
| Total | 0 | 100 |

hybridizing species with overlapping variability. The method consists in providing a series of states for each character in which the two species differ. The typical condition in species $A$ is always designated as 0, the typical condition in species $B$ most frequently as 2 and that of intermediate specimens as 1. For 12 characters, essentially typical specimens of species $A$ may vary between 0 and 3, typical specimens of species $B$ between 20 and 24, and hybrids between 8 and 16. Two refinements are possible. The first is to allow for more intermediate states in the case of important characters so that specimens may score anywhere between 0 and, let us say, 6 for such a character. The second refinement is to arrange the scale of values in such a way that the maximal score adds up to 100. In this way it is possible to express the similarity of intermediate specimens as percentages. In order to illustrate the variation, it is advantageous to graph it in the form of histograms. Hatheway (1962) shows how to construct a weighted hybrid index (Fig. 8-13).

A far more precise method is the multivariate method of *discriminant functions* (Fisher, 1936, 1938). It is based on the fact that in the discrimination between two taxa, the greater the distance between the means and the smaller the combined standard deviations of this character, the greater will be the contribution made by a character. The method then consists in calculating a factor ($b1$, $b2$, . . . $bx$) which, when multiplied by the numerical value of the character, will result in a maximal value. The calculation of the $b$ values on a desk calculator is laborious, but there are now good computer programs which calculate the discriminant function from the raw data. "Stepwise" programs are now available which permit the elimination of characters that do not significantly contribute to the discrimination.

The method of discriminant functions has been used increasingly in taxonomy in recent years. It is particularly useful in the identification of doubtful specimens and occasionally in the establishment of a well-defined class of intermediate hybrids. Recent applications were made in studies of *Drosophila* (Carson and Stalker), fish (Stone), birds (Storer), and mammals (Foster, 1965). Bühler (1964) shows, step by step, how a discrimination analysis is carried out and how, in his example, it permits the unequivocal assignment to the correct species of every skull belonging to two exceedingly similar species of shrews. Kim, Brown, and Cook (1966) point out how to employ discriminant functions to find the diagnostically most useful characters.

The analysis described in the preceding paragraph makes the assumption that one already knows the essential characteristics of the respective species and is mainly concerned in the assignment of dubious specimens. There is always a residue of cases where the variation is too great and discordant to lead to an unequivocal species separation. A good example is provided by spiders of the species group *Steatoda punctulata* in which different specialists at one time or another considered that one, two, four, six, or eight species were involved (Levi, 1959). In such a case, larger series from single localities are needed as well as the utilization of additional characters. Each of the two to four species of the *Steatoda punctulata* complex varies individually and geographically, and a sophisticated biometric analysis would be futile until better material permits the study of the trends of this variation.

Several (on the whole rather complicated) methods of *multivariate analysis* have been tested in recent taxonomic publications, particularly in their application to fossil material. The impression one gets from a study of these papers is that no method has been found that makes enough of a contribution to the analysis to justify the very considerable labor involved in applying it routinely. The $T^2$ test is perhaps most useful.

## 9B. COMPARISON OF ALLOPATRIC AND ALLOCHRONIC SAMPLES

The taxonomist has a well-defined interest when he compares samples drawn from different natural populations. He is looking for evidence that will facilitate the decision whether or not to include the populations in the same taxon and, if not, whether to rank them as different subspecies or species. The question is never whether or not the compared populations are completely identical. Population geneticists have demonstrated conclusively that no two natural populations in sexually reproducing animals are ever exactly alike. To find a statistically significant difference between several populations is, therefore, only of minor interest to the taxonomist; he takes it for granted. Even the lowest recognizable taxon, the subspecies, is normally composed of numerous local populations some of which differ "significantly" in gene frequencies and in the means of a number of variates.

If no statistically significant differences between the samples from two populations are found, the sampled populations must be referred to the same taxon. If a difference is established, additional considerations are required before it can be decided whether or not the populations should be considered different taxa or not, and, if different taxa, whether they should be ranked subspecies or species.

### 9B.1   DIFFERENT SUBSPECIES OR NOT?

In groups of animals in which polytypic species are generally recognized, it is frequently questioned whether or not the difference between two populations is sufficiently great to justify their recognition as two different subspecies. The most important prerequisite for making such a decision is a clear understanding of the nature of the subspecies category (see Chap. 3, also Mayr, 1963, pp. 347–351). If the samples are clearly different in one or several respects, they qualify as different subspecies (but see below). A problem arises when the ranges of variation overlap. How much overlap between two subspecies is permissible? The simplest way of determining overlap is to plot the linear overlap of the observed samples (Fig. 9-1), but this method is misleading in two ways: it gives only the overlap of the samples (which is always smaller than that of the sampled populations), and it exaggerates the importance of the end points of the range, that is, the tails of the overlapping curves. Linear overlap is therefore a very unsatisfactory way of describing the degree of difference between two populations.

Is there a more satisfactory way? As far as we know, there is no method that is not vulnerable to objections. Mayr (in Oliver, 1943) proposed a "coefficient of difference," which relates the difference between the means to the standard deviations of the samples.

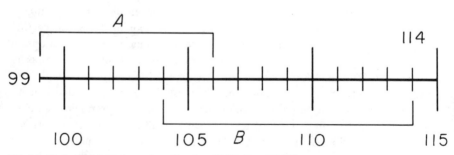

Fig. 9-1. Linear overlap of observed samples. A = 99–106, B = 104–114 mm.

**9B.1.1  Coefficient of Difference (CD).** This statistic is based on the observation that the less overlap there is between two population curves, the larger is the difference between the means M when divided by the standard deviations (SD). The formula for the coefficient of difference for populations $a$ and $b$ is ($b$ being the population with the larger mean):

$$CD = \frac{M\,b - M\,a}{SD\,a + SD\,b}$$

The CD is easy to calculate when one knows the means and standard deviations of the two populations. However, it allows only a rough approximation because it makes various assumptions that are not strictly correct, e.g. that the distributions are normal and that the sample statistics equal the population statistics. The latter is clearly not true, particularly for standard deviations of comparatively small samples. A number of authors have therefore suggested various modifications. Ride (1964) suggests calculating two CD's, one for the upper and one for the lower confidence limit. Another suggested refinement is to use the square root of the pooled variances. Géry (1962) believes that it would be simpler to base degree of difference between population samples on a $t$ test and supplies tables for this purpose. Reyment (1960, p. 28) proposes a multivariate extension of the coefficient of difference.

The reason for the uncertainties is that statisticians have until recently been mainly interested in determining presence or absence of a difference. They have paid little attention to methods establishing a degree of difference. In taxonomic work, particularly in the recognition of subspecies, so many additional considerations enter the picture that extreme accuracy is not important, while a measure that would give the order of magnitude of the overlap of two population curves is. The CD (or one of the cited alternate methods) is a simple answer for this need. However, computer methods have now become available that permit the routine calculation of measures of distance, like Mahalanobis' $D^2$, which are far more accurate. See Seal (1964) for an introduction to this method.

**9B.1.2  Degree of Difference and Subspecies Recognition.** Degree of difference is only one of a number of considerations in the recognition of subspecies (Chap. 3). A yardstick such as the CD will help to achieve more uniform standards, but other information, such as degree of isolation, presence or absence of clinal variation, presence or absence of a checkerboard type of distribution, or discordant variation of different characters, must be equally taken into consideration.

Widely different standards of subspecies recognition have been adopted by different authors. Some extreme "splitters" called every population a different subspecies that could be shown by statistical tests to be different. Some "lumpers" insist that every individual of a subspecies must be diagnostically different. Very few taxonomists hold with either extreme. A so-called 75-percent rule is widely adopted. According to this, a population is recognized as a valid subspecies if 75 percent of the individuals differ from "all" ( = 97 percent) of the individuals of a previously recognized subspecies. At the point of intersection between the two curves where this is true, about 90 percent of population $A$ will be different from about 90 percent of the individuals of population $B$ (to supply a symmetrical solution). This corresponds to a CD of about 1.28 (Table 9-2). From this table one can readily see how much overlap there is between curves at given coefficients of difference.

In the heyday of subspecies splitting in ornithology, some authors recognized subspecies on the basis of a CD of only 0.675 or even lower. There is now a tendency to demand a CD of well over 1.28, perhaps at least 1.5, in order to justify subspecies recognition. In cases involving irregular distribution patterns, position on a long cline, and discordant variation of different characters, an even higher CD is sometimes insuffi-

TABLE 9-2.    PERCENTAGE OF NONOVERLAP OF PARTIALLY OVERLAPPING CURVES ASSOCIATED WITH GIVEN VALUES OF THE COEFFICIENT OF DIFFERENCE (CD)

| Values | CD | Joint nonoverlap, percent |
|---|---|---|
| Below the level of conventional subspecific difference | 0.675 | 75 |
| | 0.84 | 80 |
| | 0.915 | 82 |
| | 0.995 | 84 |
| | 1.04 | 85 |
| | 1.08 | 86 |
| | 1.13 | 87 |
| | 1.175 | 88 |
| | 1.23 | 89 |
| Conventional level of subspecific difference | 1.28 | 90 |
| Above the level of conventional subspecific difference | 1.34 | 91 |
| | 1.405 | 92 |
| | 1.48 | 93 |
| | 1.555 | 94 |
| | 1.645 | 95 |
| | 1.75 | 96 |

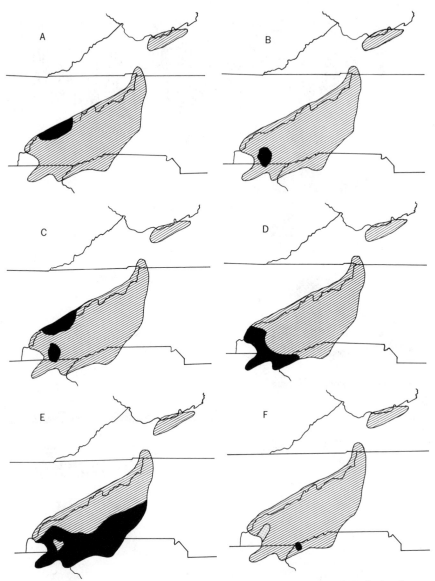

**Fig. 9-2.** Discordant geographic variation in the salamander *Plethodon jordani*. Darkened areas represent regions where more than 95% have: (A) red cheeks, (B) red legs, (C) dorsal red spots in newly hatched young, (D) lateral white spots, (E) a dark belly, (F) small dorsal brassy flecks. An area with small dorsal white spots is indicated in (F) by stippling (*from Highton, 1962*).

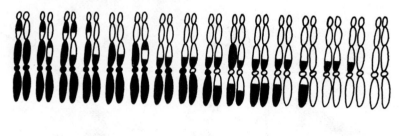

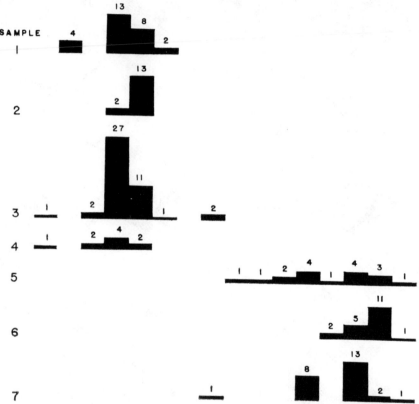

Fig. 9-3. Geographic variation in the frequency of 16 color morphs in 7 populations of the wolfspider *Geolycosa xera*. Ventral color pattern on legs 1 and 2 of females is shown above. Populations 1–4 belong to a different subspecies from populations 5–7 (*from McCrone, 1963*). Subspecies often differ from each other in the relative frequency of phena.

cient. Highton (1962) demonstrates that geographic variation in the salamander *Plethodon jordani* is too discordant to justify the recognition of formal subspecies even though the variation of each individual character shows a definite geographic trend (see Fig. 9-2). Inger (1961) gives a well-balanced discussion of various criteria to be used in the recognition of subspecies. Degree of difference, as expressed in the coefficient of difference, is only one of them. The standards for subspecies recognition are now much more rigorous than they were a generation ago (see also Chap. 3).

**9B.1.3    Subspecies Borders.** Most subspecies are geographical isolates or former isolates, and their borders are easily established. The delimitation of subspecies that are connected by primary intergradation is difficult, and their recognition usually unwise. Such subspecies are often the adaptive response to regional climatic conditions (particularly of temperature and humidity) and are no more sharply delimited than the causative climatic factors. Where substrate races are involved (black lavas, white sands or limestones, red soils, etc.), subspecies borders are sometimes remarkably sharp, particularly when reinforced by habitat selection. Lidicker (1962) plotted on a map the total character change in a kangaroo rat (*Dipodomys merriami*) per unit of distance and found that this resulted in well-defined contour lines. Bands of rapid character changes, as defined by an index of differentiation, coincided remarkably closely with the previously recorded subspecies boundaries. In other species, however, no well-defined subspecies borders seem to exist, as shown by Hagmeier (1958) for the marten (*Martes americanus*) and by Jolicoeur (1959) for the wolf (*Canis lupus*), two species in which the geographical variation of every character seems to be independent of all others.

**9B.1.4    Polytopic Subspecies.** When subspecies of a species differ only in a single diagnostic character relating to color, size, or pattern, it may happen that several unrelated and more or less widely separated populations independently acquire an identical phenotype. The evolutionist knows that such populations are not identical genetically, but since the subspecies is not an evolutionary concept, taxonomists sometimes combine such visually identical populations into a single subspecific taxon. Such a geographically heterogeneous subspecies is called a *polytopic subspecies*. The only alternative to its recognition is not to recognize any subspecies in such a species. In the absence of diagnostic differences there is no legitimate excuse for dividing a single polytopic subspecies into several subspecies merely on the basis of locality. It must be remembered that any subspecies is a heterogeneous composite, even when it consists of contiguous populations.

## 9B.2   SUBSPECIES OR ALLOPATRIC SPECIES?

Whenever the taxonomist encounters two taxonomically distinct allopatric populations, he must decide whether to consider them species or subspecies. Various types of evidence are used in making this decision. Degree of difference per se was sufficient reason for species recognition by adherents of a typological species concept. Adherents of the biological species concept look for evidence of actual or potential interbreeding and use degree of morphological difference only to draw inferences on the probability of potential interbreeding.

The word *allopatric* is essentially an antonym of sympatric and means, therefore, distribution without geographical overlap [see Smith (1965) for the proposal of a more elaborate terminology]. On the basis of geographical pattern and observed interbreeding, five kinds of allopatry can be distinguished:

KINDS OF ALLOPATRY

I    Allopatric populations $A$ and $B$ are in contact.                                          II
     Allopatric populations $A$ and $B$ are separated by an unoccupied
     area.                                                                                        V

II   Populations $A$ and $B$ intergrade or interbreed freely.                                    III
     Populations $A$ and $B$ do not interbreed, or do so only very
     occasionally.                                                                               IV

III  $A$ and $B$ intergrade clinally in a (usually fairly wide) zone of
     contact                                                                                      (a)
     $A$ and $B$ interbreed completely in a (usually rather narrow) zone
     of contact or by hybridization.                                                             (b)

IV   $A$ and $B$ meet in zone of contact where occasional hybrids occur.   (c)
     $A$ and $B$ meet in zone of contact but do not interbreed at all.       (d)

V    $A$ and $B$ do not interbreed because they are separated by a dis-
     tributional gap or natural barrier which prevents contact.               (e)

Populations that qualify under (a) and (b) are nearly always to be treated as subspecies; under (c) and (d) as species; and under (e) as species or subspecies. The following comments, designated to correspond to the key above, may be helpful:

Allopatric populations that intergrade clinally with each other belong to the same species. It depends on the degree of difference whether or not they are to be considered subspecifically different.

1. *Primary Intergradation.* There is no clear-cut distinction between intergradation and allopatric hybridization. In general, we speak of *primary intergradation* when a series of intermediate populations is intercalated between two subspecies, each population with approximately the same amount of variability as any other population of either subspecies. We speak of *allopatric hybridization* when the two subspecies meet in a well-defined zone and form there a hybrid population with greatly increased variability, often containing the entire spectrum of character combinations from subspecies $a$ to subspecies $b$. There must be evidence for random interbreeding in this zone. Allopatric hybridization is sometimes also referred to as *secondary intergradation,* because it is a secondary event, following a breakdown of previously existing extrinsic isolation of the populations.

Among North American birds the flickers (*Colaptes*), juncos (*Junco*), and towhees (*Pipilo*) furnish good examples of hybridization between widely divergent subspecies. For further details and additional examples, see Mayr (1942, pp. 263–270, and 1963, p. 118), and Sibley (1961).

2. *Secondary Intergradation.* Secondary intergradation, that is, the occurrence of zones of contact of previously isolated populations or subspecies, occurs very commonly in geographically variable species. No taxonomic difficulty arises when the zone of hybridization is narrow. However, if it is wide and if a well-defined, stabilized hybrid population with intermediate characters develops, it is sometimes convenient and justified to recognize the "hybrid" population taxonomically. It may be treated as a subspecies if it satisfies the requirements of the 75-percent rule. The taxonomic recognition of a hybrid population is not justified if it is highly variable and includes in this variation a range of phenotypes extending from one parental extreme to the other. If two taxa that were previously recognized as two allopatric species completely intergrade in a zone of secondary contact, it proves that they are not reproductively isolated and that they should be considered subspecies of a single polytypic species.

3. *Occasional Hybridization.* Allopatric forms that hybridize only occasionally in the zone of contact are full species. There are a few cases where it is difficult to decide whether the hybridization is occasional or complete. Much recent evidence indicates that hybridization has to be fairly complete in order to restore secondary intergradation. For a treatment of occasional hybridization, see 2.5 and 13.21.

More difficult to evaluate are cases where two species remain as distinct species over most of their range but form complete hybrid populations in a few areas. This happens particularly in regions where human interference in recent years has badly disturbed the natural ecological balance. It is recommended that such forms be treated as full species in spite of the occasional free hybridization under the stated conditions (see Mayr, 1963, pp. 119–124).

4. *Parapatric Species.* Allopatric populations that fail to interbreed, although in contact, are full species. Failure to interbreed indicates reproductive isolation and attainment of species rank. The absence of geographical overlap may be caused by one or the other of two opposite reasons. The zone of contact may connect two very different ecological areas (e.g. savanna and forest). If one of the two neighboring species is specialized for one of these habitats and the other species for the other, the two species cannot invade each other's range because their ecological requirements are too different.

The other possible reason for nonoverlap of full species is that their ecological requirements are so similar in every respect that they compete with each other. On one side of the zone of contact one species is slightly superior, on the other side the other. Such "competitive exclusion" (for detailed discussion see Mayr, 1963, pp. 81–82) will result in strict allopatry of full species. It is important to understand this because allopatry used to be accepted by some authors as an automatic criterion of conspecificity. Vaurie (1955) showed that in 225 species of 7 families of Palearctic songbirds, 22 good species had been treated earlier as subspecies by one or another author owing to their allopatry. See also Kohn and Orians (1962).

A careful study of the zone of contact usually reveals areas where increased habitat diversity actually permits occasional sympatry of the two species. The best evidence for species status is provided by the sharpness of phenotypic difference in the zone of contact. If two continental species display no evidence at all of any intergradation in their zone of contact (or close approach), they evidently do not exchange genes with each other and must be treated as full species, even though they are allopatric.

5. *Isolated Allopatric Populations.* Geographically isolated allopatric populations may be either species or subspecies. The most important of the species criteria, the presence or absence of reproductive isolation, cannot be used (except experimentally, and even then only with reservations) to determine the status of populations that are separated from their nearest relatives by a distributional gap. This is the reason why the classification of allopatric populations is so often subject to a considerable amount of disagreement among taxonomists. Many solutions for this dilemma have been proposed, but all of them are beset with difficulties.

Some taxonomists insist that all morphologically distinct, isolated populations be treated as full species "until it is proved that they are subspecies." This solution is of course impractical, because it is impossible in most of these cases ever to obtain clear-cut proof one way or the other. Furthermore, this solution overlooks the fact that it is just as serious an error to call a population a species when it is really only a subspecies, as vice versa.

The second solution is to treat as full species all populations that are not connected by intergradation. This procedure is founded on the correct observation that populations which are connected by intergradation are conspecific; the reverse conclusion is then drawn, namely, that populations which are not connected by intergradation are not conspecific. This however is in conflict with the rules of logic when applied to isolated allopatric populations. Geographical isolation is not an intrinsic isolating mechanism (Mayr, 1963, p. 91), and there is no guarantee that the morphological hiatus caused by the temporary stop in the gene flow is proof of the evolution of isolating mechanisms. The opposite extreme—considering all related allopatric forms to be conspecific—is equally wrong, as shown above.

A complete experimental analysis, including studies on mating preference and a cytological examination of hybrids, is usually impossible, and even where it is possible it may not be conclusive. Ecological preferences are part of the isolating mechanisms between species, and these cannot be properly evaluated in the laboratory. For example, the sympatric sibling species *Drosophila pseudoobscura* Frolova and *D. persimilis* Dobzhansky and Epling always hybridize in laboratory populations, but only a few scattered $F_1$ hybrids have ever been found in nature.

When direct proof is unavailable, it becomes necessary to decide the status of isolated populations by inference. Several kinds of probabilistic evidence are available on which to base such inferences. All rely on the observation that reproductive isolation is correlated with a certain amount of morphological difference, which is fairly constant within a given taxonomic group. The taxonomist can use this evidence to work out a yardstick which can be applied to isolated populations. There are three sets of morphological differences that can be utilized to calibrate such a scale.

1. *Degree of Difference between Sympatric Species.* Within a given genus or within a group of closely related genera, there is usually a fairly well-defined amount of morphological difference between valid sympatric species. This difference may be great, as in the case of birds of paradise, or it may be very slight, as in the case of sibling species. This amount of difference between good species can be used to determine the status of isolated populations in these same genera.

2. *Degree of Difference between Intergrading Subspecies within Widespread Species.* The amount of morphological difference between the most divergent subspecies in species of the same genus indicates how much morphological difference may evolve without acquisition of reproductive isolation.

3. *Degree of Difference between Hybridizing Populations in Related Species.* Subspecies or groups of subspecies within a species sometimes become temporarily separated from one another through the development of a geographical barrier but merge again after the breakdown of the barrier. Free interbreeding, which often occurs even after morphological difference of considerable magnitude has developed, proves conspecificity. Good examples of such free interbreeding of morphologically strongly differentiated populations are to be found in North American birds, among some of the juncos (*Junco*) and flickers (*Colaptes*) (Short, 1965). This is to be taken into consideration in the ranking of isolated populations in these genera.

Even after all these criteria have been applied, some doubtful cases remain. *It is preferable for various reasons to treat allopatric populations of doubtful rank as subspecies.* The use of trinominals conveys two important pieces of information: (1) closest relationship and (2) allopatry. Such information is very valuable, particularly in large genera. Geographical replacement suggests furthermore that either reproductive isolation or ecological compatibility has not yet been evolved. To treat such allopatric forms as separate species has few practical advantages. If further analysis shows that such a form had been erroneously reduced to subspecific rank, it can again be restored to full species status.

# Chapter 10 The Procedure of Classifying

*I*n this chapter an attempt will be made to explain the procedure of sorting species into higher taxa. The theory of biological classification, on which this procedure is based, has been treated in Chap. 4. Most of the problems of classification of higher taxa—one might designate this as macrotaxonomy—are still controversial. Compared to the extraordinary activity on the species level, there has been until quite recently very little conceptual ferment on the level of the higher categories. We are still waiting for a new systematics of macrotaxonomy, even though recent work represents significant forays into this terra incognita. What is particularly badly needed at this time is a greater interest among taxonomists in the taxonomic method as such, in the properties of the evidence on which taxonomic conclusions are based, and in the process of inference by which conclusions are derived from the evidence. The new interest in methodology, aroused by Hennig, Cain, Michener, Simpson, and the numerical pheneticists, is likely to produce eventually as much of an advance on the level of macrotaxonomy as the new systematics did on the species level.

In the past, animal taxonomists have been singularly silent about their methods and guidelines. Many of their conclusions, no matter how sound they turned out to be, were reached by "inspection," by an overall simultaneous evaluation of all readily accessible lines of evidence. Only the final result was presented to the reader, rather than the steps that led to it. The very fact that this holistic approach resulted so often in a durable product, confirmed by subsequent researches, demonstrates that it must have been based on a sound and valid method of evaluation. It would be most helpful to later workers if specialists would always record how they obtained and interpreted the evidence. One suspects that they often reached results intuitively, like the mathematician Gauss who is supposed to have said "I have the result, but I don't know yet how I can arrive at it."

The time has not yet come to present a well-balanced methodology of macrotaxonomy. The treatment in this chapter is an attempt to present a synthesis of conflicting claims and alternative procedures, and, perhaps most important, a critique of unsound approaches and assumptions. The literature cited in the various sections of this chapter will permit a deeper penetration into the field. A careful study of various recent publications will help the reader to understand the nature of current disagreements and uncertainties; some of these are Michener (1957), Inger (1958), Simpson (1961), and Mayr (1965b).

The activities of the taxonomist may be best characterized by recognizing four steps:

A. Preparatory activities
    1. The sorting of individuals into phena and these into populations (Chap. 8)
B. Genuine classification
    2. The assigning of populations to species (Chaps. 3, 9)
    3. The grouping of species into higher taxa
        a. Determination of relationship
        b. Formal delimitation of taxa
    4. The ranking of taxa in a hierarchy of categories

In the present chapter we are concerned with operations 3 and 4.

Simultaneously with operations 2, 3, and 4 names are assigned to the recognized taxa (Chaps. 12, 13).

## 10.1   THE GROUPING OF SPECIES INTO HIGHER TAXA

The above phrase suggests a single operation, but there are actually at least three involved:

1. The determination of the nearest relatives of each species
2. The searching for gaps which permit the ordering of species into clusters and groups of clusters, and the decision which of these clusters to recognize formally as genera and which others merely informally as species groups

3. The arrangement of genera into groups of higher and higher taxa, and the ranking of these in the proper categories of the taxonomic hierarchy

These three operations merge into each other and cross-affect each other, as we shall presently see. This complexity of classifying is by no means revealed by the definition given in 4.1 "Zoological classification is the ordering of animals into groups on the basis of their similarity and relationship." Before the various steps in the classifying procedure can be discussed, the meaning of the term relationship must be clarified.

*Relationship.* Unfortunately this term has been used in the taxonomic literature with three very different meanings. To the pheneticist, relationship (and its synonym affinity) simply means unweighted similarity; to the cladist, it means genealogical relationship (4.3.5), i.e. propinquity of descent; finally, to the evolutionist it means inferred genetic similarity, as determined both by distance from branching points and subsequent rate of divergence (4.8). He finds it by evaluating "weighted phenetic similarity." The use of the words affinity or relationship, when nothing but similarity is meant, leads to confusion and should be avoided (Mayr, 1965*b*).

**10.1.1  Finding the Nearest Relatives of a Species.** What is the nearest relative of a given taxon is in many cases so obvious that it requires no special investigation. Equally often the choice is not a foregone conclusion, and the question arises, Is taxon $B$ better grouped with taxon $A$ or with taxon $C$? In order to find the answer, the following procedure is usually employed. All characters likely to shed light on the relationship of $B$, or, in other words, all characters known or suspected to vary in the taxon group $A$, $B$, and $C$, are tabulated (see Table 10-1).

An analysis of Table 10-1 shows that $B$ is closer to $A$ in one set of characters $(a, b, c, d)$ while it is closer to $C$ in another set of characters $(e', f', g', h')$. There is a third class of characters $(i'', k'', l'', m'')$, in which $B$ differs from both $A$ and $C$.

Two sets of evidence are thus available: One is the relative weight of characters $a$, $b$, $c$, $d$, as compared with that of $e'$, $f'$, $g'$, $h'$. The other is the information contained in the characters peculiar to $B$ $(i'', k'', l'', m'')$. Are these characters more easily derived from the corresponding characters of $A$ or of $C$, or are they perhaps in part ancestral to the equivalent characters of $A$ or of $C$? All these questions are merely somewhat more sophisticated restatements of the traditional question: Is $B$ more similar to $A$ or to $C$? Determination of similarity then is the key operation in classification. No matter how widely the theories of classification diverge (4.3), the actual classifying operations resemble one another. Careful comparative evaluation of similarities and differences is the first step in all procedures of classification, regardless of whether the observed similarities

TABLE 10-1.    OCCURRENCE OF CHARACTERS IN THREE
RELATED TAXA

| Characters | Taxon $A$ | Taxon $B$ | Taxon $C$ |
|---|---|---|---|
| Shared by $B$ and $A$ | $a$ | $a$ | $a'$ |
| | $b$ | $b$ | $b'$ |
| | $c$ | $c$ | $c'$ |
| | $d$ | $d$ | $d'$ |
| Shared by $B$ and $C$ | $e$ | $e'$ | $e'$ |
| | $f$ | $f'$ | $f'$ |
| | $g$ | $g'$ | $g'$ |
| | $h$ | $h'$ | $h'$ |
| Peculiar to $B$ | $i$ | $i''$ | $i'$ |
| | $k$ | $k''$ | $k'$ |
| | $l$ | $l''$ | $l'$ |
| | $m$ | $m''$ | $m'$ |

are directly converted into a phenetic classification or whether they are carefully evaluated before being used as evidence for the delimitation of groups considered to have descended from common ancestors. Since in general the more closely two groups of organisms are related, the more similar they are to one another, pheneticists and evolutionists share an interest in "similarity." They differ, however, in their determination and evaluation of similarity.

10.1.2    **Similarity.** The history of taxonomy decisively refutes the assumption that similarity is self-evident and not in need of careful evaluation. It requires a great deal of knowledge and experience to be able to look through superficial similarities and to discount superficial dissimilarities. The taxonomist forever makes decisions (at a more sophisticated level) that are best illustrated by an example. Show to a biologically uneducated person pictures of a shark, a porpoise, and a cow, and ask: "Which two of these three are most similar?" and often you will get the answer: "Of course, the two fish!" In this case, the zoologist has no trouble. Yet much of his work in taxonomy consists in the elimination of spurious similarities and the approriate weighting of seeming differences. Pheneticists who are merely interested in identification schemes might be entirely satisfied with a classification that arranges the porpoise with the fishes and *mutatis mutandis* all other phenetic deviations from evolutionary groupings merely on the basis of superficial similarity.

The evolutionary taxonomist, however, treats similarity merely as evidence on which to base inferences on phyletic relationship. He attempts to penetrate the superficial aspects of the phenotype in order to determine underlying genetic similarities. He believes that taxa which are based on inferred genetic similarity have a far greater predictive value than taxa based on superficial unevaluated similarity.

Similarity established by a procedure in which the probable information content of each character is carefully weighted almost invariably gives a classification in which the taxa truly are monophyletic descendants of common ancestors. Bader (1958) points out correctly how little effect the discovery of an abundant fossil record has had on the classification of mammals. The conventional method of taxonomy, constructing classifications on the basis of carefully weighted similarity, is not often apt to lead the taxonomist astray.

**10.1.3    Causes of Similarity.** Two taxa may be similar to each other for various reasons. The different kinds of similarity have unequal weight in the construction of natural taxa, although our understanding of these differences is still incomplete. The combined similarity between two taxa may be composed of four kinds of similarities (see also Cain and Harrison, 1960).

1. Similarities resulting from joint possession of characteristics shared with a common ancestor
    a. Ancestral characters (symplesiomorphs of Hennig) shared with a remote ancestor
    b. Derived characters (synapomorphs of Hennig) shared with a more recent ancestor
2. Similarities resulting from joint possession of independently acquired phenotypic characteristics produced by a shared genotype inherited from a common ancestor (similarity through parallel evolution)
3. Similarities resulting from joint possession of independently acquired phenotypic characteristics that are not produced by a genotype inherited from a common ancestor (similarity through convergence)

As far as giving information on common descent is concerned (and hence on inferred genetic similarity), these characters must be ranked in the sequence: 1b, 2, and 1a, while 3 is to be eliminated altogether. But even within the three first groups the similarities are of unequal value (see below under weighting).

**10.1.4    Methods for the Determination of Similarity.** The many poor classifications that have been proposed in the taxonomic history of almost any group of animals prove that the establishing of natural groups is no easy matter. Operationally it requires the simultaneous consideration of numerous characters in numerous taxa to make a classification. One of the methods employed in this procedure, the ordering of characters into important and unimportant ones, often led to error. Another (simultaneously employed) method, to improve the classification by a process of trial and error leading to an ever closer approximation, though eventually successful, appeared wasteful in time and effort. These weaknesses of the traditional approach led to alternate proposals.

## 10.2    GROUPING BY UNWEIGHTED PHENETIC SIMILARITY

In order to avoid the shortcomings of traditional procedures based on single characters some taxonomists have been searching for a measure of "overall similarity." Serologists thought that protein interaction gave such an overall measure, while DNA matching (Hoyer et al., 1964) is the most recent candidate (7.4.9). So far none of these methods has fulfilled the hopes of its proponents. The pheneticists have chosen a different approach. They propose to determine overall similarity quantitatively, by compounding similarity values derived from a large number of individual character comparisons.

**10.2.1    Quantification of Similarity.** The recent efforts to quantify similarity are not the first, earlier attempts having been made nearly 100 years ago. More recently Sturtevant (1942) tabulated 33 characters of 56 species of *Drosophila* and found that the concordance between a given pair of species in the chosen characters ran from 0 to 25. When a second tabulation was made of 11 additional characters, based on features not used in the first tabulation, it was found that species that were similar on the basis of the first tabulation were also similar with respect to the second set of characters. The development of electronic computers has made the utilization of more sophisticated calculations possible. The new development started with the almost simultaneous publications of Sneath (1957), Michener and Sokal (1957), and Cain and Harrison (1958), each author proposing slightly different methods. A detailed presentation of these and other methods can be found in chaps. 6 and 7 of Sokal and Sneath (1963), *Principles of Numerical Taxonomy*. Many additional methods are now described each year. A special newsletter (*Taxometrics*) is now published by L. R. Hill which contains a current bibliography as well as short communications concerning new programs and modifications of previously published ones. Computer programs are now available for most of the suggested approaches. It cannot be our task to review in detail a field covered in an entire book and annually in more than 50 journal articles.

The field is changing so rapidly that statements made here may well be out of date within a year or two. Anyone who wants to use the computer in classification must try to keep abreast of the rapidly changing literature. The short presentation given here can serve only as a most elementary introduction to a complex field.

TABLE 10-2.  DATA MATRIX (*after Sokal and Sneath*, 1963)

| Characters | Taxa | | | | | |
|:---:|:---:|:---:|:---:|:---:|:---:|:---:|
| | A | B | C | D | E | F |
| 1 | 1 | 8 | 1 | 7 | 2 | 5 |
| 2 | 1 | 6 | 1 | 6 | 1 | 3 |
| 3 | 6 | 1 | 5 | 1 | 4 | 2 |
| 4 | 1 | 0 | 1 | 0 | 1 | 4 |
| 5 | NC | 6 | 3 | 6 | NC | 1 |
| 6 | NC | 2 | NC | 3 | 1 | 1 |
| 7 | 8 | 2 | 7 | 2 | 5 | 5 |
| 8 | 1 | 6 | 1 | 6 | 3 | 4 |
| 9 | 1 | 8 | 1 | 8 | 2 | 4 |
| 10 | 6 | 1 | 6 | 1 | 5 | 2 |
| 11 | 3 | 3 | 3 | 3 | 3 | 3 |

**10.2.2    Data Listing.** The first step is the listing of the raw data in the form of a data matrix (Table 10-2). The vertical columns list the taxa, while the horizontal rows tabulate the characters. There are two ways of dealing with characters in such a tabulation. If only the presence or absence of a character is to be indicated, plus and minus signs are all that is needed. On the other hand if different character states are to be recorded, numerals have to be used. The special definition of "characters" in computer taxonomy has to be kept in mind (7.1).

Such a matrix can be examined from two points of view. An analysis of the association of pairs of character (rows) is called the R technique; it studies the correlation among characters. Conversely, an analysis of the association of pairs of taxa is called the Q technique. This is the technique used by the pheneticists when making the simplifying assumption that all characters have equal information content. Every taxonomist realizes that this is unrealistic. When Rogers (1963) tested the contribution of various attributes to the grouping of specimens of Manihot,

> . . . it invariably happened that each group of specimens was set apart by a certain number of attributes and that the remaining attributes did not define the groups. Some of the attributes provided definitions of small groups, some of larger groups, and some were not significant at all at any level. What had happened, apparently, was that certain of the characteristics and attributes were at work in definition and characterization of the groups; others were not. Those that did indeed separate and define the taxa had more weight than those that did not.

Yet a recent study by Rohlf (1967) involving an R-type analysis did not apparently lead to an improvement of a classification previously produced by a Q-type analysis.

There are numerous ways of computing resemblance between taxa, based on the Q-type approach. Three types of coefficients of similarity, those of association, correlation, and distance have been proposed. Each has advantages and disadvantages. In many ways the most satisfactory method for expressing the difference between two taxa is to calculate total "distance" in a multidimensional space (Seal, 1964). With electronic computers available, it is possible to calculate taxonomic distance even when numerous taxa and characters are involved. This method likewise has deep-seated flaws that have not yet been overcome. Both Boyce (1964) and Minkoff (1965) have shown that different mathematical treatments of the same raw data lead to drastically different classifications.

**10.2.3    The Number of Characters.** In the old identification schemes (mislabeled classifications) taxa were characterized monothetically, that is by single characters. In contrast, empiricists in the eighteenth century already proposed using the greatest feasible number of characters. This advice is on the whole sound; however, there tends to be a difference between characterizing species and higher taxa. Species taxa can often be recognized on the basis of a single character. For instance, both the Mikado Pheasant (*Syrmaticus mikado*) of Formosa and the Ribbon-tailed Bird of Paradise (*Astrapia mayeri*) of New Guinea were named on the basis of a single tail-feather. The shape of the baculum of mammals is sometimes species-specific (Fig. 7-3), as are genitalic structures in arthropods, color patterns and songs in birds, and vocalizations in frogs, orthopterans, and cicadas. Sometimes a single character may be diagnostic of a higher taxon. However, higher taxa are groups of species, and they almost invariably require for characterization a number of traits, because each character or character complex may have its own independent evolution in each phyletic line and subline. In such a case a classification based on few characters, or a single one, will be quite misleading. For instance, on the basis of the morphology of the placenta and associated fetal membranes, one author proposed that lemurs be classified with the ungulates rather than with the higher primates, and the Megachiroptera with the rodents rather than with the regular bats (Microchiroptera). An ornithologist classified owls on the basis of the size and asymmetry of the external ears. A reexamination showed that this is a food-associated specialization closely correlated with climatic zones: tropical owls which find their prey visually have small ears without dermal flaps, while temperate-zone owls feeding during long winter nights on rodents which are acoustically located have large ears with wide dermal flaps. Altogether too many early classifications were based on a single conspicuous character complex: parasitic bees versus nonparasitic bees; songbirds with cone-shaped bills (finches) versus

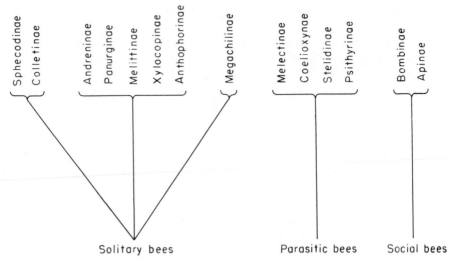

**Fig. 10-1.** A so-called "practical" classification of bees (*after Friese*) in which the parasitic species are treated as a separate group, independent of their close relatives among the solitary and social bees.

songbirds without cone-shaped bills, and so on. Every specialist can cite similar examples from the taxa with which he is most familiar.

The unreliability of single characters led to the belief that, the more characters a classification is based on, the more reliable it will be, which is not necessarily true. Even a large number of unweighted characters is not always sufficient for unequivocal classification. Rohlf's (1963) two classifications of the mosquito genus *Aedes*, one based on 77 adult and the other one on 71 larval characters, are discordant to such a degree that a very much larger number of characters would be needed even to hope for concordance. Also, there is always the problem of defining what is a character. A large-sized species may differ from a related small-sized species by hundreds of mensural characters, but all of them may be nothing more than aspects of a single trait, namely size.

Methods based on the processing of unselected and unweighted characters require large numbers of characters, preferably more than 100. These can be readily found in arthropods, particularly insects; it is not surprising that bees, mosquitoes, and mites were preferentially used to test these methods. It has not yet been determined what methods are most suitable for use in morphologically uniform groups (such as birds or lower fungi) where there is an embarrassing deficiency of useful characters. Results of principal-component analyses suggest that a relatively small number of characters might be sufficient if one only knew how to select such characters.

The only limit to the number of characters provided by a taxon is that set by the patience of the investigator. It would seem prudent not

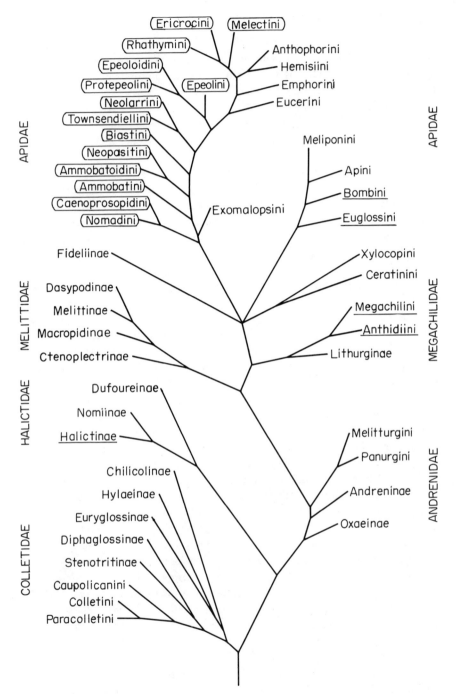

**Fig. 10-2.** A phylogenetic classification of bees (*after Michener*, 1944) in which the parasitic bees are placed according to their inferred relationships among the solitary and social bees. Wholly parasitic groups are circled, partially parasitic groups underlined.

to use more characters than are sufficient for the establishment of a sound classification. Most characters have a low information content or are completely redundant. What is important is not the number of characters but their taxonomic weight. Indeed Kendrick and Weresup (1966) and Throckmorton (1969) have demonstrated that many, if not most, characters are merely "noise" (as far as their informational value is concerned) and that their removal from the list of considered characters may increase efficiency of grouping.

When a new set of characters is utilized, it will have one of three possible effects on the existing classification. The new characters either completely confirm the traditional classification (this being what usually happens), or lead to a clear improvement of the classification, or introduce ambiguity and contradiction which can be resolved only by the application of additional new characters. At the present time it cannot be predicted which of the three will be realized in a given instance.

10.2.4   **Weaknesses of the Purely Phenetic Method.** Many recent writers have pointed out deficiencies in the methods and theory of numerical phenetics (see recent volumes of *Systematic Zoology*).

Among the theoretical weaknesses are its inability to distinguish between phena and taxa. Under its own terms this is legitimate, since the end product of the phenetic operation is an identification scheme rather than a biological theory, and this is what a good classification should be. Another theoretical weakness is the assumption that a random selection of components of the phenotype would lead to correct estimation of the properties of the genotype, provided enough characters are included in the analysis. The visible phenotype is as small a fraction of the total potential phenotype as the visible part of an iceberg is of the whole. It is not true, for this reason, that the amount of unweighted phenetic similarity necessarily runs parallel with the amount of genetic similarity. Sibling species prove conclusively that drastic genetic differences may not be reflected in the visible phenotype. The opposite extreme is demonstrated by groups like the birds of paradise, where intensive sexual selection (Mayr, 1942; Sibley, 1957, 1959) has produced enormous phenetic differences between closely related species that are still fully or largely fertile with each other. Only appropriate weighting can convert the phenetic distances between the genera of birds of paradise into a biologically meaningful classification. This proves the falseness of the assumption that each character is so polygenic that any random sample of characters will accurately reflect the properties of the genotype.

It is important here to mention computers and electronic data processing (EDP). Some users of EDP have suggested that thinking and theory become unnecessary if we merely entrust our fate to the computer. Nothing could be farther from the truth. The operations of the computer are infinitely faster than our own, and it can handle far more data at one time than we can; but objectivity is not affected by the computer, and the underlying principles are not modified in the least. The computer is defenseless against logical errors of the programmer. It is a pity that so much of the early work in computer taxonomy was based on unsound taxonomic theory (see also Ghiselin, 1966b).

Equal weighting and random selection of characters are implicitly based on the assumption that during evolution the genotype as a whole changes harmoniously and that all components of it change at approximately equal rates. Mosaic evolution and many other evolutionary phenomena, however, show that this assumption is unrealistic (10.4.5).

The phenetic method also encounters various practical difficulties. One is its inability to produce repeatable results whenever there is a slight change in the selection of characters or the methods of determining similarity (Boyce, 1964; Eades, 1965; Minkoff, 1965; Kendrick and Weresup, 1967; Michener and Sokal, 1966; Sokal and Michener, 1967).

The method is, in a way, inefficient because the programming of large numbers of taxa with about 100 (or more) characters is very time-consuming. A specialist may not have such time available, if he is the only living specialist for a group of 1,000, 5,000, or 20,000 species. In order to be able to undertake identification work, and to describe and name species rapidly, he is forced to adopt short-cut methods of classifying. He sorts numerous specimens and species into "natural groups" by scanning their total gestalt, based on an evaluation of very many characters, most of which he does not analyze or record in detail (Anderson 1954). A posteriori he determines what the most constant and most easily recognizable characters of these natural groups are.

By far the greatest practical difficulty in applying a purely phenetic procedure is the scarcity of taxonomically useful characters in most groups of organisms. Insects and other arthropods with a highly sculptured exoskeleton are the only conspicuous exceptions to this statement, but even in these there is often a "desperate need for new characters," as recently stated by a specialist in digger wasps. By the time all characters of low weight have been eliminated (see 10.4), there is often hardly anything left. In the case of most of the thirty-odd orders of birds, it is still unknown which other order is the nearest relative. The same is true for the families of songbirds (Oscines). There are several drastically different classifications of the sponges depending on the relative weight given to the very few available characters. Convergent adaptations lead to far greater similarities in such groups than does the joint possession of ancestral characters. The finches, tree creepers, titmice, flycatchers, and possibly the ratites are illustrations among birds of groups based on convergent characters. Many similar examples will occur to the specialist in rodents, snakes, urodeles, teleost fishes, bivalve mollusks, and lower invertebrates.

There are two ways to overcome the dilemma of a scarcity of useful characters. One is to look for new characters (7.4), the other is to attempt to extract more information from existing characters. Bock (1963) has shown how a deeper analysis of known characters sometimes leads to a better understanding of the phylogeny of morphologically uniform groups. A purely quantitative procedure, by which the few useful characters are diluted by a large number of useless ones, was rightly ridiculed by Adanson (Stafleu, 1965).

**10.2.5  Utility of the Phenetic Approach.** Nonweighting is least objectionable when applied to groups with immature classifications (particularly single-character classifications) and those with numerous nonredundant characters. In such taxa it has produced groupings that are clearly superior to the traditional ones. When there are several competing classifications, a phenetic analysis may be illuminating. For instance, the genera of Megascolecoid earthworms were assigned to very different families and subfamilies in three almost simultaneously published revisions. In one of these the higher taxa were based on the position and number of the calciferous glands; in the second, they were based on the number and position of the male terminalia, and in the third on the structure of the prostatic

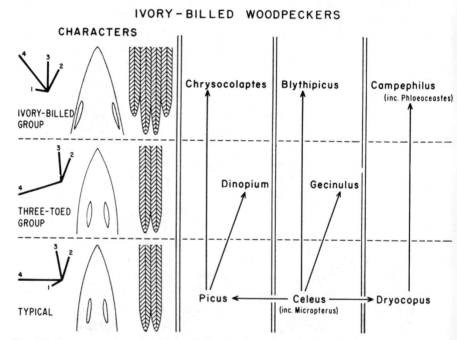

Fig. 10-3. Convergent origin of the ivory-billed woodpeckers, from three unrelated genera of typical woodpeckers. The genera in the three vertical columns have similar color patterns indicative of true relationship. The similarities in foot, bill, and tail of the ivory-bills are considered to be due to convergence (*after Bock,* 1963).

glands and the excretory system. Forty-three characters of 30 species (in 29 genera) were analyzed by Sims (1966) with the help of association coefficients, and the grouping resulting from this analysis agreed reasonably well with one of the previously proposed classifications, i.e. that based on the structure of the prostatic glands. In the mites (Acarina), also a group with an immature classification, computer techniques have likewise been very helpful (Sheals, 1965).

The phenetic method is a particularly uneconomical procedure in groups with either a mature classification or a great deficiency of taxonomically useful characters. *Weighting becomes the more important the higher the categorical level of the taxon we are attempting to place.* When we sort species within a genus, weighting is relatively unimportant, and much of it is arbitrary. But the higher in categorical rank two compared taxa are, the fewer will be the number of characters that indicate relationship. And the fewer the number of usable characters, the more carefully they must be weighted.

It is too early to pass final judgment on the phenetic method. Much of its theory (4.3.3), based on nominalism, is naive and unsuitable as an appropriate basis for work in evolutionary taxonomy (Mayr, 1965b). Yet computer methods as such have a tremendous potential. The greatest contribution the pheneticists have made is to harness the computer to the classifying procedure. The original goal, to automate taxonomic procedure and to remove from it all subjective evaluations, has not been achieved in the first ten years. "In the present state of its development, it [phenetics] is not able to provide stable classification" (Sokal and Michener, 1967). Other characters, other species in the same genera, other measures of similarity, and other clustering methods often result in different classifications. The ultimate answer will almost surely be a combination of classical and new computer methods. Nevertheless we owe the numerical taxonomists a debt of gratitude for developing new methods and for trying to find out which would give the most informative results. Methods of weighting will undoubtedly play an increasingly important role in computer taxonomy.

## 10.3  GROUPING BY CLADISTIC APPROACHES

Instead of making "overall similarity" the basis of grouping, one can also establish groups based on the inferred branching pattern of phylogeny. All methods which give branching patterns primacy in classification may be designated as cladistic approaches.

These methods share a common attempt (1) to determine the number of phyletic splits by which different taxa are separated from each other, and (2), perhaps more important, to reconstruct the sequence by which

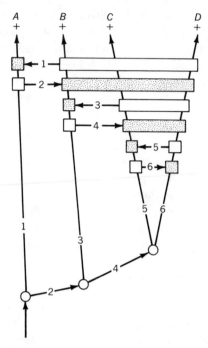

Fig. 10-4. Hennig's method of determining groups derived from the most recent common ancestor. Such groups are distinguished by the joint possession of at least one derived character (shaded rectangles) such as character 1 for taxon *A*, 2 for taxon *B + C + D*, 3 for *B*, 4 for *C + D*. This permits reconstruction of the sequence of branching ( = recency of common ancestry). Ancestral characters are designated by open circles or squares (*after Hennig*, 1966).

these branchings occurred. This strategy has long been employed by students of chromosomal evolution in order to determine the sequence in which chromosomal inversions and other chromosomal mutations must have occurred. Hennig (1950, 1966) pioneered in systematically applying this approach to taxonomy (see Fig. 10-4 for an explanation of this method). A good example of an application of this method is Wagner's (1962) classification of the Delphacidae of Europe. Wilson (1965) discusses a similar approach.

A somewhat different cladistic method was developed by Camin and Sokal (1965) for deducing phylogeny. Human geneticists (Cavalli-Sforza and Edwards, 1964) and molecular evolutionists (Fitch and Margoliash, 1967) have, largely independently, developed similar methods. All these approaches have a common objective, namely, to design a phyletic branching pattern that would result from the minimal number of evolutionary steps by which the various taxa became separated from each other.

After a phyletic line has split into two separate lines, the subsequent retention of ancestral characters and the acquisition of new, derived characters occur independently in the two lines. The basic rationale of all cladistic methods is that the more recent the common ancestry of two species (or other taxa), the more characters in common they should have. The same thought, expressed in terms of taxonomic characters, is that the occurrence of a relatively new character will be limited to the descendants of the particular species in which the new character originated. A careful study of the distribution of characters in the taxonomic hierarchy should therefore give information on the grouping and ranking of taxa. The determination of the phyletic age of characters is the key operation in this method.

**10.3.1 Ancestral Characters.** In view of the stipulation that taxa should consist of descendants from a common ancestor, a separation of characters into those similar to or identical with the characters of the ancestor and those that have more or less diverged from the ancestral condition becomes important. The terms "primitive" and "advanced" for these two classes of characters were inherited from progessionist preevolutionary theories (*scala naturae,* etc.). The term primitive implies simplicity, and it is therefore often misleading because comparative anatomy and the study of fossil series have shown that a progressive simplification is one of the most frequent trends in evolutionary lines. The terms "advanced" and "specialized" for the derived condition of a character can be misleading for the same reason. Hennig (1950) uses the terms *plesiomorph* for primitive and *apomorph* for derived characters or character states. There is a need for simple, self-explanatory terms that are not committed to any evolutionary theory. The terms *ancestral* and *derived* may serve as such simple, descriptive terms. A character is *ancestral* if it has not changed materially from the homologous character in the ancestor, while a character is *derived* if it has changed materially.

A given character may be either ancestral or derived depending on the stages in the phylogeny that are compared. Possession of wings is an ancestral character for flightless birds, but a derived character for birds as a whole when compared with reptiles. In some cases it is easy to determine the ancestral condition of a given character in a phyletic line; in other cases this is very difficult. Hennig (1950), Maslin (1952), Remane (1952), Simpson (1961), and Wagner (1962) give criteria facilitating the correct choice. The reconstruction of the entire phylogeny of the frogs (Anura) depends on decisions concerning the relative primitiveness of various characters (Inger, 1967).

In some higher taxa, e.g. the primates, many primitive taxa still survive, and it is almost possible to reconstruct the probable phylogeny by a concatenation of living forms from the lemurs to the anthropoid apes. In other groups, like the ungulates, as Simpson points out, there is no survival of the ancestral types that lived during the early Tertiary period. Among the species of a genus it is often possible to single out one that has an assortment of characters which one would expect in the ancestral species of the genus. In dendrograms such species are often placed at the point of origin of the genus, but this is somewhat misleading. Only rarely has such a species literally given rise to the others by budding off peripheral isolates. More commonly, the recent "primitive" species is that one of the many descendants of the ancestral species which happens to have retained the highest proportion of ancestral characters. Such conservative species are of extreme practical importance in helping to determine the relationship of taxa. Conclusions regarding relationships depend largely on the reliability of the determination of primitivity of characters. When a taxon is conservative in one character, it very often is also conservative in other characters, but certainly not always and rarely in all characters. *Platypus* is the most primitive of living mammals in several anatomical features, particularly in its reptilian shoulder girdle, and also its egg-laying habit. However, it is highly specialized in its dentition, bill, poison spur, and various aquatic adaptations.

Ancestral characters or character states have various properties that help in their identification (Hennig, 1950, 1966). They tend to be irregularly scattered among a broad group of related taxa, while a given derived character usually characterizes only one particular group of descendants (Fig. 10-4).

The more widely a character is distributed in the system, the more distant is the relationship which it indicates (Hennig, 1950, p. 172), and the more remote the common ancestor. The distribution of the vertebrate column in all vertebrates and the occurrence of feathers in all birds serve to illustrate this principle. Yet possessors of a given character are not necessarily more closely related to each other than they are to other taxa that lack the character. The reason is parallel evolution in the acquisition of new or in the loss of old characters. For instance, pentadactyl vertebrates (amphibians, reptiles, primates, and other mammals) are not more closely related to each other than they are to perissodactyls, artiodactyls, cetaceans, and other mammals that are not pentadactyls.

The earlier a phyletic line has acquired a character, or, as we usually say, the older a character, the more completely it seems to be integrated into the entire genotype and the more closely it is correlated with other characters. The assumption that the older a character is, the earlier it is laid down in ontogeny, is probably true but not proved. If true it would nevertheless not confirm recapitulation, because there is no evidence that an organism during ontogeny recapitulates the adult stages of its phylogeny (as postulated by some phylogenists). Indeed all the evidence refutes such an assumption.

Ancestral characters or character states are the condition from which various specialized conditions are most easily derived. For instance, cytogeneticists have shown for *Drosophila* (among other cases) that chromosomal variation among near and distant relatives can be arranged in a definite sequence of chromosomal mutations (7.4.7). Either one or the other terminal condition must have been the ancestral one. Collateral evidence (if not internal evidence) usually permits a definite choice between the two alternatives. Occasionally an intermediate condition is ancestral, giving rise to two diverging trends. Various empirical rules help in determining the ancestral condition. If one taxon is clearly derived from another taxon, as birds and mammals are derived from the reptiles, the ancestral condition may still be largely present in the ancestral taxon. Indeed, most mammalian and avian characters can be clearly derived from homologous features in the reptiles.

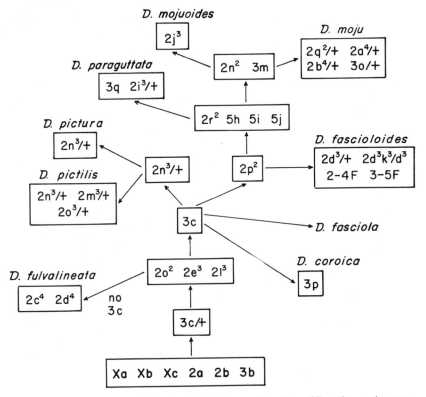

Fig. 10-5. Chromosomal phylogeny in the *fasciola* subgroup of the *Drosophila repleta* species group. Ancestral arrangement in the bottom rectangle. Arrows indicate the sequence of chromosomal rearrangements. Heterozygous inversions which occur with the standard are indicated by the inversion shown over a plus sign (*from Wasserman*, 1962).

Various phylogenetic rules such as Cope's rule, Dollo's rule, and others discussed by Rensch (1959, 1960) and Simpson (1959a), are sometimes helpful. Some of these empirical indications are definitely wrong (4.8), while others have varying numbers of exceptions. Contrary to the statement of Dollo's rule, for instance, lost ancestral structures may occasionally reappear in descendants if the potential for them was retained in the genotype. Conspicuous body setae and antennal papillae reoccur in the larvae of a number of specialized groups of Hymenoptera after being lost earlier in the evolution of this order of insects (Evans, 1965; see also Maslin, 1952). A careful evaluation of every character as to whether or not it represents an ancestral condition is a help not only in the delimitation of taxa, but even more so in ranking them as needed for the construction of a taxonomic hierarchy (10.5), and in arranging them in a linear sequence.

**10.3.2     Auxiliary Evidence.** When all the other evidence on relationship is ambiguous, one can sometimes employ two other sources of evidence.

*Parasites* evolve with their hosts but are sometimes more conservative. An existing relationship may be more evident in the parasites than in their hosts (7.4.12). The difficulty is that a parasite sometimes shifts to a new host (more or less unrelated to the old host), which deprives the parasitological evidence of some of its value. In ambiguous cases the concordance among different kinds of parasites must be analyzed. Most animal groups have several kinds among the following parasitic taxa: protozoans (particularly in the blood), cestodes, trematodes, nematodes, acanthocephalians, copepods, mites, lice, fleas, and mallophaga. Some of these parasitic taxa shift rather easily to new hosts, while others are remarkably host-specific. The relationship between man and the African anthropoids (*Pan*), for instance, is indicated by four different groups of parasites. See Baer (1957) for a discussion of various aspects of this method.

*Geographic distribution* is another important clue to relationship. Hennig (1950) presents numerous cases where the study of distribution patterns has resulted in improvements in classification. This evidence has been presented in detail in Chap. 7 (7.4.13).

During the continuous faunal turnover on all continents older elements are often forced to retreat to the southern continents (Australia and South America) or to such islands as Madagascar and New Zealand. These relicts of formerly more widespread taxa often possess exactly the ancestral characters from which the character states of more modern types can be derived. The study of relict types now found only at these peripheral locations is thus often of particular importance for the reconstruction of the inferred ancestral characters.

**10.3.3     Utility of the Cladistic Method.** The careful analysis of every taxonomic character which the cladistic method demands is of considerable heuristic value. Even taxonomists who reject the ranking criteria of cladism (see 10.5.2) often find it useful to try to infer the phyletic age of each character, since this makes such a large contribution to its taxonomic weight. In the grouping step of the classifying procedure cladistic analysis is indeed most useful, but when it comes to drawing inferences from cladistically produced evidence, one must make two reservations.

First, an operational one—none of the methods listed above has been sufficiently tested to permit judgment as to its efficiency and reliability. Second (and more important)—none of these methods allows sufficiently for mosaic evolution, for parallel evolution, and, particularly, for a combination of both. The cladistic approach has a strong typological element, demanding "yes" or "no" decisions and assuming all changes of characters to be unique events. Other difficulties are caused by evolutionary reversals and by the occasional impossibility of determining the ancestral condition. The probabilistic considerations that will have to be built into the cladistic approach to make it more reliable will inevitably bring it nearer to the phenetic approach, and it is conceivable that the best elements of both approaches can eventually be combined in a "synthetic" method. The classical method of taxonomy, of course, has essentially employed such a "synthetic" procedure.

## 10.4  INFERRING RELATIONSHIP BY WEIGHTING SIMILARITY

Experienced taxonomists have always insisted that characters differ in the contribution they make to the soundness of a classification. Darwin (1859, pp. 414–417), for instance, gives some good empirical rules concerning the usefulness of certain characters. With the failure of the phenetic and cladistic approaches to produce automatically and objectively sound classifications, there has been a renewed interest in the traditional methods of taxonomy. The weighting of similarities as evidence for relationship is the key operation in this approach. When well done, it leads to classifications of lasting value; when poorly done, it leads to much changing and controversy. Owing to their crucial importance the weighting criteria must be discussed in detail.

**10.4.1  A Priori and A Posteriori Weighting.** The Aristotelians and their successors often assigned a priori weights to certain characters. Cain (1959a) has pointed out the fallacy of this approach. Neither function nor conspicuousness nor any other known aspects of a character gives it a priori a greater weight than other characters. Indeed the very same structural difference may have high weight in one taxon and low weight in a related taxon. Nor should the taxonomist confuse weight of a character with its usefulness in a diagnostic key. Any rigid following of the dictates of a few arbitrarily chosen a priori characters leads inevitably to a classification which is not "natural," that is, which has low predictive value.

Adanson and the empirical taxonomists of the ensuing period rejected a priori weighting and replaced it by an empirical process, perhaps best called a posteriori weighting. Adanson was satisfied with merely eliminating useless and redundant characters; later authors, when the natural groups of animals were ever better understood, attempted increasingly to assess the relative merits of each character. The value of this approach, in spite of individual errors, is best substantiated by the fact that all existing good classifications are the result of such a posteriori weighting.

The acceptance of evolutionary theory after 1859 did not change this method as such. It did, however, provide scientific justification. It now became evident why some characters are better indicators of natural groups than others. Different characters contain very different amounts of information concerning the ancestry of their bearers. *Weighting, then, can be defined as a method for determining the phyletic information content of a character.* If character *a* indicates assignment of a species to genus *A* and character *b* to genus *B*, we must determine which of the two characters (*a* and *b*) has the higher information content. It is neither necessary nor even possible to give a precise numerical value to the relative weight of each character. Qualitative statements are usually more important than quantitative ones. In order to assign a species to the correct phylum, to know that it has a chorda is more important than a thousand measurements.

The scientific basis of a posteriori weighting is not entirely clear, but difference in weight somehow results from the complexity of the relationship between genotype and phenotype. Characters which appear to be the product of a major and deeply integrated portion of the genotype have a high information content concerning other characters (which are also products of this genotype) and are thus taxonomically important. Other kinds of characters, such as monogenic and oligogenic characters, as well as superficial similarities, convergences, and narrow adaptations, have low information contents concerning the remainder of the genotype and are thus of low value in the construction of a classification. What the descendants of a common ancestor share is not an aggregate of independent characters but a whole well-adapted harmonious genotype. Such a genotype has a considerable evolutionary inertia, and it appears that adaptively needed modifications can be superimposed on it without destroying it. Indeed one might speculate that the characters in a phyletic line which are most intimately tied up with the basic well-integrated genotype are the most conservative.

Simpson (1962*b*) has described good taxonomic characters as "readily observable characters that are believed to be fairly constant within taxa but different between taxa at any pertinent level." This is a good empirical description, but it requires painstaking study of large series of specimens to demonstrate that characters are "fairly constant" and a delimitation of the taxa (based on a study of characters!) before it can be shown that the characters are different between them! Nor does this description cope with the difficulties produced by the various types of similarity (10.1.2) and with certain other difficulties to be mentioned later (10.4.5). Indeed, it is precisely this problem of how to determine what is a good character, a character with high weight, that has been a main source of concern among taxonomists.

Methods of a posteriori weighting have been discussed by Hennig (1950, 1966), Remane (1952), Maslin (1952), Cain and Harrison (1960), Simpson (1961, pp. 82–106), and Throckmorton (1969). No overall treatment of this subject has yet been published, and it is still cloaked in uncertainty. There has been an increasing interest in methodology and a hope that computer programs can be developed that will facilitate a posteriori weighting. Actually, there is already a certain amount of implicit weighting in most existing methods of numerical taxonomy.

The classical approach has been to work backward from classifications that produce natural groupings and study the characters which delimit such natural groups. Tooth structure in mammals, wing venation and structure of the genitalia among insects, and the structure of the bony palate among birds are taxonomic characters that are fairly constant within groups that appear to be natural and are therefore given high weight. Each of these favored taxonomic characters fails occasionally and in certain groups frequently. The number of cervical vertebrae (seven) is a class character in mammals, while it is not even a generic character in birds; for instance, it fluctuates from 23 to 25 in the genus *Cygnus* (swans). The only reason why high weight is given to certain characters is that generations of taxonomists have found these characters reliable in permitting predictions as to association with other characters and as to the assignment of previously unknown species.

It has been said with good reason that the trial-and-error method of improving classification is ponderous and uneconomical. To undertake a successful a posteriori weighting of characters requires a thorough knowledge of the history of previous classifications of a given group and an ability to make value judgments. Yet no clearly better method has so far been found. Beginnings are being made in the development of computer methods that endeavor to select characters of special weight in classification (e.g. Throckmorton, 1969). Until such methods have been further worked out and more thoroughly tested, a survey may be presented of the traditional considerations in weighting. This will continue to remain important in all groups with a paucity of taxonomic characters.

**10.4.2 Characters with High Weight.** A few generalizations can be made about characters with high taxonomic weight. These include the following:

*Complexity.* Complex structures have greater weight than do simple structures, even when there are more of the latter. This is one of the reasons why genital armatures in arthropods are of such high taxonomic importance. They are usually highly complex and differ even in closely related species, sometimes quite drastically. The probability that species which are only distantly related would become similar in such structures by convergence is extremely low. Complex ornamentations, complex cusp patterns of teeth, complex color patterns, etc., all fall under this category.

*Joint Possession of Derived Characters.* Hennig (1950 and later) has articulated a principle followed by taxonomists for generations but apparently never before spelled out succinctly. It states: "Taxa ought to be defined on the basis of shared derived characters (synapomorphy) and not of shared ancestral ones." The strength of this principle is self-evident. If two forms share a primitive (ancestral) character, this may be due to the fact that they have not yet lost the ancestral trait rather than that they are closely related. The acquisition by two taxa of the same evolutionary novelty is almost invariably the result of close relationship, rarely of convergence.

*Constancy.* A character that is constant "throughout large groups of species" (Darwin, 1859, p. 415) has higher weight than a variable character. Farris (1966) suggests that low phenotypic variability be given very high weight. Since monogenic characters sometimes have low variability, this principle is probably valuable only with relatively complex (polygenic) characters.

*Consistency.* A character that is consistently present in one group and equally consistently absent in related groups has obviously higher weight than a character occurring sporadically in several groups and differing merely in the frequency of its occurrence.

*The Darwin Principle.* Taxonomists have long stressed the relative importance of characters that do not serve a specific ad hoc adaptation but are merely, so to speak, an indication of an underlying basic genetic similarity. As Darwin (1859, p. 414) put it, "The less any part of the organization is concerned with special habits, the more important it becomes for classification." This is why color pattern in birds, a particular configuration of cusps on molars in mammals, a particular pattern of reduction or fusion of wing veins and of elaboration of sclerotic structures in the genitalia in insects are of such high value in classification. It is possible that the special configuration of some of these structures has an adaptive significance that has not yet been discovered. In most instances it is more probable that the genotype responsible for the structural configuration was brought together by natural selection and has adaptive significance as a whole. The principle stating that a character which is the product of the general genotype is more likely to have high taxonomic weight than a character that represents an ad hoc specialization has wide application.

*Characters Not Affected by Ecological Shifts.* Most higher taxa include subtaxa (species in genera, genera in families, etc.) that have made an ecological shift. Any character not affected by such a shift has a higher weight than characters affected by it. In the mergansers (*Mergus*), to be discussed in the following section, the color pattern of the downy young, the courtship, and other characters not affected by the shift into the fish-eating food-niche have higher weight than characters affected by the shift.

*Correlated Characters.* Taxonomists are in general agreement that they rely more on correlated character complexes than on any other clue to relationship. Yet in the weighting of characters a strict distinction must be made between two kinds of correlations, one of very high and the other of very low weight. Of low weight are characters that are functionally correlated and do not deserve to be treated as separate characters because, being members of a single functional complex, they are redundant (see 10.4.3). Of high weight are characters that are not functionally correlated but are found in members of the same taxon because these characters are phenotypic manifestations of the ancestral well-integrated gene complex, let us say like the diagnostic characters of the Deuterostomia or the Chordata. One might call this kind of concordance *phyletic correlation.* When comparing the species of two well-defined higher taxa, let us say taxon $A$ and $B$, one always finds that the species within either taxon share certain functionally uncorrelated characters not displayed by the species of the other taxon and vice versa. The species of taxon $A$ may have characters $a$, $b$, $c$, and $d$, the species of taxon $B$ characters $e$, $f$, and $g$. We infer that the joint possession of characters by species of a higher taxon is the result of common descent and assign such concordant, that is, phyletically correlated, characters (like $a$, $b$, $c$, $d$ and $e$, $f$, $g$) high taxonomic weight. The character that is concordant with the greatest number of other characters has the highest weight. Each species of such a higher taxon, of course, has many additional characters that do not belong to such correlated character complexes.

Whether an association of characters represents functional or phyletic correlation is not necessarily evident at first sight. It sometimes requires a careful functional analysis. In a few cases, indeed, there is no sharp line between them. Many phyletically correlated character complexes may have originally started as a functional complex in which the genetic integration was retained even after the functional correlation had broken down owing to a change of function of individual components.

**10.4.3   Characters with Low Weight.** Every specialist knows characters which in his group are "unreliable," that is, poor indicators of relationship. They have low weight in the grouping of taxa. The characters here enumerated are usually considered as belonging to this category.

All conditions that represent the converse of characters with high weight (see 10.4.2) have low weight, other things being equal.

*High Variability.* All highly or erratically variable characters are in this category. The pattern of branching in arteries of vertebrates may be different not only in different individuals of the same populations but even in the left and the right side of the body. Differences in arterial pattern are not nearly as helpful for classification as some authors thought they were. Wing venation provides important characters for the classification of insects, but Sotavalta (1964) showed that there was far more variation in this character in the tiger moths (Arctiinae) than known and that the traditional generic arrangement of the family, based on wing venation, is thoroughly in need of revision.

Characters known to vary greatly within groups of clearly related species and to show similarly great variation in other only distantly related groups. The presence or absence of bands in snails or large body size in certain groups of mammals are such characters.

The term *variable character* may mean several things. It may mean that it is either present or absent in members of the same population or species. It may mean that it shows many different degrees of expression in the same species or population. But it also may mean that, although constant within a given species, it is either present or absent in a natural group forming a higher taxon. Variability in the characters of higher taxa may mean something very different from variability in populations.

Totally invariable characters have low weight; for instance, "possession of two eyes" in taxa belonging to the vertebrates.

Characters that are too difficult (or too time-consuming) to be determined, such as number of hairs in mammals or certain physiological constants, have low weight.

*Monogenic or Oligogenic Characters.* A simple monogenic character, like those involved in balanced polymorphism, usually varies independently of other characters. Such a character has low phyletic information content. The distribution of monogenic characters (like albinism) in the zoological system is often rather haphazard.

*Regressive (Loss) Characters.* Any kind of regressive character is usually of low taxonomic weight, because such losses may happen independently in more or less distantly related phyletic lines. Taxa based on the loss of eyes, loss of wings or wing veins in insects, loss of toes or other appendages in mammals and birds, loss of teeth in mammals, or of segments in segmented animals, are often unnatural. Trends toward simplification are often erratically realized in different phyletic lines. Losses may be the result of special environmental conditions (eyes in caves, wings on islands in stormy oceans), or they may represent tendencies in a higher taxon that are realized many times independently in subordinate lower taxa. The Sanderling (*Calidris alba*), for instance, has lost the rudimentary hind toe which is still present in related sandpipers and was for this reason placed in a special genus (*Crocethia*). Yet this bird is more closely related to certain species of the genus *Calidris* (*minutilla* and *pusillus*) than these species are to other species of *Calidris* like *C. acuminata* or *melanotos*. Michener (1949) has shown how the independent loss of characters leads to spurious similarities among saturnid moths. Neoteny (sexual reproduction in a preadult state) has also sometimes led to spurious similarities and proposals for unnatural taxa. Since the loss of a structure or feature may occur repeatedly in independent lines, taxa based on the absence of characters are often polyphyletic and unnatural.

*Narrow Specializations.* Characters reflecting a single selection pressure, such as loss of an organ, or desert coloration in desert birds and mammals, have lower taxonomic weight than their conspicuousness would suggest. As Darwin expressed it (1859, p. 414): "Nothing can be more false" than to assume "that those parts of the structure which determined the habits of life, and the general place of each being in the economy of nature, would be of very high importance in classification." (See also 10.4.2 under Darwin Principle.)

An example may illustrate this. A group of fish-eating ducks, the mergansers (*Mergus*, etc.), acquired—in connection with their fish-eating habits—a number of ad hoc adaptations, such as a streamlined body and head and a long thin bill with numerous horny teeth. Some ornithologists placed them in a separate family. Later on it was shown that the mergansers agree with the Golden-eye group (*Bucephala*) in the essentials of courtship display, color of downy young, internal anatomy, and in protein characters. The two groups also frequently hybridize with each other. The taxonomic value of the fish-eating adaptations was consequently drastically downgraded and mergansers and golden-eyes combined in a single tribe. Characters associated with shifts in the food niche are particularly susceptible to a rapid attainment of conspicuous differences or, conversely, of convergent similarities.

Structural adaptations subsequent to the invasion of a new food niche may happen very rapidly and generally have low taxonomic weight. In birds, for instance, particularly in the songbirds, the bill is highly plastic, and flycatcher bills, finch bills, warbler bills, thrush bills, and shrike bills have developed many times independently. Different subspecies of the same species may have drastically different bill types, indicating how quickly the shift from one type to the other may occur. Teeth and jaw structures

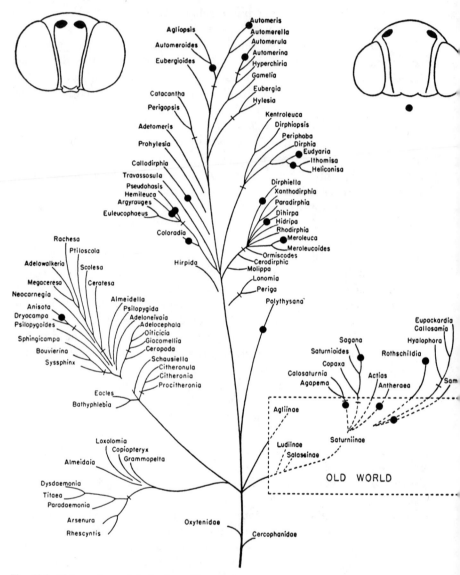

**Fig. 10-6.** Phylogenetic tree of the Saturniidae of the New World, in which all lines where eye size is greatly reduced are marked with a black spot. Normal eye upper left (*from Michener*, 1949).

of cichlid fishes in African lakes likewise illustrate the principle of the rapidity of nutritional adaptations. This is equally true for any other character difference that is the result of a single selection pressure, such as locomotory specialization or secondary sexual characters resulting from sexual selection (e.g. birds of paradise).

*Redundant Characters.* As pointed out in 10.2.3, it is not necessarily true that the more characters a classification is based on, the better it is. There is a law of diminishing returns, because different components of the phenotype ( = different characters) may be pleiotropic or functionally correlated aspects of the same information in the genotype. *A character is redundant if it is a necessary correlate of other characters.* Large size alone may make species *A* appear more similar to species *B* than to a third small species *C* even though elimination of the size factor and its universal correlated effects may reveal that *A* is actually far more closely related to *C* than to *B*. Proportions are often far more sensitive indicators of relationship, although the effects of allometry must be evaluated carefully (Gould, 1966). A tendency toward an increase in meristic elements may actually be a single factor, even though it may affect all parts of the body and all appendages that show meristic factors. To score each of them separately would grossly distort actual similarity.

The same peril of redundancy is ever-present with *functional character complexes.* Verheyen (1956), on the basis of an analysis of over 50 characters, placed the diving petrels (a family of Tubinares) in the same order as the auks (actually related to the gulls) because nearly all the characters used by him were taken from locomotory adaptations for wing-diving. The degree of inferred relationship among the anthropoid apes was distorted until quite recently because the classification was based almost entirely on arboreal locomotory and vegetarian masticatory adaptations. Bill, tongue, soft palate, bony palate, jaw muscles, and other structures in the skull of birds are all adapted to the particular food niche which a given species utilizes. Individual aspects of these structures cannot necessarily be scored as so many independent characters.

One can count and evaluate in classification only those characters that are reasonably "independent" of each other. Just exactly what "independent" is, and how this can be determined, is still controversial. To reduce the weight of characters according to their degree of correlation is part of the answer, but not the entire answer, because characters may be correlated for two entirely different reasons, function or phylogeny.

It is not yet quite clear when to consider phyletically correlated characters as redundant. A mild amount of redundancy may be useful in confirming a proposed classification. Beyond this point it will be misleading or, at best, will cause a waste of time. Various mathematical tools, such as correlation analysis and principal-component analysis, are sometimes very helpful. In his use of taxonomic characters the taxonomist must always watch for redundancy and try, instead, to sample as many independent manifestations of the genotype as possible.

**10.4.4  Summary.** The experience which taxonomists have gained from weighting can be summarized in the statement that a classification based on phyletic weighting has numerous advantages. It is the only known system that has a sound theoretical basis; it has greater predictive value than other kinds of classification; it stimulates a character-by-character comparison of organisms believed to be phylogenetically related; and it encourages the study of additional characters and character systems in order to improve the soundness of the classification, hence its information content and predictive value. Finally, it leads to the discovery of interesting evolutionary problems. Thus classifications based on phyletic weighting not only have scientific advantages but are actually best able to answer the demands of the practice by having a greater total information content than artificial systems.

**10.4.5  Difficulties in Weighting and Phyletic Grouping.** A number of difficulties in arriving at an appropriate weight for a given character were mentioned in the two preceding sections. Some additional ones will now be discussed.

*Reversal.* It is not known to what extent reversibility may cause confusion. Only few exceptions are known to the "irreversibility" rule that a structure or organ once lost is not reacquired in the same way. The few exceptions do not seriously weaken the validity of this rule. Far more frequent however are cases where, owing to a loss of specialization, an advanced form acquires secondarily a spurious similarity to a primitive ancestor. Generally such despecialization affects only a single character or character complex and does not lead to confusion with primary primitiveness.

*Convergence.* The secondary acquisition of a character by two taxa, not derived from a common ancestor showing that character, is called convergence. Does it ever happen that different taxa, through adaptation to the same mode of life, become so similar that their different origin is not recognizable? Convergence is believed by many taxonomists to be the greatest obstacle against achieving a truly phyletic classification. Actually the difficulties caused by convergence have been exaggerated. Even though convergence involving particular characters, entire organs, or general shape and proportions occurs quite frequently, convergence between the entire phenotypes of unrelated organisms is rare, if not absent. I am not aware of a single case among higher organisms. No flying mammal, reptile, or arthropod could ever be mistaken for a bird. Nor can a whale, a seal, or a manatee be mistaken for a fish. Convergence is a problem only among fairly close relatives in rather uniform groups, particularly where classificatory schemes are based on few characters.

Convergence involves almost invariably an adaptation for a similar niche utilization. That the lagomorphs (hares and relatives) were classified with the rodents was because of their similar gnawing incisors. When several kinds of apparently unrelated seed-eating songbirds were combined in the

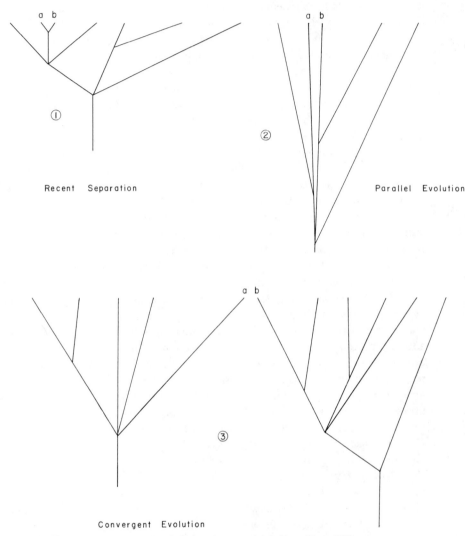

Fig. 10-7. Three alternate reasons for similarity of taxa *a* and *b* (*from Mayr, 1964*).

family Fringillidae, it was because of their cone-shaped bill. Other familiar examples are the distantly related but similar-appearing families of water beetles which share a streamlined form; the strikingly similar structure of the forelegs in mantids (Mantodea) and mantispids (Neuroptera); and the superficially similar ectoparasites of vertebrates, which belong to at least six different orders of insects. Extreme convergence in response to adaptive needs occurs also in many physiological properties, as, for instance, those of the blood pigments of marine invertebrates and of neurotoxins in certain fishes and salamanders. Most cases of convergence are restricted to a single character or functional character complex. The utilization of new characters invariably leads to the unmasking of unnatural assemblages.

*Mosaic Evolution.* The unequal rate of evolution of different features among descendants from a common ancestor, often combined with parallel evolution, is perhaps the most frequent cause of difficulties in classification. Let us assume that there are seven evolutionary trends in a major phyletic assemblage. In each subline, however, the rate at which these trends are realized will be different (Table 4-1). When this happens, one may fail to find correlated groups of characters—an almost indispensable basis for the clustering of taxa. Also, each taxon will show a different combination of primitive and advanced characters. This raises the question of which special-izing trend should be emphasized in the grouping of taxa and also makes it quite impossible to arrive at a noncontroversial sequence of taxa. Among specialists there will always be disagreement as to which of these trends of specialization has led farthest away from the ancestral condition. There is no simple solution to this dilemma.

*Incongruence among the Information Contents of Different Phena.* Separate classifications based on different phena in a species, let us say one classification based on males and another on females, or one on larvae and another on imagoes, often lead to different results. Since the different phena are either genetically identical (juveniles and adults) or differ only by sex or morph genes, such a discrepancy indicates shortcomings in the analysis of similarity. A conflict between larval and adult characters usually arises from the fact that either stage has acquired certain secondary special-izations or regressions and only a deeper analysis will give the right answer (7.4.5). If it cannot be found, new characters (biochemical, behavioral, chromosomal) will have to be employed to break the deadlock.

**10.4.6 Grouping Species by Weighted Similarity.** The long preced-ing discussion on how to determine the kind of similarity which indicates common descent places us now in the position to go back to our original question: What species shall we unite into a given group or cluster? The performance of this operation is a consequence of the fact that each higher taxon (from the genus up) consists of a cluster of species separated by minor or major gaps from other clusters. (A monotypic taxon is a "cluster" of a single species.) A species belongs to a cluster because it has had largely the same evolutionary history as the other species of that cluster and is therefore—in an evolutionary sense—more similar to them than to other species. The formal taxonomic recognition of such clusters raises a number of practical and operational problems that will be discussed in the next section.

## 10.5    CLASSIFICATION AND INFORMATION RETRIEVAL

Determination of similarity between species is the first step in the classifying procedure. It would also be the next-to-last step if this would invariably produce groups of species that are well defined and as isolated as hummingbirds, penguins, or bats. The question whether a given mammal is a bat or not never arises. Even flying squirrels or flying lemurs could not possibly be mistaken for bats. Unfortunately evolution is not so orderly as to produce only well-defined taxa, neatly arranged like the elements in the periodical table of chemistry. In almost any assemblage of species one can distinguish subgroups and supergroups of species, dense and loose clusters, and a liberal proportion of intermediate or very isolated species—in other words, one finds gaps of every width. To translate this variation optimally into a hierarchy of higher taxa is part of the art of taxonomy.

It is during this part of the classifying procedure that the taxonomist must be most keenly aware of the fact that a classification is an information storage and retrieval system. The complex relationship of species with each other and the varying rates of branching and of evolutionary divergence must be translated into a system of taxa, conveniently ranked in the appropriate categories, in such a way that it forms as efficient as possible a system of information storage. One must remember that it is possible to translate the same information about relationship into very different classifications. It is the purpose of the ensuing discussions to make the taxonomist aware of various practical and theoretical considerations that will permit him to produce a classification meeting two objectives: grouping nearest relatives near each other and facilitating information retrieval. It must be the principal concern of the classifier to delimit and rank the taxa in such a way as to meet both objectives.

**10.5.1    The Formal Recognition of Taxa.** A taxon is defined (1.2) as a group of species receiving a definite rank in the hierarchy of categories. The inevitable consequence of this definition is that the two operations, the formal recognition of taxa and their categorical ranking, are inseparable in the classifying procedure and must be treated together.

This particular step in classification, formal recognition and ranking of taxa, is the most controversial one in taxonomy and still subject to widely diverging opinions. Even though there may be a fair amount of agreement on similarity among species, the translation of this evidence into formal classifications shows wide variation. This is apparent from three lines of evidence:

1. *Historical Changes.* The genus, the family, and the other higher categories have changed their value during the history of taxonomy. The various species

recognized by Linnaeus are in most cases still listed as species today, while the status of his genera has completely changed in most cases. Most Linnaean genera of animals have since been raised to the rank of families or even higher categories.

2. *Splitters and Lumpers.* There is usually a wide difference of opinion among contemporary authors on the average size they favor for taxa (see 10.5.3) and the rank they assign to them. Contemporary treatments of the same higher taxon may show drastically different categorical assignments. For instance, in Parker and Haswell's textbook of zoology (1940) the insects are classified as a class of the phylum Arthropoda, and the Orthoptera as an order with four suborders. In Handlirsch's treatment of the insects in Kükenthal's *Handbuch der Zoologie* (1926–1936), the insects are listed as a subphylum, and the Orthoptera are arranged in two superorders and four orders.

3. *Different Standards in Different Groups.* Extremely different standards of categorical assignment are traditionally adopted in different branches of the animal kingdom. The recent birds, for instance, are classified by various authors into from 20 to 50 orders. There is perhaps less difference between some of these orders than there is between some of the families of insects or of mollusks. Likewise the families in the order Passeres (songbirds) are much less distinct than are the families in most other groups of animals. We find similar inequalities throughout the animal system. Obviously the higher categories, particularly order and family, do not have the same meaning for specialists in different areas of zoology.

The reasons for the complete absence of a uniform standard of ranking criteria is that the higher categories, in contrast to the species category, lack a nonarbitrary definition. They have only a formal definition which states that "a higher category designates the rank of a taxon which consists of a group of taxa of the next lower categorical rank and which is separated by a discontinuity from other taxa of the same categorical rank." There is no reliable criterion to tell us whether a cluster of species should be ranked in the hierarchy of categories as a genus, a tribe, a subfamily, a family, etc., and as a result hardly any two authors agree in the delimitation and the ranking of taxa.

**10.5.2  Ranking Procedures.** To get away from the traditional subjectivity and arbitrariness, two methods have been proposed by which to establish categorical rank automatically: cladistic and phenetic ranking.

*Cladistic Ranking.* The cladists (4.3.5) base categorical rank on the geological time at which the ancestor of a given taxon branched off from the parental stem (or from a sister branch). This method fails because it ignores differences in the rates of evolution and confuses genealogical with genetic relationship. As Darwin has rightly said, the degrees of modification which the different groups have undergone are very different. Even though the family Limulidae (horse-shoe crabs) is older than the mammalian order Carnivora, it does not mean that the categorical rank of these two taxa has to be reversed. Rapidly evolving taxa often branch off from slowly evolving taxa, but it would be absurd to demand that the subsequent fate of the daughter taxon should determine the categorical rank of the parent taxon.

The bird-crocodilian-reptile example (Fig. 4-4) best illustrates this problem. In almost any assemblage of related taxa, there is one or another that has started off on a new evolutionary path. Although the branching point is no more remote in time than that of related taxa, the taxon has diverged more strongly. This is the "ex-group problem" of Michener (1957). Whenever the divergent taxon has truly entered a new major adaptive zone, as birds compared to reptiles, or the hominids compared to the pongids, placing the divergent taxon in a higher category is justified. Such a separation of a single deviant evolutionary pioneer from another taxon with which it shares the nearest branching point is not in conflict with the principle of monophyly (see 4.3.5; also Mayr, 1965a). In most cases, however, the categorical separation of an aberrant species (as a separate genus) or an aberrant genus (as a separate family) leads to a fractioning of the system and lowers its information-retrieval capacity.

Cogent arguments against cladistic ranking are also provided by the chronology of parasites and their hosts (Osche, 1960). For many taxa of vertebrate parasites, for instance, it may be very probable that they originated together with their hosts, that is, at the same geological time. Should a genus of cestodes be raised to the rank of a family or order because it parasitizes a family or order of vertebrates? Should the superfamily Ascaroidea of the nematodes be given the same categorical rank as the class Cestoda because both invaded the vertebrates at the same time? For other arguments against the cladistic approach see 4.3.5 and 10.3.

*Phenetic Ranking.* The phenetic approach is to calculate a measure of overall similarity between taxa, to set arbitrary numerical levels of difference to designate categorical levels, and thus to obtain "automatically" the correct rank of each taxon. Unfortunately, this unbiological approach has numerous weaknesses.

Firstly, it is fundamentally unsound to quantify similarity in a comparison of entities as highly heterogeneous as the character complexes of different taxa (Ghiselin, 1966b). The conspicuous phenetic differences caused by some taxonomically unimportant ad hoc specializations prove this point convincingly. Striking differences in food-getting specializations of closely related species and conspicuous secondary sexual characters, as in male birds of paradise, lead to impressive phenetic differences that have very little taxonomic weight (10.4).

The second reason, in part a consequence of the first, is that genetic differences between species and higher taxa are very unevenly reflected in the phenotype (sibling species!). Also the amount of phenotypic difference between higher taxa of presumably similar genetic difference varies from order to order and from class to class. Osche (1960) points out, for instance, that the subdivisions of the tapeworms (cestodes) differ far more from each other morphologically than do the taxonomic subdivisions of the nematodes. Adoption of a uniform phenetic yardstick for the recognition of categories would lead to a badly distorted and unbalanced system.

Finally, the particular numerical level of similarity ("phenon level" of the pheneticists) depends strongly on the particular numerical method applied in a given case and is often not repeatable when a different computer program is used. The reasons why unweighted phenetic similarity is unable to reflect the underlying genetic program correctly were discussed earlier (10.2).

Nevertheless there are situations where the phenetic approach is helpful, if not indispensable. The successful application of the weighted approach (see below), in which clustering and ranking are carried out by inspection and overall evaluation, requires experience and a deep understanding of the taxon. The beginner, or someone dealing with a little-studied group, will have considerable difficulties in determining the proper weighting criteria. He should by all means employ, as a first step, one of the computer clustering methods. However, he must remember that an automated method is a short cut and that the subsequent decisions (size and ranking of taxa to be formally recognized) must be based on a careful weighting of the manifold considerations discussed below (10.5.3).

It has not yet been determined which computer clustering method gives the best results. Sokal and Sneath (1963) favor the nearest-neighbor method. Wirth et al. (1966) are experimenting with "graph clustering" (also based on the nearest-neighbor approach), while Watson, Williams, and Lance (1967) prefer "centroid sorting." Further testing will soon reveal which method is most advantageous under what circumstances. The relation between grouping, in the classical sense, and clustering, in the sense of numerical taxonomy, is not yet clear.

*Weighted Ranking.* Owing to the deficiencies of the automated approaches, most taxonomists continue to favor the classical approach of a careful weighting of numerous separate factors and considerations. The experienced taxonomist knows that every classification is a compromise between different requirements, some of which are occasionally in conflict with each other.

The most frequent conflict is between the practical objectives of any information retrieval system and our scientific understanding of relationships. For instance, specialists are often able to determine the probable relationship of species down to the level of very small groupings. Yet to establish a separate genus for each of these groups, in extreme cases for each well-defined species, would completely destroy the value of the classification as an information retrieval system.

**10.5.3   Criteria for Delimitation and Ranking.** Scrutinizing the classifications of successful taxonomists, it seems that they base the recognition of taxa on the relative merits of five considerations:

1. Distinctness (size of gap)
2. Evolutionary role (uniqueness of adaptive zone)
3. Degree of difference
4. Size of taxon
5. Equivalence of ranking in related taxa

All five factors must be weighted before each decision (see also Michener, 1957). When a taxonomist, during the classifying procedure, encounters a group of species that would seem to merit recognition as a new higher taxon, he will ask questions concerning the stated five points, such as: Is the new taxon sufficiently different from the taxon with which it is most nearly related? Is the new taxon of a size that is convenient for information retrieval? Only after all these questions have been answered in the affirmative should the taxon be formally recognized.

*Distinctness (Size of Gap).* The greater the gap between two clusters of species, the greater the justification for recognizing both clusters as separate taxa. Size of the gap is measured not merely in terms of phenetic distance but, more important, in terms of the biological significance of the difference (see following section on the evolutionary role).

Concerning the evaluation of the gap, there is a profound difference between the recognition of taxa on the level of the species and on that of the higher taxa. The mere existence of a very special kind of gap ("reproductive isolation") is the necessary and sufficient criterion of taxonomic recognition on the species level. No other condition has to be met, as proved by the recognition of sibling species. Above the species level the presence of a gap is only one of several requirements, since not every species is placed in a higher taxon of its own.

The gaps between taxa are the result of evolution. There is no factual or conceptual conflict between evolutionary continuity and the existence of these gaps. Numerous evolutionary processes generate discontinuities in spite of the complete continuity of populations in time (as far as they are in an ancestor-descendant relationship). Speciation, extinction, adaptive radiation, unequal rates of evolution, and other evolutionary phenomena are responsible for the existence as well as for the unequal size of the gaps separating higher taxa. It must be emphasized that not only the gaps but also the clusters of species separated by the gaps are realities of nature.

When there is an appreciable gap between two taxa, the terms "degree of difference" and "size of gap" mean essentially the same thing. However, when clusters of species are large and heterogeneous and not separated by a clear-cut gap, their mean values may be quite different and yet they may not be very distinct. It is not justifiable to select two extreme species in a widely scattered array of species and make them the types of genera, if there is not some sort of discontinuity between them. Indeed, the discontinuity should be reasonably wide and well-defined, in order to preclude excessive splitting. Michener (1957; 1963, p. 153) discusses the problem of gap evaluation.

*Evolutionary Role (Nature of Adaptive Zone).* There has been an increasing tendency in systematic zoology to give weight in ranking to the evolutionary potential of a taxon. Almost any prosperous higher taxon is descended from a founder species (or species group) that succeeded in shifting to a promising new adaptive zone (birds, beetles, whales, etc.). Consequently any taxon that has succeeded in entering a new adaptive niche or zone usually receives higher rank than a taxon that lacks such ecological significance. It is one of the aims of the new systematics to investigate the ecological significance of taxa and to use this information in the weighting of the evidence (5.3.2).

Well-delimited higher taxa almost invariably have a definite ecological meaning. Cats, dogs, horses, woodpeckers, etc., fill well-defined ecological niches or adaptive zones in nature. A consideration of its adaptive and evolutionary role is thus an important element in the categorical ranking of a higher taxon (7.4.11; see also Gisin, 1964).

The relationship between the hominids and the anthropoids, and its translation into ranked taxa, is a further illustration of the importance of evolutionary considerations. The study of chromosomes, of numerous biochemical characteristics, and of their parasites has revealed so much similarity between Man and the African apes *(Pan)* that some authors have suggested placing them in a single family. Yet Man has entered so unique and strikingly distinct an adaptive zone that Huxley even suggested recognizing for him a separate kingdom (Psychozoa). This would seem to go too far, but the evolutionary distinctness of Man surely justifies recognition of a separate family.

Evolutionary considerations are also very important for the ranking of taxa along a single phyletic line (without intervening branching). Evidence is mounting that the polytypic species *Australopithecus africanus* *(sensu lato)* is the immediate ancestor of *Homo*. Yet, the man ape *africanus* with a mean brain size of 450 cc filled such an entirely different niche from *sapiens* with a brain of $\pm1,500$ cc that generic separation is abundantly justified.

It is the difference in the utilization of the environment, the difference in the occupation of an adaptive zone, that is responsible for the width and the sharpness of the gap between taxa.

*Degree of Difference.* When speaking of the degree of difference between two clusters of species, the taxonomist usually thinks of the "distance" between the means of the two groups of species. This value can be employed only in conjunction with two other data, the scatter of the cluster and the amount of discontinuity (gap) between the cluster and others. The denser and more uniform a cluster of species, the more justified one usually feels in recognizing it formally. The greater the difference between the most distant species of a cluster, the greater is the justification for a splitting of the taxon, other things being equal. This is an important consideration in the evaluation of the relative weight of cluster and of gap characteristics (Fig. 10-8). In the upper two clusters, the mean difference $(A–B)$ between the clusters is greater than in the lower clusters $(A'–B')$, but the size of the gap ($a–b$ vs. $a'–b'$) is smaller.

*Optimal Size of Taxon.* The number of species included in a taxon constitutes its size. A genus with many species is called large, one with few species small. Since a classification functions like a filing system, its subdivisions (the taxa) should fulfill the demands of greatest possible efficiency, that is, they should ideally be of approximately equal size in order to facilitate information retrieval. The amount of material included under one heading should be neither too voluminous nor too scanty. As a taxonomist once remarked, all one can say about the genus is that it should be neither too large nor too small. There are a number of biological reasons that make it impossible to achieve this ideal.

The fact that the two processes of phylogeny, branching and divergence, are independent of each other is responsible for the observed inequality

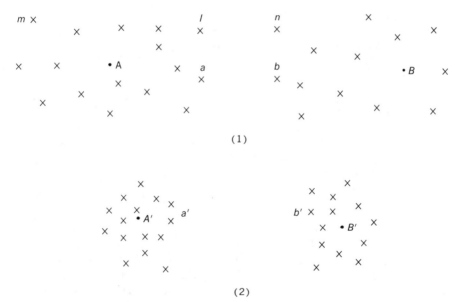

Fig. 10-8. Mean distance of clusters and size of gap are independent. Above the mean distance between the clusters $A$ and $B$ and their diameters is great, but the gap is narrow. Below $(A',B')$ the clusters are nearer and more compact, but the gap ($a'–b'$) is greater.

in the size of taxa. Continuous speciation (branching) without visible divergence results in the production of large taxa. Divergence without branching leads to the occurrence of monotypic taxa, in an extreme case to a series of superimposed monotypic higher taxa. When the line which includes the ancestor of the Aardvark (*Orycteropus afer*) split off from its parental line (Condylarthra?), it diverged steadily. Whatever branching (speciation) occurred in former times has since been obliterated by extinction. This genus became sufficiently different to be ranked as a separate family (Orycteropodidae) and eventually deserved to be ranked as a separate order (Tubulidentata). The higher taxa in this line (genus, family, and order) satisfy the definition of "classes" in the language of logic, each with a single member, and in spite of their different rank, each with the same member, *Orycteropus afer*.

The taxonomist would say that the distinctness of the species Aardvark is of such an order of magnitude that recognition of the separate order Tubulidentata is justified (and with it the subordinate categories family and genus), while the logician would say that the Aardvark has a series of properties which would qualify successively as defining the names *Orycteropus,* Orycteropodidae, and Tubulidentata (Buck and Hull, 1966).

The independence of branching and divergence results in a frequency distribution of genera and species usually referred to as "hollow curve" (Fig. 10-9; see Mayr, 1942, p. 288). When the number of species per genus is plotted for a family or an order, it is found that many genera have only one or two species, while a few genera contain a large or very large number of species. Excessively large genera as well as an excessive number of monotypic genera reduce the usefulness of a classification for information retrieval. This is the reason for the recommendation that the size of the gap justifying the separation of a higher taxon should be inversely correlated to the size of the taxa (5.4, 5.5). The recommendation to strive for taxa of equal size unquestionably leads to value judgments, therefore it is opposed by those who want to automate classifying.

Most higher taxa of organisms contain some 20 to 30 percent monotypic genera. The Okapi, the Ostrich, and many other well-known animals are monotypic and in many cases the only representatives of an entire family or order. Yet the taxonomist must make every effort to keep the number of monotypic taxa as low as possible. Slightly aberrant species should whenever possible be included in the same genus as the most nearly related species. The same advice applies to higher taxa. General similarity should be given far greater weight than conspicuous difference in a single character. A highly unbalanced classification of the African plantain eaters (Musophagidae) resulted from the recognition of three monotypic genera based almost entirely on slight differences in the shape of the nostril (Moreau, 1959; Fig. 10-10). Mayr (1942, pp. 280–289) has pointed out

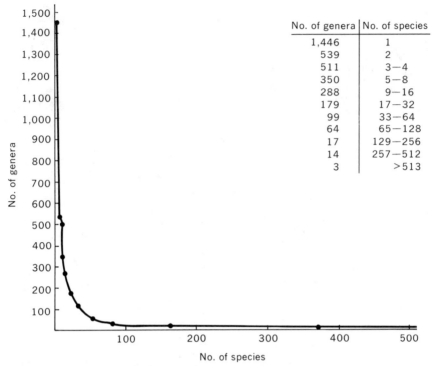

| No. of genera | No. of species |
|---|---|
| 1,446 | 1 |
| 539 | 2 |
| 511 | 3—4 |
| 350 | 5—8 |
| 288 | 9—16 |
| 179 | 17—32 |
| 99 | 33—64 |
| 64 | 65—128 |
| 17 | 129—256 |
| 14 | 257—512 |
| 3 | >513 |

Fig. 10-9. Example of a hollow curve: The number of species in 3,510 genera of weevils (Curculionidae) (*after Kissinger*, 1963).

how unhelpful for information retrieval are classifications that contain too many monotypic taxa.

The converse is equally true. There are some well-defined natural genera with over 1,000 or even over 2,000 species. Among well-known large taxa are the North American freshwater fish genus *Notropis* with about 120 good species, certain genera of weevils (Curculionidae) with more than 1,000 species each, in addition to the entire family of weevils with perhaps more than 100,000 species, and the genus *Drosophila* with more than 1,000 species (Fig. 1-1).

A consideration of size becomes important whenever there is a question whether or not to split a taxon. Splitting a genus of more than 1,000 species requires much less of an excuse than splitting a small genus. The larger a cluster of species, the smaller the permissible gap between it and other clusters. For instance, in a taxon with 1,000 species a narrow gap separating 400 species from 600 other species may be a better justification for a split than a much larger gap in a species cluster separating a single species from three others.

Instead of breaking up such a well-known entity as the genus *Droso-*

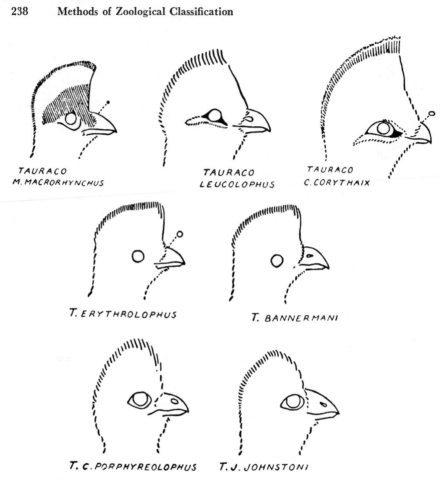

TAURACO
M. MACRORHYNCHUS

TAURACO
LEUCOLOPHUS

TAURACO
C. CORYTHAIX

T. ERYTHROLOPHUS

T. BANNERMANI

T. C. PORPHYREOLOPHUS     T. J. JOHNSTONI

Fig. 10-10. Types of beak and nostril in tauracos (Musophagidae). Nostrils may be concealed or open, round, oval, or slitlike (*from Moreau, 1959*).

*phila*, it is more useful to introduce subdivisions. The subgenus is available under the genus, but the liberal use of the informal category *species group* is an even more useful device, as a relief for one's memory. *Drosophila* specialists have adopted a few subgenera, but on the whole they get along extremely well by speaking of the *repleta* group, the *willistoni* group, the *melanogaster* group, etc. Nothing would be gained by chipping off a few aberrant species from such a large group and making them the types of monotypic taxa. The species group is the most neglected category in many areas of systematic zoology.

*Splitters and Lumpers.* No nonarbitrary method is known by which the proper size of taxa can be determined unequivocally, except in the case of very isolated species clusters. This is the reason for the eternal argument between *splitters* and *lumpers,* two types of taxonomists found among the specialists of almost any group of organisms. It is the ideal of the splitter to express every shade of difference and degree of relationship through the formal recognition of separate taxa and their elaborate categorical ranking. It is the ideal of the lumper to express in the classification the fact that the higher taxa and categories express relationship and that too fine a division of taxa leads to an intolerable burden on the memory. The author of this volume believes in broad taxa and sides, therefore, with the lumpers.

Among the reasons for splitting are the following. The splitter is aiming at the ideal of recognizing only homogeneous groups. However, evolution being disorderly, it is virtually impossible to find a polytypic taxon in which some species do not differ more from each other than do others. Removal of these species into separate taxa leads to an ever finer pulverization of the system until finally most taxa become monotypic. Splitters almost without exception classify characters rather than groups of organisms. Whenever a splitter finds a species with an aberrant character, he establishes a separate genus for such a species. He fails to remember that since the time of Linnaeus it has been realized that it is the taxon that gives the character and not the reverse (5.4.2). Roewer's splitting of the Opiliones is a notorious case. He placed 1,700 species in 500 genera, of which more than half (300) contain only one or two species. Subsequent work has shown that most of these "genera" are meaningless biologically, being sometimes based even on different individuals of the same population. In the Hominidae no less than 30 generic names have been proposed for what at most are three taxa: *Homo, Australopithecus,* and *Paranthropus.*

Splitting is particularly deleterious on the generic level. The generic name is part of the scientific name of an organism and can therefore be employed more advantageously to indicate affinity than can the name of any of the other higher categories. McKevan (1961) and Sailer (1961) have described particularly well the disadvantages of excessive splitting. The specialist must have a regard for the general zoologist; he must always remember that he is not writing merely for three or four cospecialists.

The difference in the attitude of lumpers and splitters may be illustrated with a concrete example. Figure 10-11 represents a dendrogram (10.7) of the 36 living species of river ducks of the genus *Anas (sensu lato)*. There is comparatively little difference of opinion among ornithologists as to the grouping of these species. It is agreed that there are several more closely bunched clusters of species, each of which forms a natural group, such as species 16–21 (mallard group) and species 30–36 (blue-winged teal–shoveler group). It is also agreed that certain species, such as 1, 2, 7, 22, and 23, are rather isolated, and that some of the species and species groups are somewhat intermediate between others. Furthermore, ornithologists agree that the whole group is fairly well isolated from the

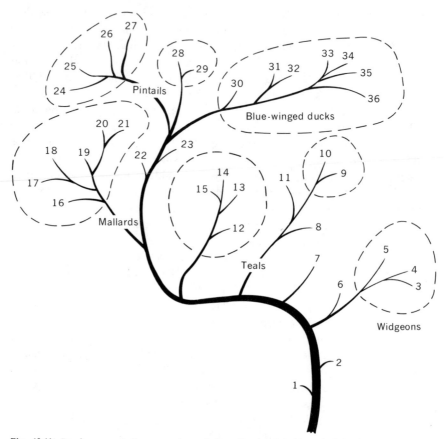

Fig. 10-11. Dendrogram of the genus *Anas* of the tribe Anatini (river ducks) (*in part based on Johnsgard, 1965*). 1 = *waigiuensis*, 2 = *sparsa*, 3 = *penelope*, 4 = *americana*, 5 = *sibilatrix*, 6 = *falcata*, 7 = *strepera*, 8 = *formosa*, 9 = *crecca*, 10 = *flavirostris*, 11 = *capensis*, 12 = *aucklandica*, 13 = *gibberifrons*, 14 = *castanea*, 15 = *bernieri*, 16 = *melleri*, 17 = *rubripes*, 18 = *platyrhynchos*, 19 = *undulata*, 20 = *poecilorhyncha*, 21 = *luzonica*, 22 = *specularis*, 23 = *specularoides*, 24 = *acuta*, 25 = *georgica*, 26 = *bahamensis*, 27 = *erythrorhyncha*, 28 = *versicolor*, 29 = *punctata*, 30 = *querquedula*, 31 = *discors*, 32 = *cyanoptera*, 33 = *platalea*, 34 = *smithi*, 35 = *rhynchotis*, 36 = *clypeata*.

other genera of the tribe Anatini. The congeneric status of this group of 36 species is supported by their biochemical and anatomical similarity, by a remarkable interspecific hybrid fertility and by the joint possession of many behavioral characteristics (Johnsgard, 1965, p. 131).

The "splitters" interpret this evidence as indicating that this group of species should be divided into at least 12 genera, many of them monotypic. They insist that this is the only way of indicating the existence of the various groups of species. The lumpers, on the contrary, feel that relationship within the duck family, and a balanced presentation of the relative distinctness of other genera of ducks, is best accomplished by including all 36 species in a single genus.

Lumping, however, can also be carried too far. When lumping becomes fashionable in a previously oversplit group, a few authors invariably go too far in their zeal and combine elements of such heterogeneity that their genera become almost as useless as the virtually monotypic genera of the splitters. Excessive lumping tends to produce a good deal of subjective (secondary) homonymy which is difficult to handle nomenclatorially. Finally, it results in very long and rather heterogeneous lists of synonyms. These drawbacks of lumping would be very minor if it were not for the preceding creation of useless names by the splitters.

Every branch of systematic zoology seems to go through a cycle. As more and more species are discovered, the taxa are split more and more finely, and a steady process of escalation of the categories to higher rank seems to occur. Finally, a saturation point is reached in the splitting, and a vigorous period of lumping follows. Let us take the case of birds. Linnaeus arranged the 554 species of birds known to him in 63 genera (8.8 species per genus). Some ornithologists in the 1920s allowed no less than 7,000 to 8,000 genera for the 8,600 species of birds now known. In the 1930s these were reduced to about 2,300 genera, and now perhaps 1,700 genera are recognized (5 species per genus).

*Equivalence of Ranking in Related Taxa.* A very important consideration in ranking is *balance.* Categorical rank in related taxa should be equivalent. For instance, the genus should designate a similar category in the Carnivora and Rodentia—in fact, as far as this is possible, even in birds and reptiles. The family (if possible) should have a similar significance in flies and beetles, and so forth. To achieve such equivalence Crowson (1958) has proposed that norms be adopted for each taxonomic group. One might, for instance, adopt the Rodentia as yardstick for a mammalian order. No other group of species of mammals should be recognized as an order unless it has a combination of qualities which make it equivalent to the rodents. The time has not yet come to implement this suggestion, but it seems to come far closer to a realistic solution of the problem of ranking than the adoption of an arbitrary numerical value of phenetic similarity.

In the past, splitters have too often concentrated their attention on a given taxon, raised it in rank, and then subdivided it finely. As a result such a taxon is completely out of step with all equivalent related taxa (not similarly treated by the splitter). Arkell and Moy-Thomas (1940) cite some typical examples of this ill-advised procedure:

> Examples of flagrant disregard of this rule [a uniform scale of values] are Buckman's innumerable genera made by splitting up contemporary species of the single good Liassic ammonite genus *Dactylioceras,* and by Heinz's pulverization of the Cretaceous lamellibranch genus *Inoceramus.* Out of what was originally a single genus *Inoceramus* Heinz created a whole systematic hierarchy, comprising 2 families, 24 subfamilies, 63 genera, and

27 subgenera; and even so he did not take the Jurassic forms into account. [Even if this minute subdivision were taxonomically justified] it should be carried out downwards in the scale, starting with the genus *Inoceramus* and proceeding through subgenera to groups and sections. All the advantages of minute subdivisions can thus be achieved without disturbance to the whole classification of Mollusca.

In order to determine how many categorical levels are needed to express degrees of relationship, a careful analysis of morphological similarity and also of behavioral and ecological information is needed. To reach the proper compromise between too little or too much is not always easy. For instance, there are only two formal categories (suborder and super-family) between order and family. Yet recently acquired biological information indicates that there are three levels of difference between the order Orthoptera and the included families: the Saltatoria and Cursoria within the Orthoptera, the Ensifera and Caelifera within the Saltatoria, and the Tettigonioidea and the Grylloidea within the Ensifera. The Tettigonioidea, in turn, are divided into several families, including the Tettigoniidae and Gryllacrididae. Utilization of the category "infraorder" is indicated in such a case.

The same levels of (genetic) relationship are frequently correlated in different higher taxa with different degrees of morphological change. Thus in the Nymphalidae, a family of butterflies, the subfamilies are separated by smaller gaps and show less morphological diversity than do the subfamilies of another family, the Papilionidae. In order to achieve equal standards of morphological difference the splitter would be inclined to raise the subfamilies of Papilionidae to family rank and thereby generate a categorical escalation throughout the system of the Lepidoptera. If families and genera of cestodes were based on the same amount of differences as in the nematodes, it would lead to a great deal of splitting of the former.

In short, the recognition of taxa and their ranking must be undertaken in such a manner as not to upset the equivalence of categories in related taxa.

**10.5.4 Special Problems.** A number of special problems require separate discussion.

*Intermediate Species.* If a species is intermediate between two genera (or other higher taxa), it should be referred to that genus which includes its nearest relatives. In practice this rarely causes difficulties. What does occur, however, is that a species $l$ of genus $A$ may be more different from the most distant species $m$ of its own genus [Fig. 10-8,(1)] than from the nearest species $n$ in genus $B$ even though separated from it by a decided gap. Nevertheless, since the most distant species of genus $A$ are connected by a whole array of intermediate species, while genera $A$ and $B$ are separated by a well-defined gap, there is no contradiction.

Very large species groups may be recognized as separate taxa even if there are a few intermediate species. For instance, the characteristic families of Chrysomelidae (leaf beetles) and Cerambycidae (long-horn beetles) are connected by a few intermediate genera. Yet the two groups are so large and the connecting element so small in comparison with the divergent element that it would completely defeat the objectives of sound classification to combine the two families into one (Michener, 1957). In such a case one may have to assign the intermediate species somewhat arbitrarily to one or the other of the related higher taxa.

*Parallel Evolution.* A particular difficulty is caused by parallel evolution. "*Parallelism is the development of similar characters separately in two or more lineages of common ancestry and on the basis of, or channeled by, characteristics of that ancestry*" (Simpson, 1961). The readiness with which taxa in different (but closely related) phyletic lines respond to the same selection pressure with similar phenotypic changes permits important inferences on the degree of similarity of their genotypes.

Parallel evolution is extremely widespread, since most of the genotype remains untouched during speciation and phyletic branching. When orangs (*Pongo*) and gibbons (*Hylobates*) acquired certain structural similarities in connection with a brachiating arboreal mode of life, this was largely an activation of dormant tendencies already present in the genotype inherited from the common ancestor. Nasute soldiers evolved twice independently in the termite subfamily Nasutitermitinae; strongly asymmetrical soldier mandibles evolved in three genera of the subfamily Termitinae; phragmotic heads (a defensive adaptation for plugging holes in the nest) evolved in three different genera of the family Kalotermitidae. A secondary jaw articulation originated at least 14 times in the class Aves (Bock, 1963).

The innate potential is often realized in a rather haphazard manner. Hennig (1950, p. 176) cites stalked eyes as an example; these occur widely among the acalyptrate dipterans in scattered species and genera, but nowhere else in the large order Diptera. The potential for acquiring such a character indicates relationship, but this does not justify combining into a taxon those species in which the potential manifests itself in the visible phenotype.

The phenomenon of parallel evolutionary changes in related groups is particularly often encountered by paleontologists. The emergence of mammals from various lines of therapsid reptiles, the repeated emergence of teleosts from holostean fishes, the independent evolution of similar functional types from separate lines of rodents, and many cases described by invertebrate paleontologists (e.g. Arkell and Moy-Thomas, 1940), are familiar illustrations.

The possibility exists in such cases that the morphological changes in related lineages parallel each other to such an extent that they will go through the same evolutionary stages or *grades,* as Huxley (1958) has called them. When such grades of different lineages resemble each other far more closely than do stages in a single phyletic line, the taxonomist faces a dilemma. Should he adopt a "horizontal" classification by combining similar grades in a single taxon? There is no easy answer. The redefinition of monophyly (4.3.5) and the realization that we aim to classify on the basis of inferred genetic similarity make us increasingly inclined to favor a certain amount of horizontal classification in such cases (quite independently of additional practical advantages). The delimitation of the mammals against the therapsid reptiles is an illustration. The experienced taxonomist must use his judgment as to what particular balance between horizontal and vertical classification he wants to adopt (see MacIntyre, 1966). In view of the important function of a classification as an information retrieval system, we recommend that genetic relationship resulting from parallel evolution be given the appropriate weight in the delimitation of higher taxa. Horizontal classifications are permissible only when based on all considerations, not just on one single character complex. A particularly strenuous effort must be made to exclude all similarities caused by convergence.

*Evolutionary Continuity.* If the fossil record were complete, a delimitation of taxa against each other would be a near impossibility (2.5.1). Fortunately for the taxonomist (as regrettable as this is for the evolutionist), the fossil record is so full of gaps as to permit a clear-cut delimitation of taxa in most cases. However, the record is beginning to be sufficiently complete to raise some real difficulties. The delimitation of the fossil Creodontia from the derived fissipede carnivores is such a case (MacIntyre, 1966). The separation of the mammals from the mammal-like reptiles is another case. *Eohippus* is about as good an ancestor for tapirs and rhinoceroses as for the Equidae. Again, the taxonomist must base his ultimate decision on a careful evaluation of cladistic and phenetic considerations as well as on the question of what classification would yield the most information.

**Summary.** The discussion of the many criteria affecting decisions on the delimitation and ranking of taxa should have made clear why there is so much disagreement among specialists. It depends on tradition, on the maturity of the existing classification in a given area, on the richness of species in a group, on the degree to which the phenotype reflects (or exaggerates) changes of the genotype, and on the number and scope of the specialists, among other factors. Presumably there will always be specialists who tend toward splitting and others who tend toward lumping. Some want to express degree of difference in their classification, others want to emphasize relationship. Also, there will be those who stress recency of common ancestry (cladistic aspects) and those who simply rely on degree of "similarity." Nevertheless, there has been an unmistakable trend in recent decades toward more uniform standards. These have already developed within restricted taxa (birds, Coleoptera, isopods), and the inequality of standards is, on the whole, not a severe impediment to communication.

## 10.6  IMPROVEMENT OF ESTABLISHED CLASSIFICATIONS

Discussions of taxonomic procedure tend to create the impression that the act of classifying is always a *de novo* operation, as if the taxonomist always had to classify huge piles of unassorted species. Actually, the greater part of our zoological system is mature, sound, and stable. The student of mammals does not have to discover the differences between marsupials and placentals or between primates and rodents. The entomologist has no trouble in delimiting Odonata, Lepidoptera, or Coleoptera; there has been no difficulty about this since long before Linnaeus. What then are the real problems of classification in modern taxonomy? The answer depends on the particular group being studied.

Most acts of classification in recent organisms are minor rearrangements of existing classifications. For instance $A(a + b + c)$ and $B(d + e)$ are reclassified into $A'(a + b)$ and $B'(c + d + e)$ or into $A''(a + e)$ and $B''(b + c + d)$. The recent classification of the echinoderms (Table 10-3) may be given as a case in which the primary classifying criterion was replaced. Such rearrangements of taxa occur at all levels of the hierarchy of categories. An entire breaking-up of existing taxa into much smaller pieces and their reassortment into a totally new set of higher taxa happen rarely and only in poorly known groups. The reclassification of the rhabdocoels (Turbellaria), or some proposals for a reclassification of the sponges (Porifera) discussed by Lévi (1957) and Hartman (1958), are examples of rather drastic reclassifications.

In the older classifications the turbellarians were classified entirely on the basis of the presence and configuration of the intestinal tract. However, the single feature of body size has a dominant influence on the manifestation of the intestines. They are often absent in very small species but highly elaborate in the larger species (to keep the ratio of intestinal surface to body volume reasonably constant). The Rhabdocoela of the older literature, turbellarians with tube-shaped intestines, were a rather heterogeneous assemblage. The replacement of this single-character classification by another one, yolk content of the eggs and associated mode of cleavage, has not been a full success. This specialization is indicative of a grade and may well have been acquired repeatedly. Which subclass one adopts for the orders of turbellarians depends entirely on the primary character chosen. The reconstruction of the ancestral characters and of the common ancestor as a whole is still controversial (Ax, 1961).

TABLE 10-3.  EFFECT OF ALTERNATE CLASSIFYING CRITERIA ON THE
CLASSIFICATION OF THE ECHINODERMATA
(*from Fell* 1965)

| Conventional classification based on growth form | Classification based on additional characters, unrelated to growth habits |
|---|---|
| Subphylum Pelmatozoa:<br>  Class Carpoidea | Subphylum Homalozoa:<br>  Class Carpoidea<br>Subphylum Crinozoa: |
| Class Cystoidea<br>Class Eocrinoidea<br>Class Paracrinoidea<br>Class Blastoidea<br>Class Edrioblastoidea<br>Class Crinoidea | Class Cystoidea<br>Class Eocrinoidea<br>Class Paracrinoidea<br>Class Blastoidea<br>Class Edrioblastoidea<br>Class Crinoidea<br>Subphylum Echinozoa: |
| ————* | Class Helicoplacoidea* |
| Class Edrioasteroidea<br>Subphylum Eleutherozoa:<br>Class Ophiocistoidea<br>Class Echinoidea<br>Class Holothuroidea | Class Edrioasteroidea<br><br>Class Ophiocistioidea<br>Class Echinoidea<br>Class Holothuroidea<br>Subphylum Asterozoa: |
| Class Asteroidea<br>Class Ophiuroidea | Class Stelleroidea (including<br>  the subclasses Somasteroidea,<br>  Asteroidea and Ophiuroidea) |

* Helicoplacoidea unknown before 1963.

TABLE 10-4.  INCONCLUSIVENESS OF CLASSIFYING CHARACTERS IN TWO
GENERA OF CALCAREOUS SPONGES (*from Vacelet*, 1961*)

| Taxon | Nucleus of choanocyte | Larva | Triradiate spicules |
|---|---|---|---|
| Subclass Calcaronea | Apical | Amphiblastula | Sagittal |
| Genus *Murrayona* | *Basal* | ? | Sagittal |
| Genus *Minchinella* | ? Filling cell | *Parenchymula* | Sagittal |
| Subclass Calcinea | *Basal* | *Parenchymula* | *Equiangular* |

* *Syst. Zool.*, 10:45–47.

The genera *Murrayona* and *Minchinella* of the Pharetronida seem to have a mixture of the characters of the two subclasses of the Calcarea (Calcaronea and Calcinea).

**10.6.1  Classificatory Activities.** The problems of classification can perhaps be tabulated as follows (without claiming to exhaust all possibilities) :

1. The assignment of a newly discovered species to the right genus by asking these questions:

   a. Is it possible to include it in an established genus?

   b. Is it necessary to erect for it a new genus and possibly for this genus a still higher new taxon?

2. The transfer of an incorrectly placed taxon to its proper position. This includes the shift of a species to a different genus, the shift of a genus to a different subfamily or family, and so forth.

3. The splitting of a taxon into several of the same rank (genus, family, etc.), either by cleaving a heterogeneous assemblage of species into several smaller, more homogeneous ones, or by removing an alien element from an otherwise homogeneous taxon.

   When breaking up too large a taxon, certain rules must be obeyed in the ranking and naming of the resulting new taxa.

   a. The rank of the original taxon is to be maintained if at all possible. Finer discrimination is possible by the elaboration of subtaxa. For instance, it is usually less desirable to raise a heterogeneous family to the rank of superfamily and then recognize as families all the previously recognized subfamilies, than to develop a finer subdivision into tribes and genus-groups.

   b. No taxon should fall out of step with its sister-groups. The classification of fossil man by certain anthropologists is a warning illustration of an unbalanced classification (see 10.5.3).

   c. A minimal number of names is desirable. By adopting informal groupings such as "species-groups" (instead of new genus or new subgenus) and "genus-group" (instead of new family, new subfamily, or new tribe), the same information can be conveyed without a burden on the memory and without disturbing the balance of the hierarchy of categories (10.5.3).

   d. Subdivide an inconveniently large taxon only if it can be "cleaved," that is, if it can be divided into taxa of approximately equal size. Splitting off a number of monotypic genera from a genus with 500 species would only impede information retrieval. This advice does not concern the removal of clearly alien elements from currently recognized taxa.

4. The raising in rank of an existing taxon, a genus to a subfamily, a subfamily to a family, etc.

5. The fusion of several taxa of the same rank and the synonymizing of those with junior names.

6. The reduction in rank of a taxon as for instance that of a genus to a subgenus, a family to a subfamily, etc.

   There is very little doubt that certain groups of animals are badly oversplit and that natural taxa in these groups are ranked in higher categories than necessary. This is true of both birds and fishes. Even the specialists concerned admit that there is little justification for having 412 families of fishes and 171 families of birds. But which of these should be reduced to the rank of subfamilies? There is no easy answer.

For instance, there is a well-defined group of songbirds in the Old World tropics, the drongos or crow shrikes. In spite of their wide distribution and numerical abundance, they consist of only 2 genera and about 20 species. Up to now, not a single good morphological character diagnostic for this group has been found, and yet in general habitus and in behavior they stand reasonably well apart from all other songbirds. Ornithologists would be perfectly willing to consider the drongos a subfamily or perhaps a tribe of some other songbird family, but as yet no character is known that would help in finding that family. Among the families that have been suggested are the Campephagidae (cuckoo-shrikes), the shrikes (Laniidae), the Muscicapidae (flycatchers), the Paradisaeidae (birds of paradise), and the Sturnidae (starlings). In desperation ornithologists finally raised the drongos to the rank of a family, the Dicruridae, while perfectly willing and ready to reduce this rank as soon as additional information becomes available. The same is true for at least twenty other families of birds.

7. The creation of a new higher taxon not by raising the rank of a taxon (e.g. a family to superfamily rank) but by making an entirely new grouping of taxa of the next lower rank. The proposal of a new superfamily for a number of existing families or a new order for a series of families illustrates this procedure.

8. The search for the nearest relative of an isolated taxon and, if it is successful, the study of the question whether a new taxon of higher rank should be created for the newly established group of relatives. For instance, behavioral and anatomical researches indicate that the Tubinares (shearwaters, etc.) are the nearest relatives of the penguins (Impennes). Should one establish a superorder for these two orders?

**10.6.2 Stability.** During such minor improvement activities a determined effort must be made to disturb the stability of the currently prevailing classification as little as possible and to maintain, if not improve, its information retrieval qualities. The usefulness of a classification as a communication system stands in direct relation to its stability, which is one of the basic prerequisites of any such system. The names for the higher taxa serve as convenient labels for the purpose of information retrieval. Terms like Coleoptera or Papilionidae must mean the same to all zoologists in order to have a maximum of usefulness. This is even more true for the genus, the name of which is included in the scientific name (13.1).

When a well-established taxon is found to be somewhat heterogeneous, it is often inadvisable to split it into several taxa of the same rank if the components are each other's nearest relatives.

For example, if cytological or biochemical investigation reveals that a certain genus consists of two groups of species, each with a different series of chromosome numbers or biochemical reactions, it is not necessary, in order to record and study this phenomenon, for the genus to be split into two separate genera, even if certain characters are correlated with the two groups. Instead, one can deal quite satisfactorily with the situation by leaving the limits and nomenclature of the genus intact, and by making a special classification of its species into the two groups, giving them ad hoc designations such as group *A* and group *B* [or species group *A* and species group *B*], thus enabling one to refer to them and study them just as satisfactorily as if one had split the genus into two—with the great advantage of leaving the general classification intact, stable, and perfectly serviceable for its purpose. (Gilmour, 1961, p. 37).

The currently adopted zoological system contains numerous taxa suspected of being polyphyletic owing to convergence. Among the birds there are many such groups, particularly among the songbirds (titmice, warblers, babblers, flycatchers, shrikes, finches, etc.). As long as it remains unknown what the nearest relatives of the components of such unnatural groups are, it is far better to retain these groups provisionally, for ease of reference. Such provisional classifications must be abandoned, however, as soon as the true relationship of the components is established.

## 10.7   PRESENTATION OF A CLASSIFICATION

When publishing the classification which has resulted from one's taxonomic studies, one must present it either as a printed list or as a diagram. Both methods of presentation raise problems.

**10.7.1   The Printed Sequence.** The technology of printing enforces a linear, one-dimensional sequence upon any printed classification. One species will have to come first, another species last, while all others will have to be listed sequentially between the first and the last. How can we determine the simplest, most convenient sequence of species?

When the classification of a group is still entirely obscure and catalogs consist merely of lists of nominal species, an alphabetical sequence is often most convenient for information retrieval. However, an alphabetical listing lacks the heuristic value of a classification arranged according to inferred genetic relationship. By not placing closest relatives near each other, such a listing makes it difficult to undertake evolutionary studies. Finally, there is no stability because it necessitates a change in the sequence every time the name (synonymy!) or the rank (e.g. shift from species to subspecies) of a taxon changes (Mayr, 1965c). Species in all better-known groups should be listed according to their relationship with each other. However, this raises various difficulties.

The multidimensional phylogenetic tree with the dimensions of time, space (longitude and latitude), and adaptational divergence must be converted into a single dimension. In order to do this the taxonomist inevitably must compromise between various considerations. Most important among these are the following three:

1. *Continuity.* Each species is to be listed as near as possible to its closest relatives.

2. *Progression.* Each series of species or higher taxa should begin with the one closest to the ancestral condition ("the most primitive one") to be followed by derived taxa which deviate increasingly from the ancestral state.

3. *Stability.* One should not make "experimental" changes from previously adopted sequences, unless the latter are proved unequivocally wrong. A classification is a reference system, and it greatly helps, particularly in a comparison of faunal lists, if different authors use the same sequence.

The three principles are often in conflict with each other (particularly 1 and 2). It is sometimes possible to establish a well-defined morphological sequence without being able to state which end of the sequence is the more primitive. In other cases, there is a dual progression from a group of primitive species toward two specialized extremes (Fig. 10-12). Instead of dividing the closely related primitive species into two groups, one leading to one extreme and the other to the other extreme, as might be demanded by the progression principle, it usually results in greater continuity if one starts at one specialized end and establishes a single sequence by first descending to the most primitive species and then ascending again to the

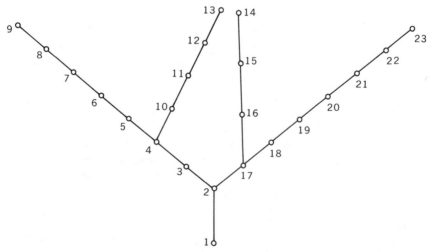

Fig. 10-12. Arranging 23 species on a dendrogram in a linear sequence. Other alternatives are: 1–13, 17, 16, 15, 14, 18–23, or 1–4, 10–13, 5–9, 17–14, 18–23.

other extreme. This avoids a more or less arbitrary split through the middle of the group of primitive species.

Because of mosaic evolution, most groups show several trends of specialization at one and the same time. In such cases, the decision regarding which specialization is considered most advanced may be entirely arbitrary. Among birds, for instance, we find four particularly conspicuous specializations of the wing:

1. Increased flight efficiency: (a) primaries reach functional peak (swifts, hummingbirds), (b) secondaries reach functional peak (albatrosses)
2. Loss of flight: (a) wing changed to swimming paddle (penguins), (b) locomotory function of wing lost in flightless birds (e.g. Ratites).

It is entirely a matter of convention which of these four groups of taxa is considered "high" or "low" in the system. Most of the so-called primitive animals, for instance the monotremes among the mammals, are highly specialized in certain aspects. Consideration of the totality of characters usually permits a decision as to the most logical or convenient sequence, but in highly uniform groups such as birds no clearly superior sequence is obvious. The taxon that is highest in wing development may be lowest in the development of the central nervous system, and vice versa.

**10.7.2  Graphic Representations.** Species living at the present time are the current end points of innumerable phyletic lines. A linear listing of species, genera, and families cannot even begin to convey an impression of the various lines of descent in a complex phylogeny. Nor can it show to what extent the grouping of species is the result of divergent, convergent, and parallel evolution. It is this deficiency of the printed sequence which has induced taxonomists to attempt to express their conclusions on relationship in the form of diagrams.

Up to the eighteenth century naturalists and philosophers arranged the diversity of nature in the form of a single *scala naturae* ascending from inanimate matter in an undivided continuous line up to man. The continued emphasis on "lower" and "higher" in our classifications is a remnant of this kind of thinking. As more and more types of organisms became known, the linearity became increasingly suspect. Pallas, Buffon, and Lamarck (Voss, 1952) suggested ramifications or anastomoses of the *scala naturae,* to satisfy the principle of plenitude. Cuvier's recognition of four major branches of the animal kingdom, each totally independent of the others, completed its demise. Darwin (1859, opposite p. 117) proposed a purely theoretical diagram of genealogy, but it was Haeckel (1866) who made the first successful attempt at presenting the relationships of all animals phylogenetically (Fig. 4-3). Since that time it has been traditional among taxonomists to present their concepts of relationships in the form of phylogenetic trees or similar diagrams (Jepsen, 1944). In spite of their

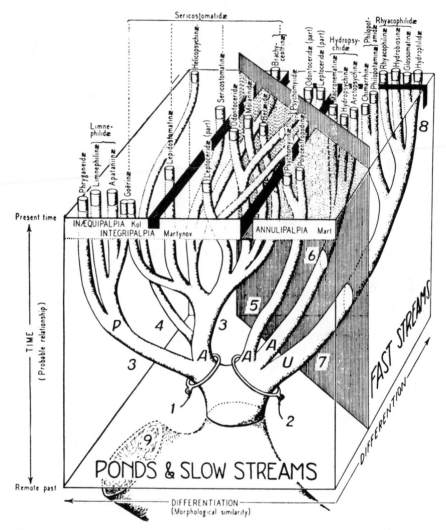

Fig. 10-13. A three-dimensional dendrogram representing the evolution of caddisworm case construction (*Milne and Milne*, 1939).

obvious shortcomings such diagrams are useful summaries of taxonomic knowledge and permit an easily accessible presentation of the author's ideas on the evolutionary history of a group. Such diagrams have great heuristic value because they tempt later students to test them on the basis of new characters and subsequently discovered taxa.

Haeckel's original phylogenetic trees were indeed drawn as trees. Many subsequent constructions have attempted to convey the impression of a three-dimensional structure (see Fig. 10-13). In most cases so-called phylogenetic trees are far more diagrammatic and often concentrate on two dimensions representing time and difference, with special attention to the branching points. A diagram of the phylogeny of the Equidae (Fig. 10-14) is a typical example. In this case the diagram is to a large extent based on the rich fossil record of this family of ungulates. Even in this

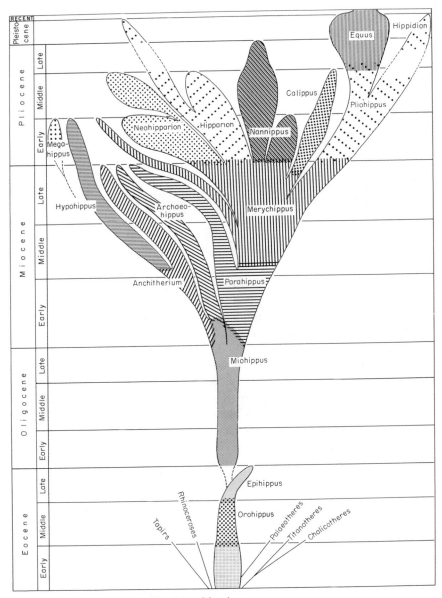

Fig. 10-14. Phylogeny of the Equidae (*after Stirton*).

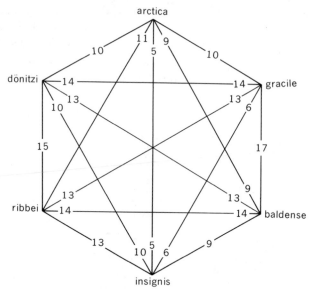

**Fig. 10-15.** Star-type phenogram of six species of ground beetles of the genus *Miscodera*. Figures indicate number of characters (in a total of 20) shared by each pair of species (*from Lindroth*, 1961).

case the interpretation is largely a matter of inference. It is true that the fossil record gives substance to phylogenetic trees, but the gaps in the record are still sufficiently large even in the best-known groups to require much conjecture.

*Diagrams of Relationship.* Numerous methods of representing graphically the similarity or relationship of taxa have been proposed by taxonomists over the years. They differ by the kind and amount of information which they convey and whether or not they include inferred phylogeny. There are various criteria on the basis of which these diagrams can be classified. The theory of classification underlying the various schemes is perhaps best brought out in the following arrangement:

*Phenograms.* Any diagram which expresses "affinity" as indicated by degree of similarity but does not aim to convey any other information is a phenogram. Serologists have long used such phenograms to indicate relationships as indicated by cross reactivity. Plant geneticists have used them to indicate degrees of interfertility between species, taxonomists to express morphological similarities (Lindroth, 1961, Fig. 10-15). This method has two virtues—it gives the degree of similarity between each species and all the other species considered, and it does not make any inferences as to probable lines of descent.

In contrast to these topologically uncommitted phenograms is the phenogram adopted by the pheneticists. It has the form of a phyletic dendrogram, with a regular branching of lines just as in a phylogenetic tree (Fig. 10-16). As shown by Boyce (1964) and Minkoff (1965), the branching points are very sensitive to changes in the methods of computation, so that each method of computation may result in a different phenogram. Rapidly diverging lines have to be inserted below the branching points of slowly evolving lines even though the phyletic branching took place later in time. For all these reasons the adoption of the dendrogram pattern for phenograms would seem inefficient and misleading (Mayr, 1965b).

*Cladogram.* In a cladogram (Fig. 10-17) the ordinate gives the estimated time, while the abscissa gives degree of difference. Degree of relationship is determined by the position of the branching points. *B*, for instance, is said to be more closely related to *C* and *D* than to *A*, because *B*, *C*, and *D* share branching point $T_2$ while *A* branched off earlier (at $T_1$). The different amounts of evolution between $T_1$ and $T_2$ as against $T_2$ and $T_3$ are ignored (see cladism, Chap. 4). Since the fossil record is rarely sufficiently complete to indicate the actual position of the branching points in space and time, they are reconstructed through a careful evaluation of the similarities and differences of taxa (Hennig, 1960).

*Phylogram.* In an orthodox phyletic dendrogram (Fig. 10-18) three kinds of information are conveyed, degree of difference in the abscissa, geological time in the ordinate, and degree of divergence by the angle

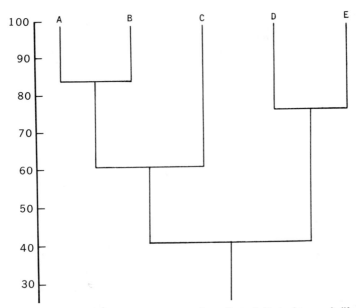

**Fig. 10-16.** A phenogram. The figures on the ordinate indicate degrees of difference (or similarity) between the taxa listed (at arbitrary distances) on the abscissa. The horizontal connecting lines of the taxa do not indicate phylogenetic branching but merely levels of phenetic difference.

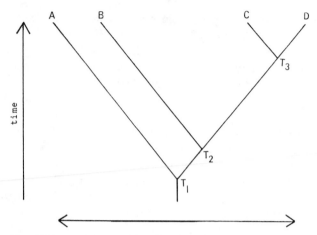

Fig. 10-17. A cladogram. The branching points at $T_1$, $T_2$, and $T_3$ determine the classification of the taxa $A$, $B$, $C$, and $D$.

of divergence. This diagram presents a maximum of information, but two of the three sets of information (branching time and rate of divergence) are inferred. Although this diagram shares with the phenogram the emphasis on degree of observed similarity, it can cope better with problems of convergence and parallelism and particularly with problems produced by different rates of evolution.

The phylogeny represented in a phylogram is only inferred and is unlikely to be an exact representation of the real phylogeny. Likewise, mosaic evolution, that is, different rates of evolution for different compo-

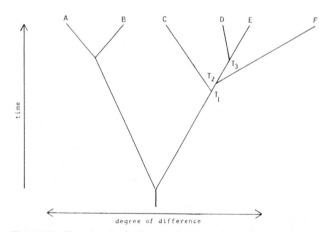

Fig. 10-18. The phyletic dendrogram (phylogram) of the evolutionary taxonomist. The slower a line evolves, the more it approaches the vertical; the more rapidly it evolves (as $T_2$ to $F$), the more it approaches the horizontal.

nents of the phenotype, makes it virtually impossible to represent phylogeny adequately in any diagram.

Curiously, such diagrams are particularly helpful where the classification must deviate from the ideal of a purely cladistic classification. It aids understanding in cases where parallel evolution requires horizontal classification (Chap. 4) or, as in the birds-crocodiles case, where a particularly rapid evolution of a single line has resulted in the evolution of a higher taxon with all other branches remaining at a lower categorical level. Finally, such multidimensional diagrams are a useful supplement to the linear sequence of the printed classification.

# Chapter *11* *Taxonomic Publication*

$N$o taxonomic study is complete until its results have been published. The advances in any area of science are made known through publication, an inseparable part of research. This is particularly true for a vast field like systematic zoology where the number of workers is so small. Applied biology is to a large extent dependent on the advances made in the correct classification of organisms.

*The Strategy of Publishing.* A research worker in any branch of science must ask himself at regular intervals whether or not he is pursuing the most productive line of research (1.4.4). Should he shift his emphasis, should he switch to a different group of organisms? What sort of research is most needed in the particular group of organisms for which he is a specialist? Should he concentrate on alpha, beta, or gamma taxonomy? The answer in most cases is that a judicious balance of various approaches is the most rewarding strategy.

A major taxonomic revision is very time-consuming and may require years of work prior to publication. The question inevitably arises whether there should be preliminary papers and whether the specialist should prepare at the same time a series of shorter papers, perhaps dealing with different taxa, in order to maintain a publication schedule. Obviously there is no excuse for duplication in publication for the sake of increasing the number of titles in one's bibliography. On the other hand the field work and research required for the preparation of a comprehensive revision may result in biological, evolutionary, and biogeographical findings which are best published separately.

Some taxonomists find it difficult to complete their manuscripts. They should realize that there is a law of diminishing returns and that no research is ever "complete." There will always be areas that have not yet been collected and collections in remote museums that have not yet been studied. The same is true in experimental research, where results are published even though more and perhaps slightly different experiments might well have been conducted. The number of specialists is very limited, and they have a real obligation to publish their results as soon as a reasonably reliable summary of their researches is possible.

*Documentation and Information Retrieval.* The names of 8,000 contemporary animal taxonomists are included in a recent listing (Blackwelder and Blackwelder, 1961), but many are inactive. Every year many new authors start publishing. To keep track of their output has become a formidable task. Sooner or later some system for automatic information retrieval will be introduced. The animal taxonomist is fortunate in having a relatively complete listing of all new names and taxonomic revisions in the *Zoological Record.*

Every effort should be made by the taxonomist to facilitate the task of information retrieval and to avoid anything that might increase its difficulty. The latter includes name changing in violation of the principles laid down in the Preamble of the Code; and unnecessary splitting or lumping on the generic level; any change in a widely adopted taxonomic sequence without compelling reason; and, finally, the proposal of new species not connected with complete revisions.

## 11A. KINDS OF PUBLICATIONS

Taxonomic publications range all the way from the short description of a new taxon, covering only part of a page, to lengthy monographs and/or handbooks that may consist of several volumes. This includes works primarily useful for the purpose of identification as well as revisionary works and new classifications. In some works the nomenclatural aspects are stressed, in others biology, distribution, or illustrative material. There is a time in the history of the study of each group of organisms in which one or the other kind of publication is most useful. A specialist must fully understand the function of each kind of publication and must select one of them as the form of his own next publication in the field. The following comments on these different types of taxonomic publication may be helpful to the beginner.

Many different titles are used for more comprehensive taxonomic publications. Unfortunately, the meaning of such titles is not always the same for different groups of organisms. A checklist in ornithology is a very careful, critical revision, usually with extensive synonymy and a detailed elaboration of the geographic distribution of each species and subspecies. A checklist in entomology is generally only a list of names, often in alphabetical sequence. Words like handbook and review likewise have different meanings in different fields. The subsequent discussion must be regarded in the light of this heterogeneity of usage. Many published works combine features of several of the kinds of publications described below.

**11A.1  Description of New Taxa.** The isolated description of new subspecies, species, and genera, divorced from revisional or monographic work, is the least desirable form of taxonomic contribution, except in well-known groups. The preparation of such a description does not permit as careful a comparison of all related species as does a revision; in many cases it results merely in the addition of still another name to the already long list of nominal species. The specialist who eventually prepares the revision must search out the type, which is often quite inaccessible, and lose much valuable time in getting it properly identified. In many (if not in most) cases the description of isolated new species in poorly known groups of animals is a handicap rather than a help to subsequent workers. Such isolated descriptions are justified only when names are needed in connection with biological or economic work or when a group has been recently revised and the new species can be readily fitted into the classification.

Theoretically, in order to provide an adequate description of even a single species an author must undertake much of the work basic to the preparation of a revision, and he could carry his work to that point with a little extra effort. All too frequently, however, the isolated description results from a rather superficial acquaintance with previous work, and a far greater percentage of synonyms is created through isolated descriptions than in more substantial revisions.

**11A.2  Synopses and Reviews.** Synopses and reviews are brief summaries of current knowledge of a group, and the inclusion of new material or new interpretations is not necessarily implied. They serve the utilitarian purpose of bringing scattered information on a group together in one place, perhaps as a basis for some future revisional or monographic study. Examples of a synopsis and a review are as follows:

LA RIVERS, IRA. 1947. A synopsis of the genus Endrodes (Coleoptera: Tenebrionidae). *Ann. Ent. Soc. Amer.,* **40**:318–328.
Ross, H. H. 1946. A review of the Nearctic Lepidostomatidae (Trichoptera). *Ann. Ent. Soc. Amer.,* **39**:265–291, 37 figs.

**11A.3  Revisions.** Revisions are presentations of new material or new interpretations that have been integrated with previous knowledge through summary and reevaluation. They vary greatly in completeness of treatment.

Some revisions are monographic in approach yet fall short of being monographs because of inadequate material. Others are limited to a new arrangement of a group. Most of the important current taxonomic contributions in groups where new species are still constantly being discovered fall in this category. Such revisions may deal with a whole family (or part of one), with a genus, or with a species group. Generic revisions, illustrated by the following example, are the most common type of such work.

SOMMERMANN, K. M. 1946. A revision of the genus Lachesilla north of Mexico (Corrodentia: Caeciliidae). *Ann. Ent. Soc. Amer.*, **39**:627–657, 4 plates.

**11A.4   Monographs.** Monographs are complete systematic publications. They involve full systematic treatment of all species, subspecies, and other taxonomic units and a thorough knowledge on the author's part of the comparative anatomy of the group, the biology of the species and subspecies included, the immature stages in groups exhibiting metamorphosis, and detailed distributional data. For the student of evolution such monographic treatises are the most rewarding type of taxonomic publication. They permit a detailed treatment of geographic variation, of relationships, and of distributional history. Generalizations on the structure of species, modes of speciation, nature of taxonomic categories, and the like are based on such monographs. They have the disadvantage that they require more complete material than other kinds of taxonomic papers. However, with the growth of collections in the museums of the world, it is becoming increasingly feasible to prepare monographs. Some papers which fully qualify as monographs are published as revisions or under some other title. Unfortunately, in the present state of our knowledge of many groups, especially among the invertebrates, few taxonomic papers can justify the title monograph. Monographs are more frequently possible among vertebrates. Three fairly typical examples of monographs are:

HUBBELL, T. H. 1936. A monographic revision of the genus Ceuthophilus (Orthoptera, Gryllacrididae, Rhaphidophorinae). *Fla. Univ. Pub., Biol. Sci. Ser.*, vol. II, no. 1, 551 pp., 38 plates.
LINSLEY, E. G. 1961ff. The Cerambycidae of North America, *Univ. Calif. Publ. Ent.*, vol. 18 (97 pp., 35 pl.), vol. 19 (102 pp., 1 pl.), to be continued.
USINGER, R. L. 1966. Monograph of Cimicidae (Hemiptera-Heteroptera). *Thomas Say Foundation Publications*, vol. 7, xi + 585 pp.

**11A.5   Atlases.** In recent times the need has been felt for complete illustrations of the species of various taxonomic groups. This is a reflection of the inadequacy of the printed word as a means of conveying a mental picture of the general facies of an animal. The idea of an atlas grew also out of the need for taxonomic data which are strictly comparable from one species to another. Since the purpose of an atlas is purely taxonomic, semidiagrammatic drawings are commonly used, though full halftones or colored plates have been employed when dealing with such groups as butterflies and birds.

Examples of this type of treatment are as follows:

FERRIS, G. F. 1937–1950. Atlas of the scale insects of North America. Stanford University Press, Stanford University, Calif., 5 vols.

ROSS, E. S., and H. R. ROBERTS. 1943. Mosquito atlas. American Entomological Society, Philadelphia; part 1, 44 pp., part 2, 44 pp.

**11A.6   Faunal Works.** In faunal works the scope is dictated by the limits of a geographic region rather than by the limits of a taxon. There are different kinds of faunal works of varying usefulness. On the whole, the broader the area covered the more useful is the faunal work.

Examples are:

Fauna of British India. Taylor and Francis, London. Many volumes covering most groups of animals, published from 1888 on.

Biologia Centrali-Americana, 1879–1915, parts 1–215. Dulau and Co., London.

Faune de France. 1921–1967 et seq., vols. 1–68. Office Centrale de Faunistique Paris.

Fauna SSSR. Zoologicheskii Institut Akademii Nauk SSSR (some 90 vols. published).

An example of a *local list* is:

BROWN, H. E. 1939. An annotated list of the species of Jassinae known to occur in Indiana (Homoptera, Cicadellidae). *Amer. Midland Nat.,* 21:663–673.

Reports of expeditions and voyages also belong to this category. These were useful publications in the early history of taxonomy, but such reports on individual collections are no longer looked at with favor. They never permit the careful revisionary work necessary for the exact placement of specimens, and they often lead to the proposal of new names later found to be synonyms. Faunistic reports on the whole are among the most inefficient kinds of taxonomic papers. The time required to run down all the locality records in a local fauna can be spent far more productively in preparing a revision or a manual.

A separate class of publications consists of identification works devoted to a local fauna (see below). Local faunas, in the case of well-known organisms, give the local naturalist an opportunity to provide ecological and behavioral information, often based on many years of field work. Such publications are possible only when the knowledge of the fauna has passed the level of alpha taxonomy.

**11A.7   Field Guides and Manuals.** Certain works, although taxonomic, are designed primarily or exclusively for field identification. In such cases new species are expressly excluded, and emphasis is placed on clear-cut key characters or recognition characters. Examples of this are cited in Chap. 6.

**11A.8   Handbooks.** The term handbook is sometimes used for field identification guides but nowadays more commonly for comprehensive works on a group of organisms with particular emphasis on their biology and distribution. An example is:

PALMER, R. S. (ed.). 1962. Handbook of North American birds, vol. 1. Yale University Press, New Haven, 567 pp.

More broadly the term handbook (traité) is used for surveys of large areas of zoology, like Kükenthal's *Handbuch der Zoologie* or Grassé's *Traité de Zoologie*.

The word treatise is also sometimes used for such handbooks as MOORE, R. C. (ed.). 1953. Treatise on invertebrate paleontology. Geological Society of America and University of Kansas Press.

**11A.9 Catalogs and Checklists.** A catalog is essentially an index to published taxa arranged in such a manner as to provide a complete series of references for both zoological and nomenclatural purposes. Species and genera are often listed in an alphabetical sequence, because most catalogs are nothing but uncritical (prerevisionary) listings of nominal species. The value of a catalog depends on its degree of completeness, and its preparation is therefore a highly technical task requiring great patience and an intimate knowledge of bibliographic resources and methods.

A checklist is an abbreviated synopsis. It provides a skeleton classification of a group and serves as a convenient source of reference for the naming of specimens and the arrangement of collections. Checklists vary greatly in their elaborateness. A list of names deserves the name checklist (in contradistinction to catalog) only if a careful distinction is made therein between valid names and synonyms. The name of each species is usually accompanied by a more or less detailed indication of the range of each species. Checklists are most useful in the better-known groups of animals. One in which species are critically evaluated and carefully distinguished from synonyms qualifies as "primary zoological literature" (13.24), whereas an uncritical listing of names or nominal species does not.

STONE, A. et al. 1965. A catalog of the Diptera of America north of Mexico. Agricultural Research Service, U.S. Dept. of Agriculture, Washington, D.C., 1696 pp.

McDUNNOUGH, J. 1938–1939. Checklist of the Lepidoptera of Canada and the United States of America. Part 1, Macrolepidoptera; part 2, Microlepidoptera. South. Calif. Acad. Sci. Mem., 1, 1–275; 2(1), 1–171.

MAYR, E. 1941. List of New Guinea birds. American Museum of Natural History, New York, 260 pp.

**11A.10 Revisions of Higher Taxa.** As a consequence of the increasing interest in macrotaxonomy, numerous works have appeared in recent years which deal exclusively with genera and still higher taxa. Frequently they contain an evaluation of the characters which have been used by the author to infer relationship and phylogeny and to determine ranking. Such taxonomic works are:

CORLISS, T. O. 1961. The ciliated protozoa: Characterization, classification, and guide to the literature. Pergamon Press, New York.

EDMUNDS, G. F. 1962. The principles applied in determining the hierarchic level of the higher categories of Ephemeroptera. *Syst. Zool.*, **11**:22–31.

EHRLICH, P. R. 1958. The comparative morphology, phylogeny, and higher classification of the butterflies (Lepidoptera: Papilionoidea). *Univ. Kansas Sci. Bull.*, **39**:305–378.

FELL, H. B. 1965.The early evolution of the Echinozoa. *Breviora*, **219**:1–17.

GREENWOOD, P. H., D. E. ROSEN, S. H. WEITZMAN, and G. S. MYERS. 1966. Phyletic studies of Teleostean fishes with a provisional classification of living forms. *Bull. Amer. Mus. Nat. Hist.*, **131**:341–455.

HONIGBERG, B. M., et al. 1964. A revised classification of the phylum Protozoa. *J. Protozool.*, **11**:7–20.

MARCUS, E. 1958. On the evolution of animal phyla. *Quart. Rev. Biol.*, **33**:24–58.

MARX, H., and G. B. RABB. 1965. Relationships and zoogeography of the viperine snakes (Family Viperidae). *Fieldiana-Zool.*, **44**:161–206.

MAYR, E., and D. AMADON. 1951. A classification of recent birds. *Amer. Mus. Novitates*, **1496**:1–42.

MICHENER, C. D. 1952. The Saturniidae (Lepidoptera) of the Western Hemisphere. *Bull. Amer. Mus. Nat. Hist.*, **98**:339–501.

NEWELL, N. D. 1965. Classification of the Bivalvia. *Amer. Mus. Novitates*, **2206**: 1–25.

ROSEN, D. E., and R. M. BAILEY. 1963. The Poeciliid fishes (Cyprinodontiformes), their structure, zoogeography, and systematics. *Bull. Amer. Mus. Nat. Hist.*, **126**:1–176.

SIMPSON, G. G. 1945. The principles of classification and the classification of mammals. *Bull. Amer. Mus. Nat. Hist.*, **85**:i–xvi, 1–350.

STUNKARD, H. W. 1963. Systematics, taxonomy and nomenclature of the Trematoda. *Quart. Rev. Biol.*, **38**:221–233.

**11A.11 Evolutionary and Biological Publications.** It is inefficient to include too much heterogeneous information in a single publication. This is particularly true for taxonomic monographs. Such works are usually read only by a few specialists; any information of broader biological interest published there will not come to the attention of the general biologist and is therefore bound to be ignored. Almost every author of a taxonomic revision makes interesting ecological, evolutionary, or zoogeographical discoveries which would be of great interest to other biologists. Instead of burying these findings in the introduction of a monograph, he should publish them in a general journal like *Ecology, Evolution,* or *Systematic Zoology,* where, in turn, these articles will draw attention to the monograph. The fact that taxonomists have increasingly adopted this policy partly explains the noticeable improvement in the general image of taxonomy in recent years. No other group of biologists can make such important contributions to all problems relating to the diversity of the organic world. The taxonomist has an obligation to make his knowledge broadly available, not to bury it in a monograph or revision.

In addition to these by-products of monographs and revisions, more and more taxonomists publish short contributions to evolutionary or behavioral biology or to ecology, these being the result of ad hoc investigations.

For instance, every volume of *Systematic Zoology* contains papers dealing with cytological or chemical attributes of species (and higher taxa), distribution patterns, behavior, taxometrics, variability, endemism, extinction, serology, geographic variation, speciation, rates of evolution, climatic rules, allometry, etc. Yet this is only one of a large number of journals publishing the results of evolutionary research conducted by taxonomists. In addition to journal articles there are such books as:

HUXLEY, JULIAN (ed.). 1940. The new systematics. Clarendon Press, Oxford.

MAYR, ERNST. 1942. Systematics and the origin of species. Columbia University Press, New York.

RENSCH, B. 1960. Evolution above the species level. Columbia University Press, New York.

**11A.12   Theory and Methods of Systematics.** Publications devoted to this area comprise a growing part of the taxonomic literature. They are cited in all chapters of this book and are listed in the terminal bibliography.

## 11B. MAJOR FEATURES OF TAXONOMIC PUBLICATIONS

Most taxonomic publications contain a definite set of major components. These form the basic substance of these publications and therefore deserve thorough discussion. Many excellent (and a few outworn) traditions have developed in the past 200 years of taxonomic research. Every taxonomist must know what they are and why he should follow certain standardized procedures if he is to facilitate quick retrieval of information.

### 11B.1   DESCRIPTIONS

The chief objective of a description is to aid subsequent recognition of the taxon involved. It was realized at an early date that different kinds of descriptions approach this goal in a different manner. Linnaeus distinguished clearly between the general *descriptio* (character naturalis) on one hand and the polynominal *differentia specifica* (character essentialis) on the other (Svenson, 1945). The latter contained "the essential characters by which the species is distinguished from its congeners." It corresponds to what is nowadays called a diagnosis.

The functions of the two kinds of descriptions, the general description and the diagnosis, are by no means identical. The diagnosis serves to distinguish the species (or whatever taxon is involved) from other known similar or closely related ones. The general description has a broader function. It should present a general picture of the described taxon. It should give information not only on characters that are diagnostic with relation to previously described species, but also characters that may distinguish the species from yet unknown species. It should also provide information that may be of interest to others besides taxonomists.

Linnaeus and many taxonomists since his time have stressed the extreme practical importance of a short, unambiguous diagnosis. It can only rarely be combined successfully with the general description. The latter, in turn, no matter how exhaustive it is, cannot always provide a substitute for a type specimen (see 13.48) or, in many cases, for illustrations.

There is still considerable confusion in the literature concerning the meaning and usage of the terms description and diagnosis. In describing animals the taxonomist should achieve two objectives, that of diagnosis and that of delimitation. Diagnosis[1] is the art and practice of distinguishing between things; delimitation is the art and practice of setting limits to things. "Both enter into taxonomy . . . they are essentially different and their complementary roles should be clearly understood" (Simpson, 1945). Although the formal diagnosis in taxonomic work sometimes assists in the delimitation of a taxon, this function is mainly performed by the general description. The two terms, diagnosis and description, may then be used as follows:

*Description.* A more or less complete statement of the characters of a taxon without special emphasis on those characters that distinguish it from coordinate units.

*Diagnosis.* A brief listing of the most important characters or character combinations that are peculiar to the given taxon and by which it can be differentiated from other similar or closely related ones.

The direct comparison of a species (or other taxon) with other specifically mentioned species (or other taxa) is usually called a *differential diagnosis.* Such a comparison with other species is of great practical help to students who have no material of the newly described form. It also forces the author of a new form to review all the evidence for and against the establishment of the new taxon (Rensch, 1934) and thus ensures that the diagnostic characters of the new form are mentioned. If the nearest relatives are rare or poorly known, it is also helpful to make a comparison with a well-known, if not so closely related, species (13.19).

**11B.1.1   Original Description.** The description published at the time of proposal of a name for a new species, genus, or other taxon is called the original description. It has two primary functions. The first, as stated above, is to facilitate subsequent recognition and identification; the second is to make the new name available by fulfilling the requirements of Arts. 11–16 of the Code (13.16).

---

[1] Ultimately from the Greek διαγιγνωσκειν, to distinguish between two (things).

The importance of preparing a proper description cannot be over-emphasized. The describer is forced to rely on words to convey his meaning. Yet words, no matter how carefully chosen, are rarely adequate to give an accurate and complete mental picture of the appearance of an organism. Nevertheless the description should enable a subsequent worker to identify specimens without reference to the type. In most cases this goal can be achieved by the careful worker, particularly when the description is properly coordinated with illustrative material.

A good description requires on the part of its author (1) a thorough knowledge of the group of organisms concerned, (2) a knowledge of structure and terminology, (3) an ability to evaluate differences and similarities, (4) an ability to select and emphasize the important, (5) a full understanding of the precise meaning of the words and the correct usage of the grammar of the language employed, and (6) a concern for the future worker. Ferris (1928) stated, "If [the describer's] work of recording the data has been properly done those data are available for re-examination and re-evaluation. His conclusions can be checked, they can be extended or modified or rejected as appears desirable, all without the necessity of recourse to his types."

In the less well-known groups much of the taxonomist's time is spent in comparing and contrasting one description with another. This task is difficult under any circumstances, but it is easier when the descriptions approximate one another in style, arrangement, and form. This does not mean that a completely standardized description is always possible, because the factors which influence the order of presentation, form, and style vary from group to group. Within a particular group, however, much can be done to standardize descriptions and thus increase their effectiveness and utility. (See also 11B.1.3.)

**11B.1.2  Style.** The style generally used in descriptions as well as in diagnoses is telegraphic and concise. It is usually characterized by the elimination of articles and verbs and by the selection of adjectives and nouns of explicit meaning. It further involves proper use of capitals and punctuation and adherence to a logical sequence of presentation. Thus the telephonic-style statement, "The head is one-third longer than it is wide, the antennae are shorter than the body, and the outer segments are serrate" becomes simply

*Head one-third longer than wide, antennae shorter than body, outer segments serrate.*

The descriptive style of the second statement has lost none of the precision or clarity of the first yet is only one-half as long and may be both read and understood more quickly.

**11B.1.3  Sequence of Characters.** The recommended sequence of characters is different in a diagnosis and a description. In a diagnosis it is customary to present characters in the order of their diagnostic importance (or what the author regards as the order of importance). This will facilitate rapid recognition. In the full description the material should be arranged in a standardized natural order—for instance, describing the body parts from anterior to the posterior, first on the dorsal and then on the ventral surface. The details may be varied to fit the groups yet still maintain a natural and readily comparable order. For instance, the sequence of presentation for a dorsoventrally flattened animal group would be different from that for either a laterally compressed or a radially symmetrical group because of the different methods of orientation during study. The standardized sequence of characters helps ensure that nothing important is being overlooked and that the description is comparative. It is very frustrating to try to use a taxonomic paper in which half a dozen species are described independently of one another, details being given, for example, of the antennae of one species, the pronotum of a second, and the elytra of a third. Such a procedure makes comparison quite impossible. Authoritative monographs usually adopt a standardized sequence of characters, and subsequent describers should follow it as far as possible.

The utility of a description may be increased by the use of devices enabling the reader to locate quickly the particular characters he is looking for. One such device is the use of paragraphs to break up the description according to main body divisions (e.g., in insects: head, thorax, abdomen, wings, genitalia, etc.). Where paragraphing is undesirable, the same effect may be gained by italicizing these same key words. If the author has followed a natural sequence of presentation, either method will permit the reader to orient himself quickly at some particular point in the description without having to read the whole description.

*Dictated Descriptions.* When preparing revisions, most modern taxonomists dictate the descriptions into an automatic recording device, in order to speed up the tedious task of describing. A typist can subsequently transcribe the dictation. When both hands are needed for viewing specimens through the microscope, dictating equipment with a foot pedal and a desk microphone is used. The describer has before him a checklist of characters in standardized sequence which he follows when dictating the descriptions of each taxon. The time saved by this method is considerable.

**11B.1.4  What to Include in a Description.** An exhaustive description of an organism would fill many volumes, as evidenced by works on the morphology (physical anthropology) and anatomy of the human species. It is therefore obvious that even the so-called "detailed description" of a taxonomic species is highly selective and in the nature of an expanded diagnosis. Furthermore, in these days of steadily increasing printing costs it becomes increasingly important not to include triva. How much subject matter should be included in a description depends on the group concerned and the state of knowledge of that group. Excessively long descriptions obscure the essential points; excessively short ones omit pertinent data. While the diagnosis serves to distinguish a species from other known species, the description should be detailed enough to anticipate possible differences from as yet undescribed species. The description should therefore be more detailed in poorly known groups, because it is impossible to predict which characters will distinguish a new species from those that are still undiscovered. It should pay particular attention to character complexes that are variable in the given group. On the other hand, subspecies in a well-known species of birds may differ from one another in so few characters that an extensive description would be a repetition of the species description. In such a case the description may not differ at all from a diagnosis—for instance, "Like subspecies *alba* but larger, upper parts blackish gray, not ash gray" (followed by a tabulation of the measurements).

As far as practicable, descriptions should include all characters, both positive and negative, which are known to be useful or potentially useful in distinguishing other taxa at the same categorical level. However, characters of higher categories should be omitted except where they are anomalous or where the rank of the taxon is in doubt. For example, the description of a subspecies of Song Sparrow should not include reference to characters that are typical for all Song Sparrows (or worse, for all sparrows!). Violation of this rule is not only uneconomical but distracts attention from the essential features of the taxon concerned.

Beyond the above generalizations, there is little to guide the describer other than his own good judgment. More than almost any other aspect of taxonomy the description provides a permanent record of the author's ability to observe accurately, record precisely, select and interpret intelligently, and express clearly and concisely the facts which are before him (see also Cain, 1959*b*).

The description should include a statement of the differences between the sexes and, if only one sex is available, a frank statement of that fact (e.g., "female unknown"). Likewise the characters of immaturity should be discussed, as well as larval stages. Available behavioral, ecological and other biological data should be presented. In the case of sibling species, such information is often more important than morphological characters.

Whether or not the description should be based exclusively on the type is a much-disputed point. Those who favor this method argue that all too often it has eventually turned out that the original material—and consequently also the description—was a composite of several species, which makes it very difficult to disentangle the characters of the various species. They argue that it is much safer to restrict the description to the type and have it followed by a discussion of the variability of the rest of the material.

Others believe that such treatment favors the erroneous typological view that the type has a special significance as far as the characters of the species are concerned. They prefer the description to be a composite drawn from a consideration of the entire material (Simpson's hypodigm) and propose to mention at the end by what characters (if any) the type specimen differs from the rest of the material.

Actually both methods agree that (1) the entire variability of the species material should be described, and (2) it is advisable to mention the special features of the holotype. Different authors may use different methods to achieve these objectives.

**11B.1.5  Description of Coloration.** Differences in coloration are among the most important diagnostic characters in many groups of animals. A detailed description of the general pattern of coloration and of the precise tone of the various colors is therefore essential in many taxonomic groups. Subspecific differences in birds, mammals, and butterflies are often largely a matter of coloration. Many attempts have therefore been made to standardize color descriptions, since words like rufous or tawny do not necessarily suggest the same shade of color to every taxonomist. It is for this reason that color keys are widely used in taxonomy. Those of Ridgway (1912), Maerz and Paul (1950), and Villalobos-Dominguez and Villalobos (1947) are specially recommended. When fine shades of color are involved, a direct comparison with topotypical material is advisable. Measuring devices (spectrophotometers) give objectivity and standardization in such comparisons and permit quantification, as needed for statistical evaluation (Lubnow and Niethammer, 1964; Selander et al., 1965).

**11B.1.6  Numerical Data.** The recording of a set of precise measurements is an integral part of a well-rounded description. If the new form differs from its relatives in its proportions, such proportions should also be recorded (see Chaps. 7 and 9). Exact data should be given of numerically variable features of structure or pattern, such as numbers of spots, spines, scales, tail feathers, and so forth. The reasons for including such data are stated in Chap. 9.

**11B.1.7  Descriptive Treatment.** A full descriptive treatment of a species may take the following form:

*Scientific Name and Its Author*

Bibliographic reference to place, date and author of original description
Type (including type locality and repository)
Synonymy (if any) (see below)
Diagnosis and differential diagnosis (brief statement of essential differences
  from nearest relatives, see above)
Description

Measurements and other numerical data
Range (geographical)
Habitat (ecological notes) and horizon (in fossils)
Discussion
List of material examined

Since the type is the name-bearer, it follows immediately after the name (and that of each synonym) and *not* after the description.

In case the species is new, the form of the description is as follows:

*X-us albus, new species*

Type (including statement of type locality and repository)
Diagnosis, etc., as in a redescription

**11B.1.8 Illustrations.** Illustrations are in most instances vastly superior to verbal descriptions. Anything that can be made clearly and sufficiently visible in a picture should be illustrated. The value of illustrations is recognized in the International Rules, since a scientific name given to a published illustration was held to be valid (prior to January 1, 1931) even when not accompanied by a single word of description [Art. 16a(i), (vii)]. Such a naming of illustrations was quite customary in the days of Linnaeus. In our day, however, sound taxonomists always present a diagnosis and full description together with the illustrations. See 11B.4 for a discussion of illustrations.

**11B.1.9 Redescriptions.** The redescription of hitherto poorly described forms is an extremely important element of revisional and other taxonomic work. In the present state of our knowledge of many animal groups, it is of greater importance than the description of new forms. Ferris (1928), in commenting on this phase of systematic entomology, has stated that:

> . . . a distressingly large percentage of the named species, in almost every group of the insects, cannot be recognized positively or even at all, on the basis of the existing literature. It is more important, for the advancement of our study, to redescribe such forms than it is to describe new species. The redescription of such forms should be regarded by the student as an essential part of his work upon any group which he may elect to study. The fact that a species has been named should make no essential difference in the way in which it is treated. . . . The proper aim is not to name species but to know them. The writer who contributes to the genuine knowledge of species is accomplishing far more than one who merely names them. The fact that the author's name accompanies the names of the new species which he describes should not be allowed to influence his activities.

With this view the author heartily concurs.

The specimen or specimens on which a redescription or illustration are based should be clearly indicated because, in the event that the species has been wrongly identified, a new species may be proposed for *X-us albus* Jones, not Smith. In such a case the type specimen of the new species is the specimen (or is selected from the specimens) on which the redescription or illustration was based.

If a good description, correctly and adequately stated, is readily available in the literature, it is wasteful to republish copies of it again and again. A bibliographic reference is sufficient.

**11B.1.10  Description of a Higher Taxon.** Much of what has been said about the description of species taxa is equally true for higher taxa. However, the description of a new higher taxon traditionally stresses diagnostic features. Citation of the type species (in case of a genus) or of the type genus (in case of a family) reduces the amount of descriptive material that needs to be included.

In the case of higher taxa of vertebrates it is good practice to utilize characters of the skeleton that would be diagnostic in fossils. In mollusks and those other groups of invertebrates that have a fossil record, the modifications of the hard parts, used for diagnosis, ar usually the same in Recent forms and in fossils.

The form of presentation is as follows:

*X-us, new genus*

Type species
Diagnosis, description, list of included lower taxa, discussion

**11B.1.11  Summary.** Recommendations on the preparation of descriptions may be summarized as follows:

1. The taxonomic characters should be treated in a standardized sequence.
2. The most easily visible characters should be featured.
3. A direct diagnostic comparsion with the nearest relative or relatives should supplement the description.
4. Since words alone can seldom give an adequate picture of the diagnostic characters of a taxon, appropriate illustrations should be provided whenever possible.
5. The description should provide quantitative data, supplemented with information on geographical range, ecology, habits, etc.
6. Species in poorly known genera should be fully described.
7. The formal description should be followed by an informal discussion of the variable characters.
8. The description should be accompanied by full information on the type specimen and other material before the author.
9. Characters that are common to all members of the next higher category should be omitted from the description.

## 11B.2  SYNONYMY

Different names given to the same taxon are synonyms (13.23). In the early taxonomic history of any group of organisms the correct establishment of synonymies is perhaps the most important task. All other tasks (such as the elaboration of a classification and the preparation of keys) depend on the correctness and completeness of the synonymies. A complete synonymy of every species and genus is therefore a necessity when a higher taxon is monographed or revised for the first time or because the previous treatment has become obsolete.

Unfortunately, the exhaustive preparation not only of a complete synonymy but also of a listing of all references to previous publications with all possible binominal combinations (in case of generic transfers) has become the misplaced ideal of scholarship for some taxonomists. The expense of printing endlessly repeated names and references (often a separate line for each reference!) is altogether out of proportion with the benefits conveyed. Friedmann (1955), illustrated in Part II of Ridgway's *Birds of North America,* what this system leads to in a taxonomic group with a rich literature. Other examples are provided by monographs on mammals and fishes.

It has now become customary in the better-known groups of animals to list in synonymies only names that were not at all or not correctly listed in the previous standard treatments. For example, in Peters' *Checklist of the Birds of the World* (1931 *et seq.*) synonyms are not listed that can be found in previous standard works, such as *Catalogue of Birds of the British Museum* (1873–1892), the *Handlist of Birds* (1896–1910), for American species Hellmayr's *Catalogue* (1918–1944), and for Palearctic species Vaurie (1959–1965). More recent checklists do not repeat synonyms correctly cited in Peters. Only genuine synonyms are listed, not mere new combinations. There is perhaps no other section of a taxonomic paper in which more economies can be achieved than in the so-called synonymies.

There are many groups of animals, particularly insects, in which only alphabetical lists of nominal species exist at the present time. Obviously the first revision in such a group must give a complete synonymy. However, it is not necessary even in poorly known groups to repeat again the synonyms correctly placed in the last previous standard treatment, even if it is otherwise out of date. Since manuscript names (*nomina nuda*) have no existence in nomenclature, their listing in synonymy is only confusing and should be avoided (13.16).

New synonymy is most usefully cited with the following sequence of data: (1) scientific name (in its original form), (2) author, (3) date of publication, (4) reference, (5) type-locality, (6) present location of type (optional). For example,

### Oncideres rhodostictus Bates

*Oncideres rhodosticta* Bates, 1885, *Biol. Cent.-Amer., Coleopt.,* **5**:367. [Lerdo, Mex.; British Mus. (Nat. Hist.)]
*Oncideres trinodatus* Casey, 1913, *Mem. Coleopt.,* **4**:352. [El Paso, Tex.; U.S. Natl. Mus.] New Synonymy.[2]

The above form is recommended for a revision of a well-cataloged group. In poorly known groups, as described above, a full synonymy (i.e., a list of scientific names, incorrect and correct) may be required. This should include all references which have nomenclatural or zoological significance, arranged chronologically under the actual name (correct or incorrect) by which the author actually referred to them.

Many authors use the convenient device of a comma inserted between the specific name and the author [*X-us albus,* Smith (not Brown)] to distinguish between a misidentification, which has no nomenclatural status, and a homonym [*X-us albus* Jones (not Brown)], which has. A full bibliographical synonymy, as used in a first revision of a poorly known genus, would appear as follows:

### Oncideres rhodostictus Bates

*Oncideres rhodosticta* Bates, 1885, *Biol. Cent.-Amer., Coleopt.,* **5**:367 [type: Lerdo, Mex.; British Mus. (Nat. Hist.)]; Linsley, 1940, *J. Econ. Entomol.,* **33**:562 synon., distr.); Linsley, 1942, *Proc. Calif. Acad. Sci.,* (4) **24**:76 (distr.); Dillon and Dillon, 1945, *Sci. Pub. Reading Mus.,* no. 5:xv (key); Dillon and Dillon, 1946, *loc. cit.,* **6**:313, 382 (revis.).
*Oncideres putator,* Horn (not Thomson, 1868), 1885, *Trans. Amer. Ent. Soc.,* **12**:195 (key, distr.); Schaeffer, 1906, *Can. Ent.,* **38**:19 (key).
*Oncideres cingulatus,* Hamilton (in part) (not Say, 1826), 1896, *Trans. Amer. Ent. Soc.,* **23**:141 (distr.).
*Oncideres trinodatus* Casey, 1913, *Mem. Coleopt.,* **4**:352 [type: El Pasco, Tex.; U.S. Natl. Mus.].
*Oncideres* sp., Craighead, 1923, *Can. Dept. Agr. Bull.* 17 (n.s.), p. 132 (larva, hosts).
*Oncideres pustulatus,* Essig (not Le Conte, 1854), 1926, *Insects of Western North America,* p. 460, fig. 368 (habits, distr.).

Note that type-locality and location of type are recorded for all genuine synonyms.

The above synonymy might appear in an abbreviated checklist as follows:

*Oncideres* Serville, 1835

1. *rhodostictus* Bates, 1885        So. Calif. to Tex.
   *trinodatus* Casey, 1913          No. Mex.
                                     L. Calif.

[2] This synonymy was published as new in *J. Econ. Entomol.,* 33:562, 1940. Its use as an example here and elsewhere in the present discussion is not to be interpreted as a nomenclatural change. The words NEW SYNONYMY are usually printed in caps or small caps to draw attention.

When a checklist contains a terminal bibliography, its usefulness may be increased by giving page references which may then be located by author, date, and page in the bibliography—thus, *rhodostictus* Bates, 1885:367. If it is desirable to indicate the various combinations under which each name has appeared, this may be accomplished by taking the oldest specific name and following it through its various combinations, then the next oldest, etc., as follows:

*Megacyllene antennata* (White)

*Clytus antennatus* White, 1855, *Cat. Coleopt. Brit. Mus.*, 8:252 [type: "W coast of America"; British Mus. (Nat. Hist.)].
*Cyllene antennatus*, Horn, 1880, *Trans. Amer. Ent. Soc.*, 8:135 (descr., syn., distr.); Craighead, 1923, *Can. Dept. Agr., Bull.* 27, p. 33 (larva, biol.); Hopping, 1937, *Ann. Ent. Soc Amer.*, 30:411, pl. 1 (revis.).
*Megacyllene antennata*, Casey, 1912, *Mem. Coleopt.*, 3:348, 351 (descr.).
*Arhopalus eurystethus* LeConte, 1858, *Proc. Acad. Nat. Sci. Phila.*, 1858:82 [type: Sonora, Mex.; Mus. Comp. Zool., Harvard]; LeConte, 1859, in Thomson, *Arcana Naturae*, p. 127, pl. 13, fig. 9.

In the above example, the comma between the specific combination and the author's name has again been used, this time to distinguish between a new combination (*Cyllene antennatus*, Horn, 1880) and an original combination (*Clytus antennatus* White, 1855).

*Generic synonymy* is handled in much the same way as a specific synonymy, except that in the case of new synonymy or full bibliographic treatment, the generic type (and its designator, if any) is cited in place of the type locality and type location. The synonymy of the genus *Dicrurus*, as cited in Vaurie's (1949) revision of the Dicruridae, may be listed as an example.

*Dicrurus* Vieillot, April 14, 1816, Analyse d'une nouvelle ornithologie élémentaire, p. 41. Type, by subsequent designation, *Corvus balicassius* Linnaeus (G. R. Gray, 1841, A list of the genera of birds, ed. 2, p. 47).
*Edolius* Cuvier, Dec. 7, 1816, Le règne animal, vol. 1, p. 350. Type, by subsequent designation, *Lanius forficatus* Linnaeus (G. R. Gray, 1855, Catalogue of the genera and subgenera of birds, p. 58).
*Drongo* Tickell, 1833, *Jour. Asiatic Soc. Bengal*, vol. 2, p. 573. Type, by monotypy, *Drongo caerulescens* Tickell = *Lanius caerulescens* Linnaeus.
*Chibia* Hodgson, 1836, *India Rev.*, vol. 1, p. 324. Type, by subsequent designation, *Edolius barbatus* J. E. Gray = *Corvus hottentottus* Linnaeus (G. R. Gray, 1841, A list of the genera of birds, ed. 2, p. 47).
*Bhringa* Hodgson, 1836, *India Rev.*, vol. 1, p. 325. Type, by original designation and monotypy, *Bhringa tectirostris* Hodgson.
*Bhuchanga* Hodgson, 1836, *India Rev.*, vol. 1, p. 326. Type, by subsequent designation, *Bhuchanga albirictus* Hodgson (Sharpe, 1877, Catalogue of birds in the British Museum, vol. 3, p. 245).
*Chaptia* Hodgson, 1836, *India Rev.* vol. 1, p. 326. Type, by monotypy, *Chaptia muscipetoides* Hodgson = *Dicrurus aeneus* Vieillot.
*Dissemurus* Gloger, 1841, Gemeinnütziges Hand- und Hilfsbuch der Naturgeschichte, p. 347. Type, by monotypy, *Cuculus paradiseus* Linnaeus.

*Musicus* Reichenbach, 1850, Avium systema naturale, pl. 88, fig. 9. Figure of generic details, no species included, cf. Bonaparte, 1854, *Compt. Rend. Acad. Sci. Paris,* vol. 38, p. 450. Type, by tautonymy, *Dicrurus musicus* Vieillot = *Corvus adsimilis* Bechstein.

*Dicranostreptus* Reichenbach, 1850, Avium systema naturale, pl. 88, fig. 12. Figure of generic details, no species included. Type, by subsequent designation, *Edolius megarhynchus* Quoy and Gaimard (G. R. Gray, 1855, Catalogue of the genera and subgenera of birds, p. 58).

## 11B.3  KEYS

The purpose of a key is to facilitate identification of a specimen (4.3.1 and 7.3). This goal is achieved by presenting appropriate diagnostic characters in a series of alternative choices. The worker finds the correct name of his specimen by making the appropriate choice in a series of consecutive steps.

The procedure involved is somewhat analogous to that of the physician who, by means of a series of questions and examinations, arrives, by a process of elimination and confirmation, at the diagnosis of the ills of a patient, or to the elimination method in culture identification of bacteria.

The use of keys in identification is old indeed. Much of Aristotle's classification of animals was presented in the form of simple dichotomous alternatives ("bloodless versus with blood," etc.). Voss (1952) gives an interesting history of the development of keys in systematic biology. Metcalf (1954) provides some hints on constructing keys. In the Aristotelian procedure keys were an instrument of classifying logic. Typologically an object was either *A* or not-*A*. To use a key typologically for classifying purposes is misleading in view of the polythetic nature of many taxa. On the other hand, if a key is used as a purely pragmatic device, the taxonomist can cope with the existence of a polythetic taxon quite easily, by keying it out repeatedly. This would be fatal for a classification but does not constitute weakness in an identification procedure.

Keys are also a tool for taxonomic analysis since in their preparation one must select, evaluate, and arrange taxonomic characters. In this sense keys are an integral part of taxonomic procedure, as well as a means of presenting findings.

The construction of keys is a laborious and time-consuming task, involving the selection and sifting of the most useful and most clearly diagnostic characters. Ideal key characters apply equally to all individuals of the population (regardless of age and sex); are absolute (two scutellar bristles versus one scutellar bristle); are external, so that they can be observed directly and without special equipment; and are relatively constant (without excessive individual variation). Unsuitable key characters include those that require a knowledge of all ages and stages of a species (e.g., "sexual dimorphism present" versus "sexual dimorphism absent"; "male larger than female" versus "male smaller than female"; "fall molt complete" versus "fall molt partial"), relative characters without absolute standard (e.g., "darker" versus "lighter," "larger" versus "smaller"), and overlapping characters ("larger, wing 152 to 162" versus "smaller, wing 148 to 158").

In most cases the data will permit the choice of several characters for the various primary and secondary divisions of the key. It is here that the writer is called upon to exercise his best judgment in order to select the most satisfactory characters at the various levels. If he is torn between a phylogenetic and utilitarian approach to the problem, he should remember that the primary purpose of a key is utilitarian; diagrams, lists, numbers, or order of subsequent treatment will take care of phylogeny. However, when making a key in a poorly known group (with many undescribed species) it is useful to arrange the key in such a manner that closely related species key out near one another. This facilitates the subsequent insertion of new species, as well as the decision whether or not a species is new. It is fortunate when the material permits the construction of a key which allows the presentation of relationship without interfering with the main function, that of ensuring identification. Key construction is a procedure for which computer programs are now available. This is one of the many possible uses of the computer in taxonomy (Kim et al., 1966).

A good key is strictly dichotomous, not offering more than two alternatives at any point.[3] Alternatives should be precise. Ideally the statements should be sufficiently definite to permit identification of a single specimen without reference to other species. In any event, identification should be possible without reference to the opposite sex or to immature stages. These should be treated in different keys when dimorphism is exhibited. In case a key character neatly separates the species of a genus into two groups, except for one or two species that are intermediate or variable, it is quite legitimate to include these variable species in both subdivisions. A given species name may thus appear repeatedly in a key. The procedure which gives the quickest and most unambiguous identification should be adopted. Ordinarily new species should not be designated as such in a key. Also, it is usually customary to omit authorities from specific names in keys unless these are not mentioned elsewhere in the article.

The style of keys is telegraphic, like that of descriptions, and the phrases are usually separated by semicolons. Even though the primary contrasting characters of each couplet may be diagnostic and definitive, supplemental characters are desirable in the event that the primary character may not be clearly discerned or the specimen may be injured or mounted in an unsatisfactory manner. One of the most satisfactory methods for assembling data for the construction of a key is shown in an example

[3] If it is impossible to work out a key that permits the identification of all species, it is advisable to indicate this clearly and to key out as groups any species that cannot be diagnosed by key characters.

TABLE 11-1.  ARRANGEMENT OF KEY CHARACTERS*

| Name of species | Wings | Antennae | Antennal color | Eyes | Tarsal segments | Leg color |
|---|---|---|---|---|---|---|
| smithi | clear | filiform | black | entire | linear | black |
| completa | opaque | serrate | black | entire | linear | black |
| emarginata | opaque | serrate | black | emarginate | linear | black |
| rufipes | opaque | filiform | black | entire | linear | red |
| nigripes | opaque | filiform | black | entire | linear | black |
| flavicornis | clear | filiform | yellow | entire | bilobed | black |
| ruficornis | clear | filiform | red | entire | linear | black |
| californica | clear | filiform | black | entire | bilobed | black |

* Characters used in examples are italicized.

of the method and the subsequent analysis given in Table 11-1. This example is oversimplified in order to demonstrate the method more clearly.

Several types of key are used in taxonomic papers, all being dichotomous and based on a series of choices. By far the most commonly used is the dichotomous-bracket key. The other is the indented key. The latter has the advantage that the relationship of the various divisions is apparent to the eye. It has the disadvantages, especially in a long key, that the alternatives may be widely separated and that it is wasteful of space. For these reasons it is generally used only for short keys, keys to higher taxa, or comparative keys (keys which not only serve the purposes of identification but also treat the same comparative characters at each level for each group). An *indented key* based on the hypothetical data given in Table 11-1 might be as follows:

    *A.* Wings opaque
        *B.* Antennae serrate
            *C.* Eyes entire.................................... *completa*
            *CC.* Eyes emarginate.............................. *emarginata*
        *BB.* Antennae filiform
            *C.* Legs red..................................... *rufipes*
            *CC.* Legs black.................................. *nigripes*
    *AA.* Wings clear
        *B.* Tarsal segments linear
            *C.* Antennae black............................. *smithi*
            *CC.* Antennae red............................... *ruficornis*
        *BB.* Tarsal segments bilobed
            *C.* Antennae black............................. *californica*
            *CC.* Antennae yellow............................ *flavicornis*

The second type of key, which is used almost exclusively by most taxonomists, is the *bracket key*. This key has the advantage that the couplets are composed of alternatives which are side by side for ready comparison, and that it is more economical of space because it is not indented. When properly constructed it may be run forward or backward with equal facility by following the numbers, which indicate the path that the various choices follow. This is the type which best fulfills the diagnostic purpose of a key. Its main disadvantage is that the relationship of the divisions is not apparent to the eye. An example based on the same data previously used is as follows:

| | | | |
|---|---|---|---|
| 1. | | Wings opaque | 2 |
| | | Wings clear | 5 |
| 2. | (1) | Antennae serrate | 3 |
| | | Antennae filiform | 4 |
| 3. | (2) | Eyes entire | *completa* |
| | | Eyes emarginate | *emarginata* |
| 4. | (2) | Legs red | *rufipes* |
| | | Legs black | *nigripes* |
| 5. | (1) | Tarsal segments linear | 6 |
| | | Tarsal segments bilobed | 7 |
| 6. | (5) | Antennae black | *smithi* |
| | | Antennae red | *ruficornis* |
| 7. | (5) | Antennae black | *californica* |
| | | Antennae yellow | *flavicornis* |

*Pictorial keys* deserve mention among various other kinds of keys designed for special purposes. The pictorial key is of value for field identification by nonscientists. During World War II, for example, malaria crews based their control operations on the results of field identifications of anopheline mosquito larvae (Fig. 11-1). The fact that critical characters were illustrated as well as described made the keys usable by such persons as medical corpsmen and engineers as well as by entomologists. Other examples of pictorial keys are one by Corliss (1959) of the higher taxa of Ciliates and one of the Rotatorian genus *Ptygura* by Edmondson (1949). Pictorial keys have been employed also in field guides to vertebrates and flowering plants.

Those who employ them must remember at all times the utilitarian purpose of keys and of the characters used in them. The character on which a given taxon "keys out" may be of no particular biological or phylogenetic significance for this taxon. It is simply the character which gives the best assurance for correct identification. Keys are not phylogenies.

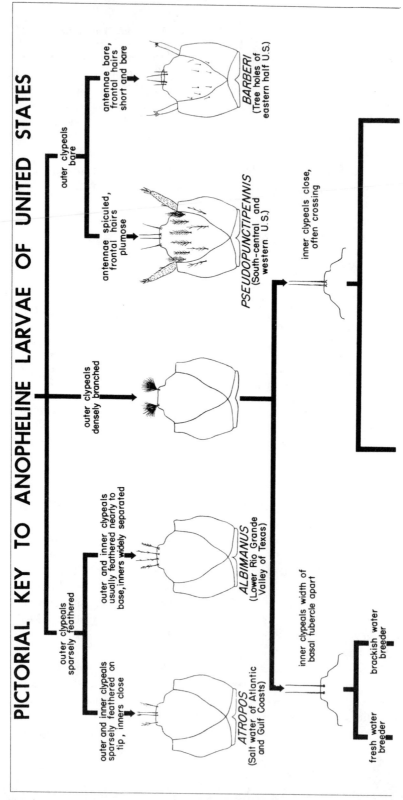

# PICTORIAL KEY TO ANOPHELINE LARVAE OF UNITED STATES

outer clypeals bare

antennae bare, frontal hairs short and bare

*BARBERI*
(Tree holes of eastern half U.S.)

antennae spiculed, frontal hairs plumose

*PSEUDOPUNCTIPENNIS*
(South-central and western U.S.)

inner clypeals close, often crossing

outer clypeals densely branched

outer clypeals sparsely feathered

outer and inner clypeals usually feathered nearly to base, inners widely separated

*ALBIMANUS*
(Lower Rio Grande Valley of Texas)

outer and inner clypeals sparsely feathered on tip, inners close

*ATROPOS*
(Salt water of Atlantic and Gulf Coasts)

inner clypeals width of basal tubercle apart

brackish water breeder

fresh water breeder

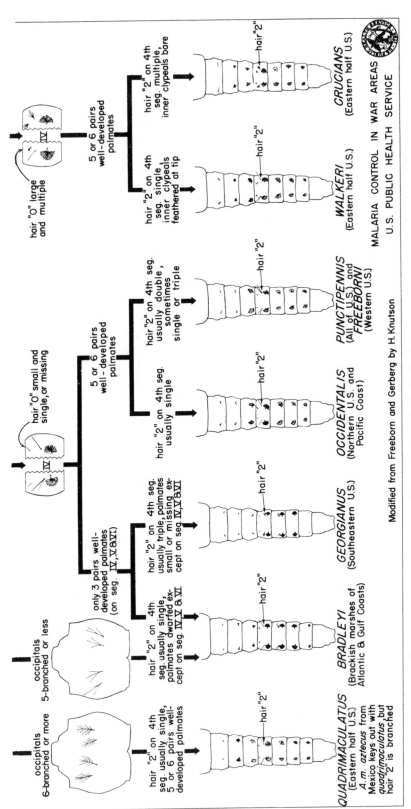

Fig. 11-1. Pictorial key to larve of anopheline mosquitoes of the United States.

### 11B.4 ILLUSTRATIONS

The old Chinese proverb says: One picture is worth a thousand words. This is all too true for illustrations in taxonomic papers. It is often quite impossible to describe adequately in words a complex structure, let us say the genital armatures of an insect or the palps of a male spider. Descriptions without illustrations are of rather limited value for many organisms.

The taxonomist must decide in each case what kind of illustration would be most useful. Except for special purposes an accurate but somewhat diagrammatical line drawing is usually the most informative illustration for a taxonomic publication. A scientist is fortunate if he happens to be endowed with talent as an artist. However, the scientist who lacks it need not be discouraged, because clear-cut diagrammatic drawings are perfectly satisfactory and, in some cases, superior to artistic drawings for scientific purposes. Ferris (1928) called this type of drawing drafting and expressed the opinion that any conscientious scientist could learn to make satisfactory drawings of this kind. Several books or manuals have been published on the subject, among which may be mentioned Kuhl (1949), Staniland (1953), Papp (1968), and particularly Zweifel (1961).

Pencil sketches should be made with a soft pencil, and bilaterally symmetrical animals should be "corrected" for symmetry by tracing one half on the other with thin semitransparent paper.

The original outline may be sketched freehand, but, at least with microscopically small organisms, it can be done more quickly and accurately by one or another mechanical means. Perhaps the most popular of these devices is the camera lucida, which, by means of prisms and a mirror, projects the microscope image on a piece of paper. With this apparatus it is possible to look in the microscope and see the specimen superimposed on the reflection of the paper. By careful adjustment of the light, one can draw an outline with specimen and pencil point both clearly in view (superimposed on each other). Another method of obtaining the outline is the direct projection of an image on a screen or paper by a microprojector attached to the microscope. Still another technique is to photograph the specimen and print an enlargement on a dull mat paper. The outline can be inked directly on the photograph, after which the photographic emulsion can be washed off. Some workers prefer to sketch freehand on a crosshatched paper, guided by a grid in the ocular of the microscope.

The type of pencils, crayons, pens, and papers suitable for various drawings are described in the cited literature. Better stationery and artists' supply stores can give advice on the suitability of various kinds of materials.

Maps, often an important kind of illustration for taxonomic papers, are best done by using available outline maps (base maps) on which transparent paper cutouts are pasted with dots, letters, figures, and crosshatchings of various kinds and sizes.

**11B.4.1    Reproduction.** In the preparation of all figures, bear in mind that their size is usually reduced (sometimes drastically) in the process of reproduction for publication. An experienced author knows that all symbols and letters will have to be sufficiently large in the original to remain easily legible after reduction. Editors and some authors have reducing lenses which show them how the illustration looks after reduction.

In the layout of the illustration due consideration must be given to format and size of type bed of the publication (book or journal) for which the illustration is intended. Proper proportions for the original drawing may be obtained by expanding on a diagonal line through a rectangle drawn to page or column size (Fig. 11-2).

Leave room at the bottom of the page for the legend. Figure numbers, letters, abbreviations, etc. should be put on neatly. In order to be legible, letters should be $1\frac{1}{2}$, $2 \times \frac{1}{16}$ or $\frac{1}{32}$ in. high, depending on the amount of reduction. Freehand lettering is rarely satisfactory. Numbers and letters may be clipped from old calendars or from standard characters printed on gummed paper, or they may be made by various mechanical lettering guides. When determining the amount of reduction the editor is limited by the size of the page, the need for captions, and other considerations, and he is often not in a position to follow the instructions of the author. This is particularly true in the case of large figures. Magnification or reduction should therefore not be stated on the figures themselves, but rather in the captions.

Where many illustrations are used, grouping is often required for economical reproduction. For zinc etchings, drawings may be assembled into plates by merely arranging and pasting on a cardboard sheet. Colorless paste or rubber cement should be used. The paper edges of individual drawings will not show. For halftones, however, trimmed edges will show, and when several drawings are to be fitted together for a plate, a mechanical paper-cutter should be used for trimming. Slight discolorations, especially when yellowish, also become conspicuous in reproduction. It is usually more satisfactory to draw numbers and letters directly on the original rather than paste them on. However, characters printed on transparent gummed paper are also available for halftone. Photographs should be mounted with smooth edges touching and symmetrical, so that the engraver can rout out neat, straight lines.

Curves and graphs are reproduced as zinc etchings and largely used as text figures. In preparing them for publication, the same instructions as to size, proportions, and lettering apply as for drawings. However,

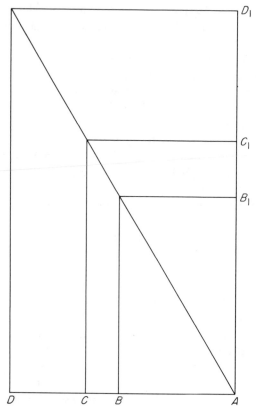

Fig. 11-2. Method of calculating proportions for enlargement or reduction of illustrations. If one side *AC* of an
illustration is enlarged to *AD* or reduced to *AB*, the
corresponding length of the other side (*AD'* or *AB'*) can
be easily determined as the point of intersection with the
diagonal.

they should be made either on white paper or blue-lined coordinate paper,
never on green-, black-, red-, or yellow-lined paper unless the coordinates
are to be reproduced.

Line drawings are usually reproduced as zinc etchings. The illustration
is photographed on a zinc plate and the background is etched away with
acid. This is the least expensive method of reproducing illustrations and
the most satisfactory for line drawings, graphs, charts, etc.

Where fine detail is needed, also the representation of delicate shades
of light and dark, as in the reproduction of most photographs, so-called
halftones have to be prepared. Here the picture is broken up into thousands
of tiny dots through a process of screening. In photolithography the illustration is photographed on gelatin and not screened, thus becoming a fulltone
process. It therefore shows even greater detail than a halftone but lacks

some of the contrast. Halftones and photogelatin plates are often printed on glossy paper or at a different printing establishment, and for this reason it is frequently easiest to assemble them together at the end of an article. Zinc etchings are usually printed on the same paper as the text and may therefore more readily be distributed through the article as plates or text figures. The latter are most economical when printed the same width as the printed page.

*Photography* for scientific publication requires special approaches and setups. Hints on equipment and lighting are given by Blaker (1965).

Colored illustrations are the most complicated and most expensive to reproduce. A screen similar to halftone screens is used, and several separate colors are used in printing, each one superimposed on the previous impressions.

Some scientific journals charge the author for copper halftone plates and for the glossy paper if halftones are to be printed in a journal which normally employs a rough eggshell-surface paper. This extra cost may include not only the paper but the hand labor involved in tipping in or pasting in the extra pages. When a journal makes page charges on a flat-rate basis, no special charges are made for illustrations.

Offset printing obviates most of these difficulties because the entire page, including printed (or typed) matter and line cuts, is photographed on a plate and then rolled onto a second roller before it is printed on the final paper. Photographs are made separately because of differences in contrast and then "stripped in" on the negative of the photolithographed page. By this method illustrations cost no more than printed matter.

It is a wise precaution to retain good, clear, photographic copies of all illustrations in the event that they are lost in the mail. A good photograph of a drawing is only slightly inferior to an original as a basis for reproduction.

## 11B.5   REFERENCES AND BIBLIOGRAPHY

References in taxonomic papers are generally treated either in footnotes, or in parentheses in the text, or in a terminal bibliography. Footnotes are useful when only a few references are involved and when repeated reference to the same bibliographical item is unnecessary. However, since they are costly to handle in typesetting and printing and may add materially to publishing costs, parenthetical references are to be preferred. When references are numerous, they are most frequently handled in a terminal bibliography. In most cases this bibliography should be as short as is consistent with its purpose, and the items included should be selected. Frequently the value of the terminal bibliography may be greatly increased by including parenthetical comments on the nature of the subject matter covered. Unverified references may be included when necessary for completeness, but they should be marked with an asterisk or some other device to indicate the fact that the author has not seen them.

Bibliographical items should receive full citation, including author, date, title, publication, volume, pages, etc. The references to the terminal bibliography may be made by enclosing an author's name and date (sometimes also page) in parentheses. Two or more references to publications by a single author in the same year may be designated by appended letters (Smith, 1940a; Smith, 1940b). The author-date system of bibliographical reference is far more satisfactory than the straight numbering system which is sometimes used. The number system tells nothing about the reference; moreover, the author-date system permits the addition of references during the preparation of the manuscript without the necessity of renumbering all references beyond the point of insertion.

The designation "Bibliography" implies completeness of coverage of the subject. "Literature Cited" indicates restriction of references and is self-explanatory. For illustrations, see the terminal bibliography of this book.

An example of footnote style[1] is given here. In the typescript, footnotes are often entered beneath a marginal line in the text wherever they occur, rather than at the bottom of the page, because in the final publication pagination is entirely different from the pagination of the typed pages.

Numerous exceptions will be encountered, especially in various government documents, but a majority of literature citations will fit into one or other of the simple styles illustrated. It is becoming standard practice to list year of publication immediately after the author's name, since this sequence agrees with that of the author-date system of reference.

If the work cited is by several authors, only the first need be reversed for alphabetical purposes. Thus,

McAtee, W. L., and J. R. Malloch. 1922. Changes in names of American Rhynchota, chiefly Emesinae. *Proc. Biol. Soc. Wash.,* 35:95–96.

The original style of capitalization and italics may be followed. However, many titles are set entirely in capitals, others entirely in lowercase, except that the first word and scientific and place names are capitalized. The full title should be given in all but the briefest of footnotes, because readers obtain valuable leads in this way. Abbreviations of journals should follow such standard works as the *World List of Scientific Periodicals.*

The citation: (Wheeler, 1889a) is specific enough because it refers to a two-page paper. On the other hand, it may be necessary to refer to a particular page of a larger work. In this case the page is indicated

[1] W. M. Wheeler, 1889. *Amer. Nat.,* 23:644–645 [*sample footnote*].

in the citation (Wheeler, 1910, p. 263, or **1910**:263), and the complete work is listed in the bibliography.

## 11C.  FORM AND STYLE OF THE TAXONOMIC ARTICLE

Much scientific literature is of a rather ephemeral nature, except for a few classical papers. This is not true in taxonomy, where publications (particularly if they contain new names) are consulted for many generations, indeed back to 1758 and earlier. The taxonomist must try to give his publications a quality which lives up to this timelessness. Innumerable details require attention, and many of them escape the beginner's attention. There are various guides to the preparation of scientific papers (Trelease, 1951; Hurt, 1949), as well as style manuals, such as *A Manual of Style* of the University of Chicago Press, the *U.S. Government Printing Office Style Manual*, the *Style Manual for Biological Journals*, published by the *A.I.B.S.*, and the *American Museum Style Manual*, the latter with particular attention to the practices of the taxonomist. There are also some useful hints in appendix *E* of the International Code (Chap. 12). Before submitting a manuscript to a journal, study the instructions to authors which nearly every journal has and the particular style it prefers for citations, bibliography, etc. This will save a great deal of retyping.

**11C.1  Title.** The title is the first part of the paper encountered by the reader, although it is often the last item to be added in the preparation of the paper. Its bibliographical prominence and significance warrant much care in its selection. The title should be long enough to be specific as to the contents of the paper but brief enough for easy indexing. Short words are preferable to long ones. The most important nouns should be near the beginning of each series of words. The title should contain key words which in indexing will classify the article. This is particularly important today when some abstracting services are based entirely on titles. Punctuation should be avoided unless essential to meaning. Among the essential elements of a title are (1) a clear indication of the field involved (taxonomy, morphology, ecology, etc.), (2) the scientific name of the taxon treated, (3) indications of the order and family either by means of scientific names (which may be in parentheses) or, rarely, by a well-known common name, and (4) the geographical area, fauna, or locality. The following are examples of good titles:

A Taxonomic Revision of American leafhoppers (Homoptera, Cicadellidae)
A Checklist of the Birds of Alabama
Geographical Variation of *Hippodamia convergens* in Southern California (Coleoptera, Coccinellidae)

Two New Species of Wood Rats (*Neotoma*) from the Rocky Mountain Region

The following are a few examples of poor titles for taxonomic papers. On the basis of the above-enumerated principles the objections to these are obvious.

New Hymenoptera
Notes on Mammals
The Western Biota
A Collecting Trip to Texas
Additions to the Fauna of Nebraska
Studies in the Mollusca
A New Acanthiza

Titles need not be as bad as these, however, to cause difficulties for catalogers, abstractors, reviewers, and other bibliographers. No author has cause for complaint that his work is being overlooked if it masquerades under an incomplete, ambiguous, or misleading title.

**11C.2   Author's Name.** The author's name follows the title. Bibliographical problems are simplified if an author always uses the same form of his name. The entomologist Laporte sometimes published under the name Laporte, sometimes as the Comte de Castelnau. The bibliographical confusion which resulted still persists in modern literature. Women taxonomists who begin publication before marriage frequently avoid confusion by continuing to publish under their maiden name or by a system of hyphenation, e.g., Dorothy McKey-Fender. It is customary in America to omit degrees and titles from the author's name, although these are used in many European journals. The author's correct postal address should follow his name in order to facilitate correspondence. The address can also be given at the end of the article or in a footnote.

When more than one author is involved, the order of names is determined by the nature of the contribution each has made. When the work has been more or less equally shared, the names are usually arranged in alphabetical order. When the work has been disproportionately divided or there is a marked discrepancy in age or experience, the name of the author to be considered "senior author" appears first. If the same coauthors publish several papers in a single year, it is advisable to change the sequence of names in order to facilitate ease of reference.

**11C.3   Introduction.** Every taxonomic paper should include an introductory paragraph stating the scope of the paper and, where pertinent, the reasons for the study, as well as the nature of the studied material. Frequently a brief historical review is appropriate. These features serve to orient the casual reader and the new student of the group, as well as to refresh the minds of other workers in the field.

11C.4    **Acknowledgments.** Acknowledgements may be included in the introduction when they can be treated as part of the natural sequence of exposition. Some authors place them in a footnote appended to the author's name. This system is regularly used in the *Annals of the Entomological Society of America* and certain other journals which are primarily taxonomic in content. Sometimes the acknowledgements are placed at the end preceding the summary.

Giving proper credit is one of the most important responsibilities of the scientist. Acknowledgment should be made of all unpublished observations, determinations, and data derived from others. If an unpublished statement is quoted, the author of the statement should be allowed to prepare it especially for that purpose. Previously published data should never be utilized in such a way as to appear original.

Acknowledgment should be made of borrowed and donated specimens studied. The methods of doing so vary with the amount of material received from any one source and the general plan of presentation of the paper. A way can be found for such acknowledgement even in complex cases.

Photographs, drawings, and other illustrative material lent or donated by others should be credited. Credit should also be given to artists and photographers for their work, whether or not they received payment for their efforts. Good drawings or good photographs are scientific contributions on a par with descriptive work and are frequently far more accurate and useful.

Full credit should be given to the collector, who, after all, is the real discoverer of the material and not the describer.

Assistance in outlining a research program (including the help of a major professor or senior colleague) should be acknowledged, as well as help in the preparation of a manuscript by means of critical reading.

Finally acknowledgment should be made of financial grants or of institutional aid, such as the use of laboratory facilities, libraries, etc. Frequently such help is a primary factor in making a particular taxonomic research project possible. Grant-giving agencies usually stipulate that the code number of the grant be recorded in all publications resulting from the grant.

11C.5    **Methods Used and Materials Studied.** In a revisional or monographic work it is desirable to include a statement about methods utilized and collections, specimens, or other materials studied. This enables the reader to evaluate conclusions and to judge the thoroughness of the work. Standard methods for measuring, mounting, staining, special preparations, etc., may be referred to by name and reference. Only new methods need to be described in detail.

11C.6    **Body of the Text.** The material comprising the body of the text will, of course, depend on the scope and objectives of the particular paper. It is perhaps sufficient to mention that a complete systematic paper includes (1) a delimitation of the highest included taxon (family, tribe, etc.), (2) a key (or keys) to all intermediate taxa treated (genera), (3) synonymies and descriptions of the taxa of intermediate rank (genera), (4) statement of the type species, (5) comparison with other genera, (6) keys to the species of each genus, (7) synonymies and descriptions of each species, and (8) statements as to type localities and to location of types, general distribution, hosts and other significant biological data, comparisons with other species, etc. (for details on preparation of descriptions and keys, see 11B.3.

**11C.7  Summary.** A summary is usually unnecessary in a strictly taxonomic paper. If required, it should be brief but not in telegraphic style. It should be written as a series of short paragraphs and should be specific, not written in broad general terms.

## 11D.  PREPARATION OF THE MANUSCRIPT FOR PUBLICATION

Aside from matters involved in the actual organization and construction of a taxonomic paper, there are some points which should be kept in mind in order to facilitate editorial handling after the paper has been submitted for publication. Editors are much more apt to accept readily (and publish quickly) papers which are in good form and require a minimum amount of editing. Most publications have special form requirements, and much editorial time can be saved by careful study of the instructions given in the journal in which the paper is to be published. The *A.I.B.S.* style manual for biological journals is very helpful (see 11C).

**11D.1  Typing.** Manuscripts submitted for publication must be typed. The original drafts may be on yellow paper, but the final copy should be on standard ($8\frac{1}{2}$ by 11 in. or 8 by $10\frac{1}{2}$ in.) white paper, entirely double-spaced (some publications require triple spacing), and with a wide margin for adding proof marks and for editing. If approximately the same number of lines is typed on each page, the editor can conveniently estimate the size of the final printed paper. However, some editors require that pages end with completed paragraphs. Pages should be numbered consecutively in the upper right-hand corner. Inserted pages are numbered alphabetically (e.g., 65$a$). Whole sheets should be used for insertions, regardless of length of inserted matter. When it becomes necessary to cut and rearrange, sheets should be assembled by pasting, not by pinning. Some journals request two copies of each manuscript, the original and one carbon.

All tabular material should be typed on separate sheets, since it is usually set in a different type from the text.

**11D.2  Underlining.** Underlining indicates that the material so marked is to be printed in italics. In a taxonomic manuscript submitted for publication, underlining should be limited to scientific names of genera and species which appear in the text. New names should not be underlined, because the editor will usually mark these with a wavy line to indicate boldface. Indications of style or sizes of type for titles, headings, subheadings, sideheadings, and the like should be left to the editor. In general, marks which the author makes merely interfere with the editor's work, though marginal notes as to the relative rank of headings may be helpful.

**11D.3 Legends and Text Citations to Illustrations.** Titles and legends should be self-explanatory. The manuscript of these titles should be typewritten, double-spaced on separate sheets (several titles on a single sheet), and assembled in numerical order at the end of the manuscript following the bibliography. A short identifying title may be placed on each plate for purposes of identification, but this title will not be printed. Usually in the process of handling, titles and legends go to the typesetter with the rest of the manuscript, whereas illustrations are sent to the engraver. The printer may never see the original drawings.

The place of insertion of the illustrations should be marked in the manuscript and also in the galley proof. Illustrations are usually numbered starting with each article, but some journals number plates consecutively throughout a volume. In any event, a new series of figure numbers or letters should be used on each plate. Many journals designate figures with Arabic numbers, plates with Roman numerals. All figures should be referred to in the text by number.

**11D.4 Revision of the Manuscript.** A few authors have sufficient mastery of the English language so that they can write directly in final form for publication. Other equally competent scientists find it necessary to revise page after page not once but many times. T. D. A. Cockerell was an example of the former type of writer, while by his own testimony Charles Darwin was an inveterate reviser.

Trelease (1951) recommends careful reading of the manuscript ten times, each time for one of the following: (1) consistency, (2) sentences, (3) clarity, (4) repetition, (5) connectives, (6) euphony, (7) punctuation, (8) style, (9) accuracy, (10) length. Authors of taxonomic papers seldom follow the details of this recommendation, but most papers would benefit from more revisions than they usually receive. It often helps to put a manuscript aside for a while before the final revision is made. It is always advisable to have other persons read a manuscript before it is submitted for publication. A fully corrected carbon copy of the manuscript should be kept by the author for use in case the original is lost.

**11D.5 Proofreading.** Most scientific journals permit the author to read proof on his papers before publication. A few journals place the entire burden of the proofreading on the author and hold him responsible for typographical or other errors which may pass undetected. In any event, where the author sees the proof, proofreading becomes a very important part of his scientific responsibility. The scientific value of his paper can be greatly lessened by unfortunate typographical errors. Such errors are sometimes obvious to the reader, others are insidious and may be wholly misleading.

In general, proof is submitted to the author to permit the elimination of printer's errors. Author's errors are his own responsibility, and many publications charge authors for corrections other than the printer's errors. Changes in proof are costly and therefore should not be made unless necessary, or unless the author is willing to assume the cost of the change.

Proofreading cannot always be done satisfactorily by one person. It is advisable to supplement the personal reading by having someone else read slowly from the original manuscript, while the proofreader (preferably the author) carefully reads the proof. Special attention should be given to punctuation, spelling of scientific names, numbers, and dates of all kinds. When corrections are necessary, they should be made according to the standard system of proofreader's marks, as given in most dictionaries and style manuals.

Most authors see only galley proofs of their papers. These are long sheets with the text continuous and not broken into pages. For most journals a galley is the equivalent of about three printed pages. Some publications also submit page proofs to the authors. In such cases proofreading cannot be restricted to individual words which were corrected in the galley proof but must include the whole line in which the correction was made. Modern linotype machines cannot change a single letter in a word but must reset the whole line. If a word was inserted, it may have been necessary to reset several lines or perhaps the remainder of the paragraph. The author should carefully check everything which has been reset. It is advisable also to read the top and bottom line of every page, because in the conversion of the galley into page proof mishaps sometimes occur in these lines. Corrected proof should be returned at once to the editor or printer in order to avoid delay in publication. The printing of an entire issue of a periodical may be held up by a single tardy author.

**11D.6　Reprints.** Reprints must be ordered at the time the proof is returned to the editor. It is advisable to order a larger than usual number of reprints of papers that deal with general principles, because requests for these will come not only from cospecialists.

A good reprint collection is particularly valuable in taxonomic research, where so many publications are only one or two pages long. With the help of modern copying methods (Xerox, etc.) an author can build up rapidly and inexpensively a specialized library of the entire literature of a taxon (except for a few very popular groups).

Letters requesting reprints from other authors should be specific. Most authors resent a request for "a set of reprints of your papers," except under unusual circumstances. Few authors have an inexhaustible supply of their papers, and the majority prefer to distribute their limited stock to those who will obtain the greatest use from them.

If the author wishes to send all his reprints, he is still free to do so. The receipt of requested reprints should be acknowledged, and in some cases, especially where the expense of shipment is significant, the recipient should offer to refund the postage involved.

# Part III Principles and Application of Zoological Nomenclature

# Chapter 12    The Rules of Zoological Nomenclature

*T*he zoologist deals with an enormous number of items; each species, genus, and higher taxon is a different item. He would not be able to refer to them if each one did not have a separate name. The term nomenclature comes from the Latin words *nomen* (name) and *calare* (to call) and means literally to call by name. It is the role of nomenclature to provide labels for taxa at all levels, in order to facilitate communication among biologists. The scientific names for species of organisms and for the higher taxa in which they are placed form a system of communication, a language; they must fulfill the same basic requirements as any other language.

It is most unfortunate that some taxonomists take a far greater interest in the names of animals than in the animals themselves. Nomenclature is an area in which one can operate even if one has only a minimal knowledge of biology. The result has been the elaboration of all sorts of complex rules and regulations, often without a clear understanding of the underlying principles. Many leading taxonomists, such as Hubbs, K. P. Schmidt, and Simpson, have admonished taxonomists not to make a game of their occupation with names, but to remember at all times that nomenclature is a means to an end and not an end in itself.

What are the important requisites of any system of communication, scientific nomenclature included? Surely there are many, but three seem to be outstanding.

*Uniqueness.* A classification is a filing system, an information retrieval system. The name of an animal, like the index number of a file, gives immediate access to all the known information about the particular taxon. Every name has to be unique because it is the key to the entire literature relating to this species or higher taxon. If several names have been given to the same taxon, there must be a clear-cut method whereby it can be determined which of them has validity. Normally (but see 13.1, 13.3) priority decides in cases of conflict.

*Universality.* Scientific communication would be made very difficult if we had only vernacular names for animals; specialists would have to learn the names of taxa in innumerable languages in order to communicate with each other. To avoid this, zoologists have adopted by international agreement a single language, a single set of names for animals, to be used on a worldwide basis.

*Stability.* As recognition symbols the names of objects would lose much of their usefulness if they were changed frequently and arbitrarily. It would surely create confusion if we were to call an object a spoon today but an apple next week. Yet this basic principle of communication has been constantly violated by zoologists. Altogether too much name-changing has occurred in zoological taxonomy during the past 200 years.

These three major objectives of the communication system of taxonomists are singled out for attention in the Preamble of the Code: *"The object of the Code is to promote stability and universality in the scientific names of animals, and to ensure that each name is unique and distinct. All its provisions are subservient to these ends."*

The Preamble stresses another vitally important principle, that none of the provisions of the Code shall "restrict the freedom of taxonomic thought or action." This principle is of relevance in Arts. 11, 17, 36, 49, 59*b*, 64, and 67–69 of the Code. It means that no taxonomist shall be forced to accept a particular classification, or the delimitation of a taxon, or be guided in his original choice of the type of a taxon by any but zoological reasons (see 13.2).

## THE INTERNATIONAL CODE

The valid rules of zoological nomenclature are contained in an authoritative document entitled the *International Code of Zoological Nomenclature.* Its most recent version, adopted by the Fifteenth International Congress of Zoology (London), came into force on the day of its publication, November 6, 1961. A few minor modifications, affecting Arts. 11, 31, 39, and 60, were adopted by the Sixteenth Congress at Washington in 1963 and are incorporated in the 1964 edition of the Code. This Code is a stage in a long history described in part in the introduction to the Code (pp. ix–xix) and by Linsley and Usinger (1959).

Linnaeus himself was the father of a set of rules of nomenclature published in the *Critica botanica* (1737) and the *Philosophia botanica* (1751). As the Linnaean authority faded out during the nineteenth century, new and local nomenclatural traditions developed and so did local sets of scientific names, owing to a lack of communication among taxonomists. As Strickland said in 1842,

> If an English zoologist, for example, visits the museums and converses with the professors of France, he finds that their *scientific* language is almost as foreign to him as their *vernacular*. Almost every specimen which he examines is labeled by a title which is unknown to him, and he feels that nothing short of continued residence in that country can make him conversant with her science. If he proceeds thence to Germany or Russia, he is again at a loss; bewildered everywhere amidst the confusion of nomenclature, he returns in despair to his own country and to the museums and books to which he is accustomed.

To stem this confusion zoologists and paleontologists in England, France, Germany, Russia, and America drafted formal rules of nomenclature. What was particularly needed, however, was an International Code. This was adopted in principle at the Fifth International Zoological Congress (Berlin, 1901) but was not formally issued until after the Sixth Congress (Bern, 1904) as the *Règles Internationales de la Nomenclature Zoologique* (Paris, 1905).

Certain of the articles of the original Code were amended by the later congresses at Boston (1907), Graz (1910), Monaco (1913), Budapest (1927), and Padua (1930). Far more extensive changes were discussed at the three postwar congresses of Paris (1948), Copenhagen (1953), and London (1958). By this time it had become apparent that a drastic reediting of the rules was necessary. The sequence of the articles in the old Code was rather arbitrary, and much nomenclatural tradition was not codified at all. A large body of case law had also accumulated in the meantime, published in the *Opinions of the International Commission on Zoological Nomenclature*. An editorial committee under Norman R. Stoll edited the version of the new Code as it had been adopted by the previous congresses, particularly the London Congress.

The new Code, like any human document, is not perfect. We shall point out in Chap. 13 some aspects that would seem to be still in need of improvement. Nevertheless, this Code is a document of extraordinary clarity and comprehensiveness. A great deal that was left to implication in the old rules is now explicitly stated in the new Code. The rules are a code of law, which makes it necessary to state matters precisely and without explanation. In order to improve understanding, Chap. 13 of this book attempts to provide explanatory comments, numbered 1–66, on all those provisions of the Code, the understanding of which can be facilitated by explanations. Some very valuable explanations are also found in the "Bradley draft" (*Bulletin of Zoological Nomenclature,* vol. 14, pp. 1–286), which formed the basis for the deliberations of the London Congress. However, the Bradley draft has no official status, and some of its provisions were rejected by the London Congress.

The relations between the Zoological Congresses and the International Commission on Zoological Nomenclature are frequently misunderstood. The Congress is a legislative body, and all provisions of the Code as well as the Constitution of the Commission were adopted by vote of the Congress. The Commission is a judicial body elected by the Congress as provided in the Constitution and By-laws of the Commission (see 13.59).

## EVOLUTION OF THE THEORY OF NOMENCLATURE

Many provisions of the Code are the results of compromise between conflicting principles. Such compromises go back to Linnaeus, whose nomenclature embodied on one hand the principles of Aristotelian logic and on the other hand some very practical considerations. The conflicts between stability and priority, between taxonomic freedom and nomenclatural rigidity, between typification through type-fixation and through type-restriction are some other examples. Equally important are certain changes in the last 150 years in the basic concepts of taxonomy. These include the concept of taxa as populations rather than types, the nature of the type as a name bearer, the categorical status of infrasubspecific names, and finally the application of some basic legal principles to the laws of nomenclature, such as the impropriety of a retroactive application of laws and the stabilizing effect of statutes of limitation.

All good law is living law. This is as true for the Code as for all other codes of law. The International Rules have prevented the development of chaos in the naming of animals and have greatly helped to standardize taxonomic procedures. Any change of these rules is upsetting and should not be undertaken lightly. Yet the rules should not become rigid, losing contact with the conceptual evolution of taxonomy. It is the rules that will have to be adjusted to the conceptual development of taxonomy and not the reverse.

An unresolved difficulty is posed by the drastically different needs of different taxonomic groups. The needs that exist in popular groups like birds or mammals, in which a limited number of species are referred to by their scientific names hundreds or thousands of times annually, are very different from the needs in an obscure group of arthropods attended to by a single specialist, with many species not mentioned in the literature more frequently than once every 30 or 50 years. Different problems exist for the parasitologist, who must make tentative assignments of stages in the life cycle of parasites (e.g. cercariae) even though the connection with the adult stages can be established only by experiment, and for the paleontologist, who may be forced to adopt form genera until the true taxonomic status of the objects placed in these form genera is fully established. Dissension and controversy will inevitably result if specialists in one group of organisms are oblivious of the needs of specialists in other groups.

## TEXT OF THE INTERNATIONAL CODE OF ZOOLOGICAL NOMENCLATURE

The 1964 Code consists of a Preamble, 86 Articles (Article 31 was repealed in 1963), five Appendices, an official Glossary, and a detailed Index, all items in parallel English and French versions. The excerpt here reprinted is limited to the English version of the preamble, articles, and recommendations, needed as the basis for the commentary in Chap. 13 (comments 1–66). Since the French version, as well as the Appendices, the official Glossary, and the official Index are indispensable for a full understanding of the Code, every zoologist, whether a taxonomist or not, is urged to acquire an official copy of the Code. It can be ordered from the International Trust for Zoological Nomenclature, 14 Belgrave Square, London S.W.1, at the cost of $3.00.

**INTERNATIONAL CODE OF ZOOLOGICAL NOMEN-**
**CLATURE ADOPTED BY THE XV INTERNATIONAL**
**CONGRESS OF ZOOLOGY, LONDON, JULY 1958.**
**Published in 1961.   Revised edition 1964**
**Excluding Preface, Introduction, Appendices, Glossary,**
**and Index.**

### PREAMBLE

The International Code of Zoological Nomenclature is the system of rules and recommendations authorized by the International Congresses of Zoology. The object of the Code is to promote stability and universality in the scientific names of animals, and to ensure that each name is unique and distinct. All its provisions are subservient to these ends, and none restricts the freedom of taxonomic thought or action.

1*

2,43

* The numbers in the margin refer to the numbered comments in Chap. 13.

Priority is the basic principle of zoological nomenclature. Its application, however, under conditions specified in the Code, may be moderated to preserve a long-accepted name in its accustomed meaning.                    3,4

When stability of nomenclature is threatened in an individual case, the strict application of the Code may under specified conditions be suspended by the International Commission on Zoological Nomenclature.                    50,60

## I. ZOOLOGICAL NOMENCLATURE

**Article 1.**  Zoological nomenclature is the system of scientific names applied     5
to taxonomic units of animals (taxa; singular: taxon) known to occur in nature, whether living or extinct. This Code is concerned with such names in the family-, genus-, and species-groups [VIII, IX, X; for work of an animal, see Art. 16*a*(viii)]. Names given to hypothetical concepts, to terato-     6,17
logical specimens or to hybrids as such, to infrasubspecific forms as such, or     21,41
names proposed for other than taxonomic use, are excluded.

**Article 2. Independence of zoological nomenclature.**  Zoological nomenclature is independent of other systems of nomenclature in that the     5
name of an animal taxon is not to be rejected merely because it is identical with the name of a taxon that does not belong to the animal kingdom.

   (*a*) **Transfer of taxa to the animal kingdom.**  If a taxon is transferred to the animal kingdom, its name or names enter into zoological nomenclature with the original date and authorship.

   (*b*) **Removal of taxa from the animal kingdom.**  If a taxon is removed from the animal kingdom, its name or names continue to compete in homonymy with names in the animal kingdom.

**Recommendation 2A.  Names already in use outside the animal kingdom.**     7
It is preferable not to propose for a genus of animals a name already in use for a genus outside the animal kingdom.

**Article 3. Starting point.**  The 10th edition of Linnaeus's *Systema Naturae*
marks the beginning of the consistent general application of binominal     8,9
nomenclature in zoology. The date 1 January 1758 is arbitrarily assigned in this Code as the date of publication of that work and as the starting point of zoological nomenclature. Any other work published in 1758 is to be treated as having been published after that edition.

## II. NUMBER OF WORDS IN ZOOLOGICAL NAMES     10,14

**Article 4. Taxa of rank above the species-group.**  The name of a taxon     11
of higher rank than the species-group consists of one word (uninominal).

**Article 5. Species and subspecies.**  The name of a species consists of two     11,12,
words (binomen) and that of a subspecies of three words (trinomen); in each     13,14
case the first word is the generic name, the second word is the specific name,     40
and the third word, when applicable, is the subspecific name.

**Article 6. Subgenus.** The name of a subgenus, when used in combination with a generic and a specific name, is placed in parentheses between those names; it is not counted as one of the words in the binominal name of a species or trinominal name of a subspecies. 11

## III: CRITERIA OF PUBLICATION 15

**Article 7. Application.** The provisions of this Chapter apply to the publication not only of a new name but also to any other information that affects nomenclature. 22

**Article 8. What constitutes publication.** To be regarded as published within the meaning of this Code, a work when first issued must 15

(1) be reproduced in ink on paper by some method that assures numerous identical copies;
(2) be issued for the purpose of scientific, public, permanent record;
(3) be obtainable by purchase or free distribution; and
(4) not be reproduced or distributed by a forbidden method (Art. 9).

**Recommendation 8A. Mimeographing and similar processes.** Zoologists are strongly urged not to use mimeographing, hectographing, or similar processes for a publication containing a new name or a statement affecting nomenclature.

**Article 9. What does not constitute publication.** None of the following acts constitutes publication within the meaning of the Code: 15

(1) distribution of microfilms, or microcards, or matter reproduced by similar methods;
(2) distribution to colleagues or students of a note, even if printed, in explanation of an accompanying illustration;
(3) distribution of proof sheets;
(4) mention at a scientific or other meeting;
(5) labelling of a specimen in a collection;
(6) mere deposit of a document in a library; or
(7) after 1950, anonymous publication.

## IV. CRITERIA OF AVAILABILITY

**Article 10. When a name becomes available.** A name becomes available, and takes date and authorship, only when it satisfies the provisions of Article 11; in addition, names published after certain dates must satisfy also the conditions of Articles 12, 13, 14 and 15. 16

17

(*a*) **Interrupted publication.** If publication of the data relating to a new nominal taxon is interrupted and continued later, the name becomes available only when it satisfies all the relevant provisions.

**Recommendation 10A. Divided description.** Editors should not permit the description of a new taxon below the family-group to be so divided that portions appear at different times.

(b) **Infrasubspecific names.** A name first established with infrasubspecific rank becomes available if the taxon in question is elevated to a rank of the species-group, and takes the date and authorship of its elevation.    41

**Article 11. General requirements.**  A name, to become available, must satisfy the following provisions:

(a) **Publication and date.**  It must have been published, in the meaning of Chapter III, after 1757.    22

(b) **Language.**  The name must be either Latin or latinized, or, if an arbitrary combination of letters, must be so constructed that it can be treated as a Latin word [VII].    28

   (i) The letters "j", "k", "w", and "y", more common in Neo-Latin, may be used in zoological names.

(c) **Binomial nomenclature.**  The author must have consistently applied the principles of binominal nomenclature [II] in the work in which the name is published.    11

   (i) Uninominal genus-group names published before 1931 without associated nominal species are accepted as consistent with the principles of binominal nomenclature, in the absence of evidence to the contrary.

   (ii) Names published before 1931 in the index to a work, if they satisfy the relevant provisions of this Article and of Articles 4, 5 and 6, are available, even if the author did not use binominal nomenclature in the body of the work, provided that there is a clear bibliographic reference to a description, indication, or figure of the animal in question, or, if it is a family-group name, provided that it is founded on an included nominal genus [Art. 16a(ii)].

(d) **Publication in synonymy.**  A name first published as a synonym is not thereby made available unless prior to 1961 it has been treated as an available name with its original date and authorship, and either adopted as the name of a taxon or used as a senior homonym.    18
    2

(e) **Names of the family-group.**  A family-group name must, when first published, be based on the name then valid for a contained genus, and must be a noun in the nominative plural.    33

   (i) The name must clearly be used to denote a suprageneric taxon, and not merely be employed as a plural noun or adjective referring to the members of a genus.

   (ii) A family-group of which the suffix is incorrect is available with its original date and authorship, but in properly emended form [Art. 29].

**Example.**  Latreille [1802–1803], proposed a family TIPULARIAE, based on *Tipula* Linnaeus, 1758. The name must be corrected to Tipulidae and attributed to Latreille [1802–1803], not to the author who first corrected the spelling.

   (iii) A family-group name published before 1900 in accordance with the above provisions of this Section, but not itself fully latinized, is available with its original date and authorship, provided that it has been latinized by later authors and that it has been generally accepted by zoologists interested in the group concerned as dating from its first publication in vernacular form.

**Example.**  The family name TETRANYCHIDAE is generally attributed to Donnadieu, 1875. He published the name as "Tétranycidés," but in view of the general acceptance of his name, it is to be attributed to his work and date, not to Murray, 1877, who first latinized it.

(*f*) **Names of the genus-group.**   A genus-group name must be a noun in the   38
nominative singular or be treated as such.

(*g*) **Names of the species-group.**
  (i) A species-group name must be a simple word of more than one   40
  letter, or a compound word, and must be or be treated as
    (1) an adjective in the nominative singular agreeing in gender
    with the generic name (e.g. *Felis marmorata*), or
    (2) a noun in the nominative singular standing in apposition to
    the generic name (e.g. *Felis leo*), or
    (3) a noun in the genitive case (e.g. *rosae, sturionis, thermopylarum,
    galliae, sanctipauli, sanctaehelenae, cuvieri, merianae, smithorum*), or
    (4) an adjective used as a substantive in the genitive case, derived
    from the specific name of an organism with which the animal
    in question is associated (e.g. *Lernaea lusci*, a copepod parasite
    on *Gadus luscus*).
  (ii) A species-group name must be published in combination with a
  genus-group name, but the latter need not be valid or even available.
  (iii) A species-group name must not consist of words related by a conjunction, nor include a sign that cannot be spelled out in Latin.
**Example.**   Expressions like *rudis planusque* and *?-album* are not admissible as
specific names.

**Article 12. Names published before 1931.**   In addition to satisfying the
provisions of Article 11, a name published before 1931 must have been ac-   16
companied by a description, definition, or indication [Art. 16].

**Article 13. Names published after 1930.**

(*a*) **Names in general.**   In addition to satisfying the provisions of Article 11,
  a name published after 1930 must be either
    (i) accompanied by a statement that purports to give characters differ-   19,38
    entiating the taxon; or   40
    (ii) accompanied by a definite bibliographic reference to such a state-   47
    ment; or
    (iii) proposed expressly as a replacement for a preexisting available   47
    name.

(*b*) **Genus-group names.**   A genus-group name published after 1930 must,   38
  in addition to satisfying the provisions of Section (*a*), be accompanied by   52
  the definite fixation of a type-species [Art. 68].
    (i) The provisions of this Section do not apply to names of collective
    groups [Art. 66].

**Article 14. Names published after 1950.**   After 1950, a new name published anonymously is not available.

**Article 15. Names published after 1960.**   After 1960, a new name proposed conditionally, or one proposed explicitly as the name of a "variety" or
"form" [Art. 45*e*], is not available.   20

**Article 16. Indications.**

(*a*) **What constitutes an indication.**   The word "indication" as used in this
  Chapter applies only to the following:
    (i) a bibliographic reference to a previously published description,
    definition, or figure;
    (ii) the inclusion of a name in an index to a work, provided that the
    provisions of Article 11*c*(ii) are satisfied;

(iii) the substitution of a new name for a previously established name;

(iv) the formation of a new family-group name from the stem of the name of a genus, which thus becomes the type-genus;

(v) the citation, in combination with a new genus-group name, of one or more available specific names;

(vi) a single combined description of a new nominal genus and a new nominal species, which provides an indication for each name;

(vii) the publication of a new genus- or species-group name in connection with an illustration; or

(viii) the description of the work of an animal, even if not accompanied by a description of the animal itself.     27

(b) **What does not constitute an indication.** The following are not "indications" in the meaning of this Chapter:     20

(i) mention of a vernacular name, type-locality, geological horizon, host, or a label or specimen in a collection; or

(ii) citation of a name in synonymy [see also Art. 11d].

## Article 17. Conditions that do not prevent availability. A name is or remains available even though

(1) it becomes a junior synonym: such a name may be re-employed if the synonymy is judged to be erroneous, or if the senior synonym is found to be invalid or unavailable; or     2,18

(2) in the case of a species-group name, it is found that the original description relates to more than one taxonomic unit, or to parts of animals belonging to more than one taxon, or to an animal and animals later found to be hybrid; or     21

(3) in the case of a species-group name, the genus-group name with which it is first combined is invalid or unavailable; or

(4) it is based only on any part of an animal, sex of a species, stage in life-history, one of several dissimilar generations, or one form of a polymorphic species; or

(5) it was originally proposed for an organism now but not then considered an animal; or

(6) it was incorrectly spelled with respect to any of the provisions of Chapter VII, in which case it is to be corrected and the correct spelling is the available one [Art. 32c]; or

(7) before 1951, it was published anonymously; or

(8) before 1961, it was proposed conditionally; or

(9) before 1961, it was proposed as a "variety" or "form."

## Article 18. Unallowable causes for rejection

(a) **Inappropriateness.** A genus- or species-group name, once established, cannot afterwards be rejected, even by its own author, because of inappropriateness.

**Examples.** Names such as *Polyodon, Apus, albus, sinensis,* etc., once published, are not to be rejected because of a claim that they denote a character or distribution not possessed by the animal in question.

(b) **Tautonymy.** A name is not to be rejected because of tautonymy, that is, because the species-group name or names are identical with the generic name.

**Examples.** *Bison bison; Apus apus apus.*

**Article 19. Status of emendations and errors.** In the meaning of the Code, an emendation, whether justified or unjustified, is an available name, but an incorrect spelling, whether original or subsequent, has no standing in nomenclature and is not an available name [Art. 32c, 33].

**Article 20. Genus-group names ending in -ites, -ytes, or -ithes given to fossils.** If an existing genus-group name has been modified by substituting -ites, -ytes, or -ithes for its original termination, the modified name if applied only to fossils is not available, except for the purposes of the Law of Homonymy [Art. 56b], unless there is clear evidence of intent to establish a distinct genus or subgenus.

**Example.** The generic names *Pectinites* and *Tellinites* Schlotheim, 1813, used to denote fossil shells thought to belong to the Recent genera *Pecten* Müller, 1767, and *Tellina* Linnaeus, 1758, are available only for the purposes of the Law of Homonymy.

## V. DATE OF PUBLICATION

**Article 21. Interpretation of date.** The date of publication of a work and of a contained name or statement affecting nomenclature is to be interpreted in accordance with the provisions of this Article.

22

(*a*) **Date specified.** The date of publication specified in a work is assumed to be correct in the absence of evidence to the contrary.

(*b*) **Date incomplete.** If the date of publication is not completely specified, it is to be interpreted as the earliest day demonstrated by evidence, but in the absence of such evidence, as
  (i) the last day of the stated month, when month and year, but not the day, are specified; and as
  (ii) the last day of the year, when only the year is specified.

(*c*) **Date incorrect.** If the date of publication specified in a work is found to be incorrect, the date is to be interpreted as the earliest demonstrated by the evidence.

(*d*) **Dates of work issued in parts.** If parts of a work were published on different days, the date of each is reckoned independently.

(*e*) **Range of dates.** If the specified date of publication contained within a work is a range of dates, the work is to be dated from the latest day within that range; but if evidence proves that one or more parts were issued before that day, it or they are to be interpreted as dating from the earliest day demonstrated by the evidence.

(*f*) **Date not stated.** In the absence of internal evidence of its date of publication, a work is to be dated in whole or in part from the earliest date demonstrated by external evidence, such as mention in another work.

**Recommendation 21A. Responsibility of editors, publishers, and authors.** Editors and publishers should not put any copy or part of a work into circulation in advance of the specified date of publication. Authors should not distribute reprints (separata) in advance of such publication.

**Recommendation 21B. Dating of publications.** Editors and publishers should take care to state the exact date of issue of each component part of a serial publication or of any work issued in parts. A completed volume containing parts brought out separately should state the exact day of publication of each part, and the exact pages, plates, maps, etc., that constitute it.

**Recommendation 21C.  Responsibility of librarians.**  If a work deals with zoology, librarians should not remove covers that bear information relative to the date of publication and content of the work or its parts, or to the dates of their receipt in the library.

**Recommendation 21D.  Information on reprints and preprints.**  Reprints (separata) should contain information sufficient for complete citation, including pagination and date of publication coinciding with the source-publication. Preprints should be definitely identified as such.

**Article 22. Citation of date.**  The date of publication of a name, if cited, follows the name of the author with a comma interposed.

**Recommendation 22A.  Method of citation.**  In citing the date of publication of a name, an author

(1)  should not enclose the date in either parentheses or square brackets if the work containing it specifies the date of publication;

(2)  should enclose the date, or a part of it, in parentheses if it is determined by evidence derived from the volume concerned other than in (1); or

(3)  should enclose the date, or a part of it, in square brackets if it is determined only from external evidence.

**Recommendation 22B.  Date in a changed combination.**  If the original date of publication is cited for a species-group name in a changed combination, it should be enclosed within the same parentheses as the name of the original author, separated by a comma [Art. 51*d*].

**Example.**  *Arion ater* (Linnaeus, 1758).


## VI.  VALIDITY OF NAMES

**Article 23. Law of Priority.**  The valid name of a taxon is the oldest available name applied to it [taking into consideration the provisions of Sections (*d*)(i) and (*e*), below], provided that the name is not invalidated by any provision of this Code or has not been suppressed by the Commission.

3
23

(*a*) **Exceptions.**  A name that is not the oldest available name is nevertheless the valid name of the taxon in question

(i) if it is conserved under Section (*b*) of this Article; or

(ii) if the Commission has expressly validated it.

(*b*) **Limitation.**  A name that has remained unused as a senior synonym in the primary zoological literature for more than fifty years is to be considered a forgotten name (nomen oblitum).

1
24

(i) After 1960, a zoologist who discovers such a name is to refer it to the Commission, to be placed on either the appropriate Official Index of Rejected Names, or, if such action better serves the stability and universality of nomenclature, on the appropriate Official List.

63

(ii) A nomen oblitum is not to be used unless the Commission so directs.

(iii) This provision does not preclude application to the Commission for the preservation of names, important in applied zoology, of which the period of general usage has been less than fifty years.

(c) **Change of rank.**   The priority of the name of a taxon in the family-,    35
genus-, or species-group is not affected by elevation or reduction in rank
within the group.

(d) **Family-group names.**    33

   (i) A family-group taxon formed by the union of two or more taxa of    25
that group takes the oldest valid family-group name among those
of its components, with change of termination if required.

   (ii) if a zoologist observes that the strict application of the Law of    73
Priority to two or more synonymous family-group names would
upset general usage, he is to request the Commission to decide    1
which name is to be accepted for the Official List of Family-Group    63
Names in Zoology.

(e) **Genus- and species-group names.**

   (i) A genus-group taxon formed by the union of two or more genus-    25
group taxa takes the oldest valid name among those of its components.

**Example.**   The valid name of a genus formed by the union of genus *A-us*, 1850,
and subgenus *B-us*, 1800, is *B-us*, 1800.

   (ii) A species-group taxon formed by the union of two or more species-    40
group taxa takes the oldest valid name among those of its components.

   (iii) If the name of a genus or species having subordinate taxa is found
to be invalid or unavailable, it must be replaced by the next oldest    23
valid name among those of the included co-ordinate taxa, including synonyms.

**Example.**   Genus *A-us*, 1850, contains the subgenera *A-us*, 1850, *C-us*, 1900,
and *D-us*, 1860. If the name *A-us* is found to be a homonym, it is replaced as the
name of the genus by *D-us*, 1860, the next oldest valid name.

(f) **Spelling.**   For the application of the Law of Priority to the spelling of
names, see Chapter VII.

## Article 24. Interpretation of oldest name    23

(a) **Names published simultaneously.**   If more than one name for a single    26
taxon, or identical names for different taxa, are published simultaneously,
whether in the same or different works, their relative priority is determined
by the action of the first reviser.

   (i) The expression "first reviser" is to be rigidly construed. In the case    4
of synonyms, an author must have cited two or more such names,
must have made it clear that he believes them to represent the same
taxonomic unit, and must have chosen one as the name of the taxon.

**Recommendation 24A.   Action of first reviser.**   In acting as "first reviser"    4
in the meaning of this Section, a zoologist should select the name that will best    26
ensure stability and universality of nomenclature. If none of the names has an
advantage of this sort, nor has any special appropriateness, he should select the
name that has precedence of position in the work in question.

(b) **Names founded on any part or form of an animal or on its work.**   The
Law of Priority applies

   (i) when any part of an animal is named before the whole animal; or

   (ii) when two or more generations, forms, stages, or sexes of a species
are named as different taxa; or

   (iii) when, before 1931, a name is founded on the work of an animal    27
before one is founded on the animal itself.

## VII.  FORMATION AND EMENDATION OF NAMES

**Article 25. Formation of names.**  Zoological names must be formed in       28
accordance with the provisions of Articles 26 to 30.

**Recommendation 25A.  Transliteration and latinization.**  Zoologists forming new names should, in the interest of proper usage and universality of nomenclature, consult Appendices B to D in order to ensure correct transliteration and
latinization.

### Article 26.  Compound names.

(a) **Acceptable compounds.**  If a name based on a compound name is published as two separate words in a work in which the author duly applied
the principles of binominal nomenclature, the component words are to be       28
united without a hyphen, and the name is to be treated as though it had
been originally published in that form. [For treatment of hyphenated compounds, see Article 32c(i).]

**Examples.**  *Coluber novae hispaniae* Gmelin is corrected to *Coluber novaehispaniae*,
*Calliphora terrae novae* Macquart to *Calliphora terraenovae*, and *Cynips quercus phellos*
Osten Sacken, based on the binominal name of the host species, to *Cynips quer-
cusphellos*.

(b) **Numerals in compounds.**  A number or numerical adjective or adverb
forming a part of a compound name is to be written in full as a word and
united with the remainder of the name (e.g. *decemlineata*, not *10-lineata*).

(c) **Latin letters in compounds.**  If the first element of a compound species-
group name is a Latin letter used to denote a character of the taxon, it is
connected to the remainder of the name by a hyphen (e.g. *c-album*).

**Article 27. Diacritic and other marks.**  No diacritic mark, apostrophe,
or diaeresis is to be used in a zoological name; the hyphen is to be used only       29
as specified in Article 26c.

**Article 28. Initial letters.**  Names of the family- and genus-groups must
be printed with a capital initial letter and names of the species-group with a
lower-case initial letter.

**Article 29. Formation of family-group names.**  A family-group name is       30
formed by the addition to the stem [see Glossary] of the name of the type-
genus, of -IDAE in the case of a family and -INAE in the case of a subfamily
[see Appendix D, Table 2, Part A].

**Recommendation 29A.  Superfamilies and tribes.**  It is recommended that
the termination -OIDEA be adopted for the names of superfamilies and -INI
for the names of tribes.

(a) **Generic name of classical origin.**  In zoological nomenclature, if the
name of a type-genus either is or ends in a Greek or Latin word, or ends in       30
a Greek or Latin suffix, the stem is found by deleting the case-ending of        9
the appropriate genitive singular [see Appendix D, VII].

(i) Where the word "Latin" is used in the Code, it includes ancient,
mediaeval, and modern Latin, but the word "Greek" refers only to
ancient Greek [see also Article 11b(i)].

(*b*) **Generic name not of classical origin.** If the name of a type-genus is or ends in a word not Greek or Latin, or is an arbitrary combination of letters, the stem is determined by the zoologist who first publishes a family-group name based on that nominal genus [see Appendix D, VII].

(*c*) **Generic name latinized from Greek.** If the name of a type-genus is or ends in a Greek word latinized with a change in termination, the stem is that appropriate for the latinized form.

**Example.** For *Leptocerus*, of which the second part is from *keras*, the stem for the formation of a family-group name is *Leptocer-*, not *Leptocerat-*.

## Article 30. Agreement in gender.

A species-group name, if an adjective in the nominative singular, must agree in gender with the generic name with which it is at any time combined, and its termination must be changed, if necessary, when the species transferred to another genus. The gender of a genus-group name is determined by the following provisions.

(*a*) **Genus-group names consisting of or ending in a Greek or Latin word or ending in a Greek or Latin suffix.**

  (i) A genus-group that is or ends in a Greek or Latin word takes the gender given for that word in the standard Greek or Latin dictionaries, unless the Commission rules otherwise.

**Examples.** Greek nouns transliterated without change into Latin as the whole or as part of a name, e.g. *Hoplites*, from ὁπλίτης, masculine; *Ichthyornis*, ending in -όρνις, masculine; *Wattonithyris*, ending in θύρις, feminine; Names ending in -*ops*, if from Greek ὄψ ("voice," or rarely "face"; but see under (2), below), -*opsis* ὄψις, -*gaster* γαστήρ, -*caris* κάρις, or -*lepis* λέπις, are feminine. *Tuba* from Latin *tuba* is feminine. Names ending in certain Latin nouns in -*us* are feminine (e.g. -*alvus*, -*humus*, -*vanus*, -*fraus*, -*laus*, -*acus*, -*colus*, -*domus*, -*tellus*). Names ending in -*ceras* (κέρας), -*soma* (σῶμα), -*stigma* (στίγμα), or -*stoma* (στόμα) are neuter.

  (1) A name is to be regarded as a Greek or Latin word of the same spelling, unless its original author states otherwise.

  (2) A noun of variable gender, masculine or feminine, is to be treated as masculine, unless its author states, when he first publishes the name, that it is feminine, or so treats it in combination with an adjectival specific name.

**Examples.** Compound Latin nouns ending in -*cola*, such as *Sylvicola*, are considered as masculine. Names ending in -*ops* derived from the Greek ὤψ ("face" or "eye"), of which the usual classical gender is masculine, are to be treated as masculine unless the author indicated otherwise or unless, failing such indication, zoologists have generally treated them as feminine. [For names derived from ὄψ, see example under (i) above.]

  (3) If a genus-group name is a Greek word latinized with a change of termination, it takes the gender appropriate to that termination.

**Examples.** Names ending in -*us*, latinized from the Greek endings -os (os) (masculine or feminine), -α (a) (neuter), or -ον (on) (neuter), are masculine, e.g. -*echinus* εχινος, echinos), -*cephalus* (κεφαλη, kephale), -*crinus* (κρινον, krinon), -*stomus* (στομα, stoma), -*somus* (σωμα, soma), -*cheilus* (and -*chilus*) (χειλος, cheilos), -*gnathus* (γνάθος, gnathos), -*rhamphus* (ῥάμφος, rhamphos), -*rhynchus* (ῥύγχος, rhynchos), or -*stethus* (στῆθος, stethos). Names ending in -*cera* (κέρας, keras) or -*metopa* (μέτωπον, metopon) are feminine.

  (4) If a genus-group name is a Latin word of which the termination has been changed, it takes the gender appropriate to the new termination.

**Example.** *Dendrocygna*, feminine, although partly formed from *cygnus*, masculine.

(ii) A genus-group name that ends in a Greek or Latin suffix, or in a letter or letters considered as such, takes the gender appropriate to its ending.

**Examples.** Names ending in *-ides*, *-istes*, *-ites*, *-odes*, or *-oides* are masculine. Names such as *Scatella* and *Oculina* are feminine because of the suffix, although derived respectively from the Greek neuter noun σκῶρ, σκάτος and the Latin masculine noun *oculus*. A name based on a word in a non-classical language or on an arbitrary combination of letters, with the addition of a Greek or Latin suffix, takes the gender appropriate to the suffix, e.g. *Buchia* (from von Buch), *Cummingella* (from Cumming), *Velletia* (from Vellet), *Dafila*, and the anagram *Solubea* are all treated as feminine, but the anagram *Daption* as neuter.

(*b*) **Genus-group names entirely of non-classical origin.**—

(i) A genus-group name that reproduces exactly a word in a modern Indo-European language having genders takes the gender of that word.

**Example.** *Pfrille*, from the German "die Pfrille," is feminine.

(ii) A genus-group name consisting of a word that is not Greek, Latin, or modern Indo-European, or that is an arbitrary combination of letters, takes the gender expressly attributed to it by its author, or implied by an originally associated species-group name. If no gender was assigned or implied, the name is to be treated as masculine, except that if the ending is clearly a natural classical feminine or neuter one, the gender is that appropriate to the ending [Art. 30*a* (ii)].

**Examples.** *Dacelo* (anagram of *Alcedo*) is feminine as treated by its author, but *Vanikoro*, *Gekko*, *Abudefduf*, and the anagram *Milax* are treated as masculine.

**Recommendation 31A.\*** **Species-group names formed from modern personal names.** A species-group name, if a noun formed from a modern personal name, should usually end in *-i* if the personal name is that of a man, *-orum* if of men or of man (men) and woman (women) together, *-ae* if of a woman, and *-arum* if of women [see Article 11*g*(i)(3) and Appendix D III].     40

## Article 32. Original spelling.     32

(*a*) **Correct original spelling.** The original spelling of a name is to be retained as the "correct original spelling," unless     40

(i) it contravenes a mandatory provision of Articles 26 to 30; or

(ii) there is in the original publication clear evidence of an inadvertent error, such as a lapsus calami, or a copyist's or printer's error (incorrect transliteration, improper latinization, and use of an inappropriate connecting vowel are not to be considered inadvertent errors); or

(iii) in the case of a family-group name, there has been a necessary correction of termination (other than necessitated by a change in the rank of the taxon), or a justified emendation [Art. 33*a*] in the stem of the name of the type-genus.     32

(*b*) **Multiple original spellings.** If a name is spelled in more than one way in the original publication, the spelling adopted by the first reviser is to be accepted as the correct original spelling, unless the adopted spelling is subject to emendation under the provisions of Articles 26 to 30.     4

(*c*) **Incorrect original spelling.** An original spelling that does not satisfy the provisions of Articles 26 to 30, or that is an inadvertent error [Art.     32

* This recommendation replaces Art. 31, deleted by the XVIth Congress of Zoology, 1963.

32a (ii)], or that is one of the multiple spellings not adopted by a first reviser [Art. 32b], is an "incorrect original spelling" and is to be corrected wherever it is found; the incorrect spelling has not separate status in nomenclature, and therefore does not enter into homonymy and cannot be used as a replacement name.

    (i) A name published with a diacritic mark, apostrophe, diaeresis, or hyphen is to be corrected by the deletion of the mark concerned and any resulting parts are to be united, except for one specified use of the hyphen [Art. 26c], and except that when, in a German word, the umlaut sign is deleted from a vowel, the letter "e" is to be inserted after that vowel.

**Examples.** The name *terrae-novae* is corrected to *terraenovae*, *d'urvillei* to *durvillei*, and *nuñezi* to *nunezi;* but *mülleri* becomes *muelleri* and is not a homonym of *mulleri* [Art. 57d].

## Article 33. Subsequent spelling.

(a) **Emendations.** Any demonstrably intentional change in the original spelling of a name is an "emendation."

    (i) A "justified emendation" is the correction of an incorrect original spelling, and the name thus emended takes the date and authorship of the original spelling.

    (ii) Any other emendation is an "unjustified emendation"; the name thus emended has status in nomenclature with its own date and author, and is a junior objective synonym of the name in its original form.

(b) **Incorrect subsequent spellings.** Any change in the spelling of a name, other than an emendation, is an "incorrect subsequent spelling"; it has no status in nomenclature and therefore does not enter into homonymy and cannot be used as a replacement name.

## Article 34. Endings.

(a) **In names of the family-group.** In names of the family-group, the ending must be changed when a taxon is raised or lowered in rank.

(b) **In names of the species-group.** In names of the species-group, the ending must be changed, if necessary, to conform with the gender of the generic name with which the species-group name is at any time combined [Art. 30].

# VIII.   TAXA OF THE FAMILY-GROUP AND THEIR NAMES

## Article 35. Categories and their names.

(a) **Categories included.** The family-group includes the categories tribe, subfamily, family, and superfamily, and any supplementary categories required.

(b) **Basis.** Each taxon of the family-group is defined by reference to its type-genus [see XIII, XIV].

(c) **Names.** A family-group name is to be formed and treated in accordance with the relevant provisions of Art. 29.

**Article 36. Categories co-ordinate.**    All categories in the family-group are of co-ordinate status in nomenclature, that is, they are subject to the same rules and recommendations, and a name established for a taxon in any category in the group, and based on a given type-genus, is thereupon available with its original date and author for a taxon based on the same type-genus in each of the other categories, with appropriate change of suffix.

35

2

**Example.**    The proposal of HESPERIIDAE Latreille, 1809 (as HESPER-IDES), based on *Hesperia* Fabricius, 1793, thereupon makes available, from the year 1809, the superfamily name HESPERIOIDEA and the subfamily name HESPERINAE, even though the former was first used by Comstock, J. H. and A. B., 1904, and the latter by Watson, 1893.

**Article 37. Subordinate taxa.**    The subordinate taxon that contains the type-genus of a subdivided family-group taxon bears the same names as the latter, except for suffix, and is termed the "nominate" subordinate taxon (e.g. nominate subfamily, nominate tribe).

25,36

**Example.**    The nominal family TIPULIDAE (type-genus *Tipula* Linnaeus, 1758) is divided into a number of subfamilies, each named after its own type-genus. The subfamily TIPULINAE, containing *Tipula*, is the nominate subfamily of the TIPULIDAE.

35

**Article 38. Homonymy between family-group names.**    See Article 55.

37

**Article 39. Homonymy of the type-genus.**    The name of a taxon of the family-group is invalid if the name of its nominal type-genus is a junior homonym.

**Article 40. Synonymy of the type-genus.**    When, after 1960, a nominal type-genus is rejected as a junior synonym (objective or subjective), a family-group name based on it is not to be changed, but continues to be the valid name of the family-group taxon that contains both the senior and junior synonyms.

1

33

(*a*) **Conservation of certain names.**    If a family-group name, changed before 1961 because of such synonymy, has won general acceptance, it is to be maintained in the interests of stability.

1

   (i) In the event of divergent interpretation of the expression "general acceptance," reference is to be made to the Commission.

(*b*) **Date of conserved name.**    A name adopted by virtue of the provisions of Section (*a*) takes the date of the rejected name, of which it is to be considered the senior synonym.

**Example.**    If an original type-genus *A-us* is a junior synonym of *B-us:*

(1) When before 1961, A-IDAE has not already been replaced by B-IDAE, or B-IDAE has not yet won general acceptance, then A-IDAE is to be continued as the name for the family;

(2) But if A-IDAE has already been replaced by B-IDAE, and the latter has won general acceptance, then B-IDAE is to be accepted as the correct name, and is to considered as the senior synonym of A-IDAE, with the same date as that of A-IDAE.

**Recommendation 40A. Citation of author and date.**    If author and date are cited [see Appendix E9], a family-group name adopted under the provisions of Article 40 should be cited with its own author and date, followed by the date of the replaced name in parentheses.

**Article 41. Misidentified type-genera.** If stability and continuity in the meaning of a nominal family-group taxon are threatened by the discovery that its nominal type-genus is based on a misidentified type-species, or by the discovery of an overlooked type-designation, the case is to be submitted to the Commission [Arts. 65, 70].

1
49

# IX. TAXA OF THE GENUS-GROUP AND THEIR NAMES

## Article 42. Taxa of the genus-group

38

(a) **Categories included.** The genus-group, which is next below the family-group and next above the species-group in the hierarchy of classification, includes the categories genus and subgenus.

(b) **Basis.** Each taxon of the genus-group is objectively defined only by reference to its type-species [XV].

(c) **Collective groups.** The names of certain biological groupings known as "collective groups" [see Glossary] are to be treated as generic names in the meaning of the Code, but collective groups require no type-species.

39

**Examples.** *Agamodistomum, Agamofilaria Cysticercus, Diplostomulum, Glaucothoe, Sparganum.*

(i) Wherever the terms "taxon" or "name" are used in this Code at the level of genus, the provision in question is to apply also to a collective group or its name, unless there is a statement to the contrary, or unless such application would be inappropriate.

(d) **Subdivision of genera.** A uninominal name proposed for a primary subdivision of a genus, even if the subdivision is designated by a term such as "section" or "division," has the status in nomenclature of a subgeneric name, provided the name satisfies the relevant provisions of Chapter IV.

**Article 43. Categories co-ordinate.** The categories in the genus-group are of co-ordinate status in nomenclature, that is, they are subject to the same rules and recommendations, and a name established for a taxon in either category in the group, and based on a given type-species, is thereupon available with its original date and author for a taxon based on the same type-species in the other category.

35

## Article 44. Subordinate taxa.

(a) **Nominate subgenus.** The subgenus that contains the type-species of a subdivided genus bears the same name as the genus and is termed the "nominate" subgenus.

25,38

(b) **Change of nominate subgenus.** When the invalid name of a subdivided genus is replaced by the name of a different subgenus [Art. 23e (iii)], the latter then becomes the nominate subgenus.

(c) **Citation of the subgenus.** See Art. 6.

**Recommendation 44A. Citation of synonyms.** In order to avoid misunderstanding, a synonym, or any term other than subgenus, should never be cited between the generic and specific elements of a binomen.

## X.  TAXA OF THE SPECIES-GROUP AND THEIR NAMES

**Article 45.  Taxa of the species-group.**                               40

(*a*) **Categories included.**  The species-group, for the purposes of this Code, includes the categories species and subspecies.

(*b*) **Basis.**  Each taxon of the species-group is objectively defined only by     48
reference to its type-specimen.                                              51

(*c*) **Infrasubspecific forms.**  Infrasubspecific forms are excluded from the species-group and the provisions of this Code do not apply to them [Art. 1;     41,6
see also Art. 10 (*b*)].

(*d*) **Determination of subspecific or infrasubspecific status.**  The original     41
status of any name of a taxon of lower rank than species is determined as

    (i) subspecific, if the author, when originally establishing the name, either clearly stated it to apply to a subspecies or, before 1961, did not clearly state its rank [see also Art. 45(*e*) below], and as

    (ii) subspecific, if the author, when originally establishing the name, stated the taxon to be characteristic of a particular geographical area or geological horizon and did not expressly refer it to an infrasubspecific category; but as

    (iii) infrasubspecific, if the author, when originally establishing the name, either expressly referred the taxon to an infrasubspecific rank, or, after 1960, did not clearly state that it was a subspecies.

(*e*) **Interpretation of the terms "variety" and "form"**                     41

    (i) Before 1961, the use of either of the terms "variety" or "form" is not to be interpreted as an express statement of either subspecific or infrasubspecific rank.

    (ii) After 1960, a new name published as that of a "variety" or "form" is to be regarded as of infrasubspecific rank.                              6

**Article 46.  Categories co-ordinate.**  The categories in the species-group are of co-ordinate status in nomenclature, that is, they are subject to the       35
same rules and recommendations, and a name established for a taxon in either category in the group, and based on a given type-specimen, is thereupon available with its original date and author for a taxon based on the same type-specimen in the other category.

**Article 47.  Subordinate taxa.**

(*a*) **Nominate subspecies.**  The subspecies that contains the type-specimen of a subdivided species bears the same name as the species and is termed     25,36
the "nominate" subspecies.

(*b*) **Change of nominate subspecies.**  If the invalid name of a subdivided species is replaced by the name of a different subspecies [Art. 23*e*(iii)],     23,36
the latter then becomes the nominate subspecies.

**Article 48.  Binomina.**

(*a*) **Change of genus.**  After its original establishment, a specific name becomes part of another binomen whenever it is transferred to a different genus.

(*b*) **Generic name conditional.**  If a zoologist publishes a binomen, using a previously established generic name in combination with an available specific name, but at the same time conditionally proposes a new generic name in combination with the specific name, he is considered to have established two binomina, of which the former has priority.

**Example.** Lowe in 1843 published a new species, *gracilis*, in the genus *Seriola* Cuvier, 1816, but at the same time conditionally proposed a new genus *Cubiceps* for that species. He is considered to have established first the binomen *Seriola gracilis*, and second the binomen *Cubiceps gracilis*.

**Article 49. Misidentifications.** The specific name used in an erroneous specific identification cannot be retained for the species to which the name was wrongly applied, even if the two species in question are in, or are later referred to, different genera [see Art. 70*b*].

2
49

**Example.** It is found that Smith, 1850, had recorded as "*A-us b-us* Dupont, 1800" a species different from that actually so named by Dupont. The specific name *b-us* cannot be used to denote the species that was before Smith, even if placed in a different genus from the true *b-us* Dupont.

## XI. AUTHORSHIP

42

**Article 50. Author of a name.** The author (authors) of a scientific name is (are) the person (persons) who first publish(es) it [III] in a way that satisfies the criteria of availability [IV], unless it is clear from the contents of the publication that only one (or some) of the joint authors, or some other person (or persons), is alone responsible both for the name and the conditions that make it available.

(*a*) **Exception for names in minutes.** If the name of a taxon is established by publication in the minutes of a meeting, the person responsible for the name, not the secretary or other reporter of the meeting, is the author.

**Recommendation 50A. Information in minutes.** Secretaries and other reporters of meetings should not include in their published reports new names of taxa or any information affecting nomenclature.

(*b*) **Change in rank.** Change in rank of a taxon within the family-, genus-, or species-group does not affect the authorship of the nominal taxon.

(*c*) **Justified emendation.** A justified emendation is attributed to the original author of the name [Art. 33*a*].

(*d*) **Unjustified emendation.** An unjustified emendation is attributed to the author who published it [Art. 33*a*].

**Article 51. Citation of name of author.**

42

(*a*) **Optional use.** The name of the author does not form part of the name of a taxon and its citation is optional.

(*b*) **Form of citation.** The original author's name, when cited, follows the scientific name without any intervening marks of punctuation, except as noted in Section (*d*) and Recommendation 51A.

  (i) The name of a subsequent user of a scientific name, if cited, is to be separated from it in some distinctive manner, other than by a comma.

**Example.**   Reference to *Cancer pagurus* Linnaeus as used by Latreille may be cited as

> *Cancer pagurus* Linnaeus sensu Latreille,
> *Cancer pagurus:* Latreille,

or in some other distinctive manner, but not as

> *Cancer pagurus* Latreille, nor as
> *Cancer pagurus,* Latreille.

**Recommendation 51A.   Anonymous authors.**   If the name of a taxon was published anonymously, but its author is known, his name, if cited, should be enclosed in square brackets to show the original anonymity.

(*c*) **Citation of contributors.**   If a scientific name and its validating conditions are the responsibility not of the author(s) of the publication containing them, but only of one (or some) of the authors, or of some other zoologist, the name of the author(s), if cited, is to be cited as "B" in "A" or "B" in "A & B," or whatever combination is appropriate.

(*d*) **Use of parentheses in new combinations.**   If a species-group taxon was described in a given genus and later transferred to another, the name of the author of the species-group name, if cited, is to be enclosed in parentheses.

**Example.**   *Taenia diminuta* Rudolphi, when transferred to the genus *Hymenolepis*, is cited as *Hymenolepis diminuta* (Rudolphi).

(i) The use of parentheses here applies only to transfers from one nominal genus to another, and is not affected by the presence of a subgeneric name, or by any shifts of rank or position within the same genus.

**Recommendation 51B.   Citation of author of new combination.**   If it is desired to cite the names both of the original author of a species-group name and of the reviser who transferred it to another genus, the name of the reviser should follow the parentheses that enclose the name of the original author.

**Example.**   *Limnatis nilotica* (Savigny) Moquin-Tandon.


## XII.   HOMONYMY                                                       43


**Article 52. Homonymy.**   In the meaning of the Code, homonymy is the identity in spelling of available names denoting different species-group taxa within the same genus, or objectively different taxa within the genus-group or within the family-group.

**Article 53. Law of Homonymy.**   Any name that is a junior homonym of   16
an available name must be rejected and replaced.

**Article 54. Names that do not enter into homonymy.**   The following names do not enter into homonymy:

(1) names that are unavailable in the meaning of the Code, except as noted in Arts. 20 and 56*b*;
(2) names that have never been used for a taxon in the animal kingdom;
(3) names that are excluded from zoological nomenclature [Art. 1]; and
(4) incorrect spellings, both original and subsequent.

**Article 55. Family-group names.**   Family-group names that are based   33
on different type-genera and that are identical, or differ only in suffix, are   37
homonyms.

(*a*) **Homonymy from similar generic names.**  If homonymy between names
in the family-group results from similarity but not identity of the names of
their type-genera, the case is to be referred to the Commission.

**Example.**  Two generic names, *Merope* (Insecta) and *Merops* (Aves), each
resulted in the family name MEROPIDAE. To avoid the homonymy, the Commission ruled that *Merope* should form the family name MEROPEIDAE (Opinion 140).

(*b*) **Homonymy from incorrect spelling.**  A family-group name is not to be
rejected as a junior homonym if the homonymy results from the incorrect
spelling of the earlier name.

**Example.**  PSILOPINAE Cresson, 1925, based on *Psilopa* Fallén, is not to be rejected as a homonym of PSILOPINAE Schiner, 1868, which was based on *Psilopus*
Meigen and should have been written PSILOPODINAE.

**Article 56. Genus-group names.**  The Law of Homonymy applies to all
names in the genus-group, including those of collective groups.

(*a*) **One-letter difference.**  Even if the difference between two genus-group
names is due to only one letter, these two names are not to be considered
homonyms.

**Example.**  Two genera of Diptera, *Microchaetina* Van der Wulp, 1891, and
*Microchaetona* Townsend, 1919, are not homonyms; but see Appendix D3.

(*b*) **Names ending in ites, ytes, or ithes given to fossils.**  A genus-group name
formed for use in paleontology by substituting -*ites*, -*ytes*, or -*ithes* for the
original termination of a generic name, and applied only to fossils, enters
into homonymy [Art. 20].

(*c*) **Precedence of genus over subgenus.**  Of two homonymous genus-group
names of identical date, one proposed for a genus takes precedence over
one proposed for a subgenus.

**Article 57. Species-group names.**  The Law of Homonymy applies to
species-group names originally published in (primary homonyms), or later
brought together in (secondary homonyms) the same genus or collective
group, except as noted in Art. 59*c*.

**Example.**  As separate proposals of new names, the following nominal taxa
called *intermedius* are primary homonyms of one another:

> *A-us intermedius* Pavlov,
> *A-us intermedius* Dupont,
> *A-us albus intermedius* Black, and
> *A-us concolor intermedius* Schmidt.

(*a*) **Subgeneric name.**  The presence of a subgeneric name does not affect
homonymy between species-group names within the same genus.

**Example.**  *A-us* (*B-us*) *intermedius* Pavlov and *A-us* (*C-us*) *intermedius* Dupont are
primary homonyms, but *A-us* (*B-us*) *intermedius* Pavlov is not a primary homonym
of *B-us intermedius* Black.

(*b*) **Differences in spelling.**  Species-group homonymy within a given nominal genus is not obviated by any emendation or incorrect spelling of the
generic name.

    (i) Differences in termination that are due solely to gender are to be
disregarded in determining whether adjectival species-group names
are homonyms.

(*c*) **Identical names in homonymous genera.**  Homonymy does not exist

between two identical species-group names originally or subsequently    44
placed in different genera that bear homonymous names.

**Example.** *Noctua* (Insecta) and *Noctua* (Aves) are homonyms, but *Noctua variegata* Jung (Insecta) and *Noctua variegata* Quoy and Gaimard (Aves) are not.

(*d*) **One-letter difference.** Except as specified in Art. 58, the difference of a single letter is sufficient to prevent homonymy.

**Examples.** *Raphidia londinensis* and *Raphidia londonensis* (derived from Londinium and London, words of the same origin and meaning), and *Chrysops calidus* and *Chrysops callidus* (derived from words of different origin and meaning) are not homonyms; but see Appendix D5.

(*e*) **Precedence of species over subspecies.** Of two homonymous species-    45
group names of identical data, one proposed for a species takes precedence over one proposed for a subspecies.

**Article 58. Variable spelling.** Two or more species-group names of the same origin and meaning and cited in the same nominal genus or collective group are to be considered homonyms if the only difference in spelling consists of any of the following (for diacritic and other marks, see Arts. 27 and 32*c*):

(1) the use of *ae, oe,* or *e* (e.g. *caeruleus, coeruleus, ceruleus*);
(2) the use of *ei, i,* or *y* (e.g. *cheiropus, chiropus, chyropus*);
(3) the use of *c* or *k* (e.g. *microdon, mikrodon*);
(4) the aspiration or non-aspiration of a consonant (e.g. *oxyrhynchus, oxyryncus*);
(5) the presence or absence of *c* before *t* (e.g. *auctumnalis, autumnalis*);
(6) the use of a single or double consonant (e.g. *litoralis, littoralis*);
(7) the use of *f* or *ph* (e.g. *sulfureus, sulphureus*);
(8) the use of different connecting vowels in compound words (e.g. *nigricinctus, nigrocinctus*);
(9) the transcription of the semi-vowel *i* as *y, ei, ej,* or *ij;*
(10) the termination *-i* or *-ii* in a patronymic genitive (e.g. *smithi, smithii*);
(11) the suffix *-ensis* or *-iensis* in a geographical name (e.g. *timorensis, timoriensis*); and
(12) three pairs of names treated as special cases: *saghalinensis* and *sakhalinensis; sibericus* and *sibiricus; tianschanicus* and *tianshanicus.*

**Article 59. Primary and secondary homonyms** [Art. 57].    43

(*a*) **Primary homonyms.** A species-group name that is a junior primary    16
homonym must be permanently rejected.

(*b*) **Secondary homonyms.** A species-group name that is a junior secondary    2
homonym must be rejected by any zoologist who believes that the two    44
species-group taxa in question are congeneric.

(*c*) **Revival of secondary homonyms.** A name rejected after 1960 as a    9
secondary homonym is to be restored as the valid name whenever a zoolo-    43
gist believes that the two species-group taxa in question are not congeneric, unless it is invalid for other reasons.

(i) In such a case the name proposed in replacement of the secondary homonym becomes a junior objective synonym of the latter.

**Example.** *A-us niger* Smith, 1960, is believed to be congeneric with *B-us niger* Dupont, 1950, and when transferred to *B-us* becomes a junior secondary homonym, and is renamed *B-us ater* Jones, 1970. If subsequently *A-us niger* Smith, 1960 is no longer believed to be congeneric with *B-us niger* Dupont, 1950, the former specific name is again to be used. *B-us ater* Jones, 1970, then becomes a junior objective synonym of *A-us niger* Smith, 1960.

**Article 60. Replacement of rejected homonyms.** A rejected homonym must be replaced by an existing available name or, for lack of such a name, by a new name.

    (*a*) **Junior homonyms with synonyms.** If the rejected homonym has one or more available synonym(s), the oldest of these must be adopted, with its own authorship and date.

        (i) A subjective synonym retains eligibility as a replacement name only so long as it is regarded as a synonym of the rejected name.

    (*b*) **Junior homonyms without synonyms.** If the rejected homonym has no known available synonym, it must be replaced by a new name which will then compete in priority with any synonym recognized later.

## XIII. THE TYPE-CONCEPT

**Article 61. Relationship of the type to the taxon.** The "type" affords the standard of reference that determines the application of a scientific name. Nucleus of a taxon and foundation of its name, the type is objective and does not change, whereas the limits of the taxon are subjective and liable to change. The type of a nominal species is a specimen, that of a nominal genus is a nominal species, and that of a nominal family is a nominal genus. Each taxon has, actually or potentially, its type. The type of any taxon, once fixed in conformity with the provisions of the Code, is not subject to change except by exercise of the plenary powers of the Commission [Art. 79], or, exceptionally in species-group taxa, under the provisions of Art. 75.

    (*a*) **Types of nominate subordinate taxa.** The type of a taxon is also the type of its nominate subordinate taxon, if there is one, and vice-versa. Therefore, the designation of one implies the designation of the other.

        (i) If different types are designated simultaneously for a nominal taxon and for its nominate subordinate taxon, the designation for the former takes precedence.

    (*b*) **Types and synonymy.** If two taxa are based on the same type, their names are objective synonyms. If two taxa with different types are subjectively united into a single taxonomic unit, their names are subjective synonyms.

## XIV. TYPES IN THE FAMILY-GROUP

**Article 62. Application.** The provisions of this Chapter apply equally to all categories in the family-group.

**Article 63. Types of family-group taxa.** The type of each taxon of the family-group is that nominal genus upon [the name of] which the family-group name is based [Arts. 35*b*, 39, 40].

**Article 64. Choice of type-genus.**   A zoologist establishing a new family-group taxon is free to choose as type-genus any included nominal genus, not necessarily that bearing the oldest name.

2
35

**Recommendation 64A.   Type-genus should be well-known.**   So far as possible, a zoologist who establishes a family-group taxon should base it on a genus that is both well known and representative of the family.

## Article 65. Identification of the type-genus.

- (*a*) **Correct identification assumed.**   It is to be assumed that an author publishing a new family-group name has correctly identified the nominal type-genus of the taxon in question.

49

- (*b*) **Misidentification or altered concept.**   If the nominal type-genus of a family-group taxon is found to be based on a misidentified type-species [Art. 70], or if a subsequent fixation of the type-species of a nominal type-genus has confused the accepted meaning of a family-group name, the case is to be referred to the Commission.

49

## XV.   TYPES IN THE GENUS-GROUP

**Article 66. Application.**   The provisions of this Chapter apply equally to the categories genus and subgenus but not to collective groups, which require no type-species [Art. 42*c*].

## Article 67. General provisions.

- (*a*) **Types of genus-group taxa.**   The type of each nominal genus is a nominal species known as the "type-species" [Art. 42*b*].

**Recommendation 67A.   Terminology.**   Only the term "type-species" or a strictly equivalent term in another language should be used in referring to the type of a genus. The term "genotype" should never be used for this purpose.

- (*b*) **Kinds of type-designation.**   The type-species of a nominal genus is termed "type by original designation" if it is definitely designated in the original publication [Art. 68*a*], "type by indication" if determined by the application of provisions (*b*) to (*d*) in Art. 68, and "type by subsequent designation" if designated after the establishment of the nominal genus [Art. 69].

56

- (*c*) **Designation.**   The term "designation" in relation to the fixation of a type-species must be rigidly construed; a designation made in an ambiguous or qualified manner is invalid.
    - (i) Mention of a species as an example of a genus does not constitute a type-designation.
    - (ii) Reference to a particular structure as "type" or "typical" of a genus does not constitute a type-designation.

**Examples.**   A statement such as any of the following is not to be regarded as a type-designation in the meaning of this Section: "*A-us b-us* may possibly be regarded as the type of *A-us*"; "*A-us b-us* is a typical example of the genus *A-us*"; "the venation of the anterior wings of *A-us b-us* is typical of the genus *A-us*."

- (*d*) **Types of nominate subgenera.**   See Art. 61*a*.
- (*e*) **Objective synonymy of the type-species.**   If a nominal species, type of

a genus, is found to be a junior objective synonym, the senior synonym is to be cited as the name of the type-species of the genus in question.

**Example.** *Astacus marinus* Fabricius, 1775, one of the species originally included in the genus *Homarus* Weber, 1795, was designated as the type-species of *Homarus*. However, *Astacus marinus* is a junior objective synonym of *Cancer gammarus* Linnaeus, 1758, which is therefore to be cited as the type-species of *Homarus*.

(*f*) **Actions of original author.** Only the statements or other actions of the the original author when establishing a new nominal genus are relevant in deciding

 (i) whether the type-species has been designated or indicated in conformity with provisions (*a*) to (*d*) of Art. 68, and

 (ii) which species are the originally included species in the meaning of Art. 69*a*.

(*g*) **Incorrect reference to establishment of genus.** If, in designating the type-species for a nominal genus, an author refers the generic name to an author or date other than those denoting the first establishment of the genus or the first express reference of nominal species to it, he is nevertheless to be considered, if the species was eligible, to have designated the type-species correctly.

**Example.** *A-us* Dupont, 1790, established without a designated or indicated type-species, is best known from the work of a later author, Smith, 1810. If subsequently *b-us* is designated as the type-species of "*A-us* Smith, 1810," that designation is to be accepted as valid for *A-us* Dupont, 1790, if *b-us* was eligible for designation as type-species of the latter.

(*h*) **Exclusions.** A nominal species that was not included, or that was cited as a species inquirenda or species incertae sedis when a new nominal genus was established, cannot be validly designated or indicated as the type-species of that genus.

(*i*) **Replacement names.** If a zoologist proposes a new generic name expressly as a replacement for a prior name, both nominal genera must have the same type-species, and, subject to (i) below, type-fixation for either applies also to the other, despite any statement to the contrary.

**Example.** *B-us* Schmidt, 1890, is proposed expressly as a replacement name for a junior homonym, *A-us* Medina, 1880, non Dupont, 1860. If *x-us* is the designated type-species of *A-us*, it is ipso facto the type-species of *B-us*.

 (i) The type-species must be a species eligible for fixation as the type of the earlier nominal genus.

 (ii) An emendation of a generic name, whether justified or unjustified, is an objective synonym of the original name and therefore has the same type-species.

(*j*) **Misidentified type-species.** If a designated or indicated type-species is later found to have been misidentified, the provisions of Art. 70 apply.

(*k*) **Union of genera.** The union of two or more nominal genera to form a single taxonomic genus does not change the type-species of any nominal genus involved [XIII], and the type-species of the combined genus is that of the senior component nominal genus.

## Article 68. Type-species fixed in the original publication.

The provisions of this Article apply in the following order of precedence.

(*a*) **Original designation.** If one nominal species is definitely designated as the type-species of a new nominal genus when the latter is established, that species is the type-species, regardless of any other consideration (type by original designation).

  (i) The formula "gen.n., sp.n.," or its exact equivalent, applied before 1931 to only one of the new nominal species included in a newly established nominal genus, is to be interpreted as original designation if no other type-species was designated.

(*b*) **Use of typicus or typus.** If, when a new nominal genus is established, one of the included new species is named *typicus* or *typus*, that species is the type-species.

(*c*) **Monotypy.** A genus originally established with a single nominal species takes that species as its type, regardless of whether the author considered the genus to contain other species that he did not name, and regardless of cited synonyms, subspecies, unavailable names, and species that are doubtfully included or identified (type by monotypy).

(*d*) **Tautonymy.** If a newly established nominal genus contains among its originally included nominal species one possessing the generic name as its specific or subspecific name, either as the valid name or as a cited synonym, that nominal species is ipso facto the type-species (type by absolute tautonymy).

  (i) If, in the synonymy of only one of the species originally included in a nominal genus established before 1931, there is cited a pre-1758 name of one word identical with the new generic name, that nominal species is construed to be the type-species (type by Linnean tautonymy).

## Article 69. Type-species not fixed in the original publication. The

provisions of this Article apply in the following order of precedence only to nominal genera that were established before 1931 without an originally designated or indicated type-species.

(*a*) **Subsequent designation.** If an author established a nominal genus but did not designate or indicate its type-species, any zoologist may subsequently designate as the type-species one of the originally included nominal species, or, if there were no original nominal species, one of those first subsequently referred to the genus (type by subsequent designation).

  (i) In the meaning of this provision, the "originally included species" comprise only those actually cited by name in the newly established nominal genus, either as valid names (including subspecies, varieties, and forms), as synonyms, or as stated misidentifications of previously established species [Art. 70(*b*)].

  (ii) If no nominal species were included at the time the genus was established, the nominal species-group taxa that were first subsequently and expressly referred to it are to be treated as the only originally included species.

    (1) Mere reference to a publication containing the names of species does not by itself constitute the inclusion of species in a nominal genus.

    (2) If only one nominal species was first subsequently referred to a genus, it is ipso facto the type-species, by subsequent monotypy.

    (3) If two or more nominal species were simultaneously referred to a nominal genus, all are equally eligible for subsequent type-designation.

(iii) In the absence of a prior valid type-designation for a nominal genus, an author is considered to have designated one of the originally included nominal species as type-species, if he states that it is the type (or type-species), for whatever reason, right or wrong, and if it is clear that he himself accepts it as the type-species.

(iv) If an author designates (or accepts another's designation) as type-species a nominal species that was not originally included, and if, but only if, at the same time he synonymizes that species with one of the originally included species, his act constitutes designation of the latter as type-species of the genus.

(v) A nominal species is not rendered ineligible for designation as a type-species by reason of being the type-species of another genus.

(vi) A subsequent designation first made in a literature-recording publication is acceptable, if valid in all other respects.

**Recommendation 69A. Preference for figured species.** In designating a type-species for a genus, a zoologist should give preference to a species that is adequately figured.

**Recommendation 69B. Other considerations in designating type-species.** In the subsequent designation of a type-species a zoologist should guide himself by the following precepts, listed in order of precedence:

(1) In the case of Linnean genera he should designate the most common species or one of medical importance (Linnean aphorism 246, *Critica Botanica*, 1737).

(2) If the name or a synonym of one of the originally included nominal species is virtually the same as the generic name, or is of the same origin or meaning, that species should be designated as the type-species (type by virtual tautonymy), unless such designation is strongly contra-indicated by other factors.

**Examples.** *Bos taurus, Equus caballus, Ovies aries, Scomber scombrus, Sphaerostoma globiporum;* contra-indicated in *Dipetalonema dipetalum* because only one sex was described, based on a single specimen not studied in detail.

(3) If some of the originally included nominal species have been removed to other genera, preference should be given to a remaining species, if any such are suitable (choice by elimination).

(4) A species based on a sexually mature specimen is generally preferable to one based on a larval or otherwise immature specimen.

(5) Preference should be given to a species named *communis, vulgaris, medicinalis,* or *officinalis.*

(6) Preference should be given to the best described, best figured, best known, or most easily obtainable species, or to one of which a type-specimen is accessible.

(7) If more than one group of species is recognized in a genus, preference should be given to a species that belongs to as large a group of species as possible (de Candolle's rule).

(8) In genera of parasites, preference should be given to a species that parasitizes man, or an animal of economic importance, or a common and widespread host-species.

(9) All other things being equal, preference should be given to a species well known to the author of the nominal genus, prior to publishing the generic name.

(10) If an author habitually placed a leading or typical species first as "chef de file," and described others by comparison with it, that fact should be considered in the designation of a type-species.

(11) If an author is known to have denoted type-species by their position

("first species rule"), the first nominal species cited by him should be designated as the type-species.

(12) All other things being equal, preference should be given to the species cited first in the work, page, or line (position precedence).

**Recommendation 69C. Citation of type-species.** When designating a type-species for a nominal genus established before 1931, a zoologist should cite the name of that species first in its original binomen and then in its current binomen, if this is different. He should give a bibliographic reference to the work where the species was established.

**Article 70. Identification of the type-species.** It is to be assumed that an author correctly identifies the nominal species that he either (1) refers to a new genus when he establishes it, or (2) designates as the type-species of a new or of an established genus.

    (*a*) **Misidentified type-species.** If a zoologist considers that such a species was misidentified, he is to refer the case to the Commission to designate as the type-species (by use of its plenary powers if necessary [Art. 79]) whichever species will in its judgment best serve stability and uniformity of nomenclature, either

        (i) the nominal species actually involved, which was wrongly named in the type-designation; or

        (ii) if the identity of that species is doubtful, a species chosen in conformity with the usage of the generic name prevailing at the time the misidentification is discovered; or

        (iii) the species named by the designator, regardless of the misidentification.

    (*b*) **Deliberate use of misidentification.** If the type designated for a new nominal genus is a previously established species, but the designator states that he employs its specific name in accordance with the wrong usage of a previous author, the type-species is to be interpreted as the one actually before the designator, not the one that correctly bears the name.

        (i) In such a case, the author of the new nominal genus is considered to have established also a new nominal species, with the same specific name as the misidentified species, in the new nominal genus.

**Example.** If Jones, 1900, designated as type-species of *C-us*, gen. n., a species that he cites in some such manner as *A-us b-us* Dupont sensu Schmidt, 1870, the type-species of *C-us* is that which was before Jones, not that named by Dupont, and its name is to be cited as *C-us b-us* Jones, 1900.

## XVI.  TYPES IN THE SPECIES-GROUP

**Article 71. Application.** The provisions of this Chapter apply equally to all categories in the species-group.

**Article 72. General provisions.**

    (*a*) **Types of species-group taxa.** The type of each taxon of the species-group is a single specimen, either the only original specimen or one designated from the type-series (holotype, lectotype), or a neotype [Art. 45*b*].

    (*b*) **Type-series.** The type-series of a species consists of all the specimens on which its author bases the species, except any that he refers to as variants, or doubtfully associates with the nominal species, or expressly excludes from it.

(c) **Specimens that are already types.**  The fact that a specimen is already the type of one nominal species does not prevent its designation as the type of another.

(d) **Types of replacement nominal species.**  If an author proposes a new specific name expressly as a replacement for a prior name, but at the same time applies it to particular specimens, the type of the replacement nominal species must be that of the prior nominal species, despite any contrary designation of type-specimen or different taxonomic usage of the replacement name.

(e) **Types of nominate subspecies.**  See Art. 61(a).

(f) **Value of types.**  Holotypes, syntypes, lectotypes, and neotypes are to be regarded as the property of science by all zoologists and by persons responsible for their safe-keeping.

**Recommendation 72A.  Institutional custody.**  A zoologist who designates a holotype or lectotype should deposit it in a museum or other institution where it will be safely preserved and will be accessible for purposes of research. Deposit of neotypes in a museum or other institution is mandatory [Art. 75c(6)].

**Recommendation 72B.  Labelling.**  A zoologist designating a holotype, lectotype, or neotype should unmistakably label the specimen in a way that will clearly indicate its status.

**Recommendation 72C.  Information on labels.**  When designating a holotype, lectotype, or neotype, a zoologist should publish all information that appears on the labels accompanying the specimen, so as to ensure the future recognition of the specimen.

**Recommendation 72D.  Institutional responsibility.**  Every institution in which types are deposited should
  (1) ensure that all are clearly marked so that they will be unmistakably recognized;
  (2) take all necessary steps for their safe preservation;
  (3) make them accessible for study;
  (4) publish lists of type-material in its possession or custody; and
  (5) so far as possible communicate information concerning types when requested by zoologists.

**Recommendation 72E.  Type-localities.**  An author who either designates or restricts a type-locality should base his action on one or more of the following criteria:
  (1) the original description of the taxon;
  (2) data accompanying the original material;
  (3) collector's notes, itineraries, or personal communications; and
  (4) as a last resort, localities within the known range of the species or from which specimens identified with the species have been taken.
If a type-locality was erroneously designated or restricted, it should be corrected.

## Article 73.  Holotypes and syntypes.

(a) **Single specimen.**  If a nominal species is based on a single specimen, that specimen is the "holotype."

(b) **Specified type.**  If an author states in the description of a new nominal species that one specimen and only one is "the type" or uses some equivalent expression, that specimen is the holotype.

(c) **Syntypes.**  If a new nominal species has no holotype under the provision of (*a*) and (*b*), all the specimens of the type-series are "syntypes," of equal value in nomenclature.

> (i) Syntypes may include specimens labelled "cotype" (in the meaning of syntype), "type," or by some other term, or with no identifying label, or specimens not seen by the author but which were the bases of previously published descriptions or figures upon which he founded his taxon in whole or in part.

**Recommendation 73A. Original designation.** A zoologist when describing a new species should clearly designate a single specimen as its holotype.

**Recommendation 73B. Procedure.** If a zoologist, in basing a new nominal species on specimens before him, subjectively associates with it specimens that he believes to have been misidentified by another author, he should designate his holotype from the former.

**Recommendation 73C. Data on the holotype.** A zoologist in establishing a new species should publish at least the following data concerning its holotype, in so far as they are relevant and known to him:

> (1) the size;
> (2) the full locality, date, and other data on the labels accompanying the holotype;
> (3) the sex, if the sexes are separate;
> (4) the developmental stage, and the caste, if the species includes more than one caste;
> (5) the name of the host species;
> (6) the name of the collector;
> (7) the collection in which it is situated and any collection- or register number assigned to it;
> (8) in the case of a living terrestrial species the elevation in metres above sea-level at which it was taken;
> (9) in the case of a living marine species the depth in metres below sea-level at which it was taken;
> (10) in the case of a fossil species, its geological age and stratigraphical position, stated if possible, in metres above or below a well-established plane.

52

**Recommendation 73D. Paratypes.** After the holotype has been labelled, each remaining specimen (if any) of the type-series should be conspicuously labelled "paratype," in order clearly to identify the components of the original type-series.

**Recommendation 73E. Avoidance of "cotype."** To avoid misunderstandings, a zoologist should not use the term "cotype."

## Article 74. Lectotypes.

> (*a*) **Designation of a specimen.** If a nominal species has no holotype, any zoologist may designate one of the syntypes as the "lectotype."
>
> > (i) The first published designation of a lectotype fixes the status of the specimen, but if it is proved that the designated specimen is not a syntype, the designation is invalid.
>
> (*b*) **Designation by means of a figure.** Designation of a figure as lectotype is to be treated as designation of the specimen represented by the figure; if that specimen is one of the syntypes, the designation as lectotype is valid from the nomenclatural standpoint.
>
> (*c*) **Designation to be individual.** Lectotypes must not be designated collectively by a general statement; each designation must be made specifically for an individual nominal species, and must have as its object the definition of that species.

4
54
56
57

**Example.** A published statement that, in the type-series of all species described by a particular author, the specimen bearing the author's determination label, or the only surviving syntype, is to be treated as the lectotype, is not a valid designation of lectotypes.

**Recommendation 74A. Agreement with previous restriction.** In designating a lectotype, a zoologist should in general act consistently with, and in any event should give great weight to, previous valid restrictions of the taxonomic species, in order to preserve stability of nomenclature.

**Recommendation 74B. Figured specimen.** A zoologist should choose as lectotype a syntype of which a figure has been published, if such exists.

**Recommendation 74C. Data on the lectotype.** A zoologist who designates a lectotype should publish the data listed in Recommendation 73C, besides describing any individual characteristics by which it can be recognized.

**Recommendation 74D. Syntypes in several collections.** When possible, a lectotype should be chosen from syntypes in the collection of a public institution, preferably of the institution containing the largest number of syntypes of the species, or containing the collection upon which the author of the nominal species worked, or containing the majority of his types.

**Recommendation 74E. Paralectotypes.** A zoologist who designates a lectotype should clearly label any remaining syntypes with the designation "paralectotype."

**Article 75. Neotypes.** Subject to the following limitations and conditions, a zoologist may designate another specimen to serve as the "neotype" of a species if, through loss or destruction, no holotype, lectotype, or syntype exists.

55

(*a*) **Cases admitted.** A neotype is to be designated only in connection with revisory work, and then only in exceptional circumstances when a neotype is necessary in the interests of stability of nomenclature.

    (i) The words "exceptional circumstances" refer to those cases in which a neotype is essential for solving a complex zoological problem, such as the confused or doubtful identities of closely similar species for one or more of which no holotype, lectotype, or syntype exists.

(*b*) **Cases excluded.** A neotype is not to be designated for its own sake, or as a matter of curatorial routine, or for a species of which the name is not in general use either as a valid name or as a synonym.

(*c*) **Qualifying conditions.** A neotype is validly designated only when it is published with the following particulars:

    (1) a statement of the characters that the author regards as differentiating the taxon for which the neotype is designated, or a bibliographic reference to such a statement;

    (2) data and description sufficient to ensure recognition of the specimen designated:

    (3) the author's reasons for believing all of the original type-material to be lost or destroyed, and the steps that have been taken to trace it;

    (4) evidence that the neotype is consistent with what is known of the original type-material, from its description and from other sources: however, if a nominal species is based on a sex or an immature stage that lacks good diagnostic characters, the neotype may differ in that respect from the original material;

    (5) evidence that the neotype came as nearly as practicable from the

original type-locality, and where relevant, from the same geological horizon or host-species as the original type-material;

(6) a statement that the neotype is, or immediately upon publication has become, the property of a recognized scientific or educational institution, cited by name, that maintains a research collection, with proper facilities for preserving types, and that makes them accessible for study.

(d) **Priority.** The first neotype-designation published for a given nominal species in accordance with the provisions of this Article is valid, and any subsequent designation has no validity unless the first neotype is lost or destroyed.

**Recommendation 75A.   Consultation with specialists.**   Before designating a neotype, a zoologist should satisfy himself that his proposed designation does not arouse objections from other specialists in the group in question.

(e) **Status of previously designated neotypes.**   A neotype-designation published before 1961 takes effect from the time when it fulfills all the provisions of this Article.

**Recommendation 75B.   Validation.**   A zoologist who published an invalid neotype-designation before 1961 should be given an opportunity to validate it before another zoologist designates a neotype for the same nominal taxon.

**Recommendation 75C.   Preference for earlier neotypes.**   If an invalid neotype-designation was published before 1961, the specimen then designated should be given preference when a neotype for the same nominal taxon is validly designated.

(f) **Status of rediscovered type-material.**   If, after the designation of a neotype, original type-material is found to exist, the case is to be referred to the Commission.

## XVII. THE INTERNATIONAL COMMISSION ON ZOOLOGICAL NOMENCLATURE

**Article 76. Status.**   The International Commission on Zoological Nomenclature is a permanent body which derives all its power from the International Congresses of Zoology.                                                                59

**Article 77. Duties.**   The Commission is charged with the following duties:

(1) to consider for a period of at least one year in advance of a Congress (or for such less time as the Commission may agree) any proposal for a change in the Code;

(2) to submit to the Congress recommendations for the clarification or modification of the Code;

(3) to render between successive Congresses Declarations (i.e. provisional amendments to the Code) embodying such recommendations;

(4) to render Opinions and Directions on questions of zoological nomenclature that do not involve changes in the Code;

(5) to compile the Official Lists of accepted, and the Official Indexes of rejected, names and works in zoology;                                              63

(6) to submit reports to the Congresses on its work; and

(7) to discharge such other duties as the Congresses may determine.

**Article 78. Exercise of powers.**  The Commission has the power, when an
application is referred to it by any zoologist, to interpret the provisions of
the Code and to apply such interpretation to any question of zoological
nomenclature.

59
60

(a) **Declarations.**  If a case before the Commission involves a situation that
is not properly or completely covered by the Code, the Commission is to
issue a Declaration (a provisional amendment to the Code) and to propose
to the next succeeding Congress adoption of this amendment in the manner
prescribed in Article 87.

(i) A Declaration remains in force until the next succeeding Congress
ratifies it, either in its original form or in a modified form (whereupon the Code is accordingly amended), or rejects it; the Declaration is thereupon deemed repealed for all except historical purposes.

(b) **Opinions.**  If the case in question involves the application of the Code
to a particular situation relating to an individual name, act, or publication, the Commission is to render a decision, termed an Opinion, and
either

(i) to state how the Code is to be applied or interpreted; or

(ii) acting in the interest of stability and universality, to exempt, under
its plenary powers [Art. 79], the particular case from the application
of the Code, and to state the course to be followed.

(c) **Effective date of Opinions.**  Opinions have force immediately upon
publication of the ruling of the Commission, and are to be reported to the
next succeeding Congress.

(d) **Directions.**  Decisions completing earlier rulings, and formal instruments
required under automatic provisions of the Code, are called Directions.
They have the same status as Opinions.

(e) **Construction.**  All decisions are to be rigidly construed and no conclusions other than those expressly specified are to be drawn from them.

(f) **Official Lists and Indexes.**  Names and works that are accepted or rejected in Opinions are to be entered on the relevant Official Lists or Indexes, whereupon the Opinions concerned are deemed repealed for all
except historical purposes.

63

(g) **Review by Congress of Commission's decisions.**  A motion to modify
or reject any decision of the Commission is not to be considered by a Congress until notice of at least one year (or such less period as the Commission
may agree) has been given to the Commission.

**Article 79. Plenary powers.**  The Commission is empowered to suspend,
on due notice as prescribed by its Constitution, the application of any
provisions of the Code except those in the present and the next succeeding
Chapter, if such application to a particular case would in its judgment
disturb stability or universality or cause confusion. For the purpose of pre-
venting such disturbance and of promoting a stable and universally accepted
nomenclature, it may, under these plenary powers, annul or validate any
name, type-designation, or other published nomenclatural act, or any publication, and validate or establish replacements.

1
59

(a) **Guiding principles.**  In exercising its plenary powers, the Commission
is to be guided as follows:

(i) a name suppressed so as to validate the use of the same name pub-

3

lished at a later date in another sense, is to be suppressed for the purposes both of the Law of Priority and of the Law of Homonymy;

(ii) a name suppressed so as to validate a later name given to the same taxon is to be suppressed for the purposes of the Law of Priority but not for those of the Law of Homonymy;

(iii) if the Commission refuses to use its plenary powers in a given case, the Opinion rendered is to specify the name(s) to be used in the case in question, and the action (if any) to be taken.

**Article 80. Status of case under consideration.**  When a case is under consideration by the Commission, existing usage is to be maintained until the decision of the Commission is published.     61

**Article 81. Exemption.**  The Commission is under no obligation to search out violations of the Code, or to supplement or verify information contained in applications submitted to it, or to initiate any action within its field of competence, although it may, at its discretion, do any of these things.

**Article 82. Constitution and Bylaws.**  Regulations dealing with the membership of the Commission, its governing board, elections, voting procedures, meetings, and related matters are incorporated in the Constitution and Bylaws of the Commission.     59

(*a*) **Amendments to the Constitution.**  Changes in the Constitution can be made only by the Congresses, on recommendation by the Commission, in the same manner as amendments to the Code [Art. 87].

(*b*) **Amendments to the Bylaws.**  Changes in the Bylaws can be made by the Commission under procedures set forth in its Constitution.

## XVIII.  REGULATIONS GOVERNING THIS CODE

**Article 83. Title.**  The title of these rules and recommendations is: "INTERNATIONAL CODE OF ZOOLOGICAL NOMENCLATURE ADOPTED BY THE XV INTERNATIONAL CONGRESS OF ZOOLOGY, LONDON, JULY 1958."

**Article 84. Effective date.**  This Code comes into force on the day of its publication, and all previous editions of the International Rules of Zoological Nomenclature are thereby superseded.

(*a*) **Previous decisions affecting the Code.**  All amendments affecting the Code, adopted by the Congresses prior to the XV Congress, are no longer valid unless reaffirmed herein, and then only as here expressed.

**Article 85. Language of the official texts.**  The official French and English texts of the Code are equivalent in force, meaning, and authority. If it appears that there is a difference in meaning between the two texts, the problem is to be referred to the Commission for decision, and its interpretation is final.

**Article 86. Application.**  The provisions of the Code apply to all zoological names and works, published after 1757, that affect zoological nomenclature.

(*a*) **Previous decisions of the Commission.**    No decision taken by the Commission in relation to a particular name or work, prior to the effective date of this Code, is to be set aside without the consent of the Commission.

**Article 87. Amendment to the Code.**    Amendment to this Code can be made only by an International Congress of Zoology, acting on a recommendation from the Commission presented through and approved by the Section on Nomenclature of that Congress [Arts. 77 and 78*a*].

62

# Chapter 13    Interpretation of the Rules of Nomenclature

*L*ike any code of law, the International Code presents the rules starkly and without explanation. Even the experienced specialist is sometimes in doubt how to interpret the rules, and the beginner is often totally bewildered. The difficulty of understanding individual provisions is aggravated by the unavailability of a well-articulated presentation of the basic principles, which the individual rules attempt to implement. To fill this void is one of the aims of this chapter. It is hoped that the 66 numbered comments presented here will help the taxonomist when applying the rules to his particular problems. Over and above this purely practical objective, I hope that these comments will help the taxonomist to acquire a better understanding of the entire field of nomenclature. Stress throughout will be on principles rather than on practices. These comments are numbered consecutively, and the numbers refer to the equivalent numbers in the margin of the Code in Chap. 12. MLU refers to Mayr, Linsley, and Usinger, 1953, *Methods and Principles of Systematic Zoology; BZN* to the *Bulletin of Zoological Nomenclature; SZ* to the journal *Systematic Zoology*.

Where the application of the rules to a specific situation is involved, consultation of the Index of the Code is often very helpful.

TABLE OF TOPICS DISCUSSED IN CHAPTER 13

## 1. STABILITY

The statement in the Preamble of the Code that "the object of the Code is to promote stability" is perhaps the most important provision in the entire rules of nomenclature. The International Congress of Monaco in 1913 first accepted the principle, since frequently reaffirmed, that "when stability of nomenclature is threatened in an individual case, the strict application of the Code may under specified conditions be suspended by the International Commission on Zoological Nomenclature" (Preamble). In such cases any zoologist has the right to appeal to the Commission and ask for the protection of the threatened name.

More specific stability-promoting provisions in the Code can be found in Arts. 23b (statute of limitation), 23d(ii), 40, 40a (family names), 41 (misidentified genera), 70a (misidentified type-species), and 79 (plenary powers, for all actions on behalf of stability, including the designation of neotypes). The wise application of these provisions has greatly contributed to the stabilization of zoological nomenclature and to the refutation of the claim that taxonomists indulge in reckless and irresponsible name-changing.

Even if all nomenclatural name changes were eliminated (see 13.3), and even if the binominal system were replaced (see 13.14), certain name changes are inevitable. These include cases where:

1. Several species (e.g. sibling species) had been referred to by a single name prior to a more penetrating analysis
2. Several intraspecific phena were named in the belief that they represented different species
3. Several authors (especially in different countries) had named the same taxon, unbeknown to each other

Since the names of animals are part of a system of communication, total stability is the obvious ideal. Inevitable shortcomings in taxonomic analysis will make it impossible to achieve this goal at the present time, but every effort should be made to minimize interference with the stated ideal.

I shall give only a single example (supplied by Corliss, *BZN*, 24:155–185) for the need for stability in the biological literature. In a period of 27 years about 1,500 papers were published on the physiology, etc., of the protozoon *pyriforme* under the generic name *Tetrahymena* Furgason 1940. To replace this generic name for the sake of priority would obviously be a calamity as far as information retrieval is concerned. It was therefore quite legitimate for Corliss to ask for the suppression of ten senior synonyms of *Tetrahymena* that were totally or virtually unused in the vast biological literature of this genus.

## 2. FREEDOM OF TAXONOMIC THOUGHT

It is axiomatic that the freedom to make taxonomic decisions should not be restricted by rules of nomenclature. This is now expressly guaranteed in the Preamble, which states that none of the provisions of the Code "restricts the freedom of taxonomic thought or action." This basic principle is applied to specific situations throughout the Code. In Art. 11*d* it acknowledges that the subjective action of an author in placing a name in synonymy should not determine the nomenclature to be used by others; in Art. 59*b* it permits a zoologist to continue using a name considered by some other author to be a junior homonym; in Art. 64 it authorizes the zoologist to choose (as the type of a new family) that genus which he considers the most typical or characteristic genus of that family but "not necessarily that bearing the oldest name." The same is true for type designation on the generic level (Arts. 67–69).

Conversely, taxonomic mistakes should not affect nomenclature except where this is expressly authorized by the Code. For instance, the erroneous assignment of a species to a genus does not constitute the proposal of a new specific name in that genus (Art. 49); the naming of individual variants that do not form a taxon has no validity in nomenclature (Art. 45*c*); the erroneous placing of a name in synonymy does not invalidate it [Art. 17(i)].

It is possible that the principle of freedom of taxonomic thought should be extended to some additional articles. For instance, the provisions of Art. 64 are not applied to the choice of a type-genus of a new superfamily which has been created by a taxonomist by uniting a number of previously described families into an entirely new taxon (see 13.35).

## 3. PRINCIPLE OF PRIORITY

The most controversial problem in nomenclature arises whenever two names have been given to the same taxon. Fairness and common sense would seem to dictate a simple solution: let the prior name prevail. As soon as the oldest name for each taxon has been found, it is often asserted, complete stability of nomenclature will have been established automatically. This seemingly persuasive argument has, however, created great difficulties in many areas of systematic zoology and has been vigorously opposed by distinguished zoologists. In order to show why there is a controversy, the arguments pro and con will be presented in detail.

Linnaeus (1753) endorsed priority in principle. In Aphorism 243 he said, "If a generic name is suitable, it is not allowable to change it, even for another which is more fitting." In practice, however, he was an inveterate name-changer, perhaps because he considered many names not "suitable." For this he was upbraided by his friends. For instance, Peter Collinson wrote him in 1754, "But my dear friend we that admire you are much concerned that you should perplex the delightful science of botany with changing names that have been well-received." Even in his own works he shifted the names of taxa from one edition to the next, and indeed the rule of the first reviser (13.4) seems to have been more important to Linnaeus and his followers than the rule of priority. Fortunately, the last (12th) Linnaean edition (1766–67) of the *Systema Naturae* was accepted by his successors as the standard of reference for the next 50 or more years and provided for great stability during the ensuing period. Some of his successors, such as Fabricius among the entomologists, exerted a similar authority, ensuring great stability. As this authority weakened and as more and more zoologists entered the field of taxonomy, it occurred increasingly often that one and the same animal was given several names. During the French Revolution and the Napoleonic wars communication among scientists was difficult, and taxonomists in one country were often unaware of the new species and genera described by taxonomists in other countries. Equally or even more annoying was the rise of a generation of pedants who changed names the formation of which they considered not classically correct or not precisely descriptive. When a bird called *capensis* (in the belief that it came from the Cape of Good Hope), was found to have come from Java, it was renamed *javensis*. If a blue-green bird was called *viridis* but a subsequent author considered it more blue than green, he would rename it *caeruleus*. The result was nomenclatural confusion, if not anarchy.

Eventually the situation became so critical that the British Association for the Advancement of Science appointed a committee to draw up a general set of rules for zoological nomenclature. The resulting code (Strickland, 1842), often referred to as the "Strickland Code," formed the basis of all future codes and was the beginning of stabilization. Priority was considered to be the best method of achieving stability, and the changing or replacing of an earlier name merely because it was incorrectly formed or misleading, or for other personal, aesthetic, or even scientific reasons, was outlawed. It is evident from much of the contemporary literature on the subject that "priority," as conceived by these authors, was a *priority of usage* rather than a *priority of publication*. Yet priority of usage is subjective, and in the subsequent codification of the rules it was replaced by priority of publication in the endeavor to achieve objectivity. Unfortunately, while gaining objectivity, the nomenclaturists abandoned one of the most important objects of nomenclature, stability. As a result of the upheaval caused by a strict application of the principle of priority of publication, zoologists soon began to rebel against "priority." As early as 1849 Darwin wrote to Strickland with regard to cirriped nomenclature, "I believe if I were to follow the strict rule of priority more harm would be done than good. . . ." And this conviction was shared by an increasing number of zoologists. MLU (pp. 215–216) presents data on the percentage of name changes resulting from priority in various taxonomic groups.

In spite of serious objections, when the first International Code was adopted, it incorporated "strict priority" as a basic law. This started an extraordinary period of searching through the earliest post-Linnaean zoological publications and resulted in the discovery of countless totally forgotten names. Worse than that, it placed an undeserved emphasis on names that had been deliberately ignored by contemporaries because the descriptions were too poor for identification, contradictory, or misleading. It led to a search for still existing specimens (often erroneously called "types") of these early authors and deflected taxonomists from biological research into bibliographic archeology. Four types of discoveries were often made as the result of these searches: earlier synonyms, earlier homonyms, earlier fixations of generic types, and misidentified type specimens. None of these discoveries advanced our knowledge of animals or their classification. As Michener (1963, *SZ, 12*:163) has rightly said: "In other sciences the work of incompetents is merely ignored, in taxonomy, because of priority, it is preserved."

A rebellion against the unrestricted application of priority soon arose. An opinion poll among Scandinavian zoologists in 1911 showed that only 2 were in favor of a strict interpretation of the rule of priority while 120 were against it. Zoologists in other countries likewise voted in favor of limitations to priority, and the plenary powers decision (Art. 79) was adopted at Monaco in 1913.

Many zoologists felt that this still did not go far enough in protecting well-established, familiar names. It was pointed out that priority is a means to an end, the end being stability. Where the end (stability) is in conflict with the means (priority), it is the end which should have primacy. The feeling of exasperation with the strict application of priority is well expressed by a statement of the Chicago group of taxonomists (Schmidt, 1950):

> The whole business of elaborate argument over rule and "validity" of names has been a disgrace to zoology and has contributed more than any other single factor to the low repute of systematics among zoologists as a whole. Any system seeking continuity rather than change would have been infinitely preferable to the elaborate search for priority that was established as an international game. . . . The fundamental requirement is a reformation in attitude. Strict following of the law of priority, regardless of the consequences, must be set aside as the guiding principle in nomenclatural procedure, both in the thinking of the systematist and in the working of the Commission. In its stead, there must be substituted a proper regard for the convenience of zoologists generally and a determination, by stabilizing names and current use, to avoid change and chaos.

The succeeding international congresses (Copenhagen and London) paid attention to these sentiments and adopted the important statement in the Preamble that even though "priority is the basic principle of zoological nomenclature, its application however, under conditions specified in the Code, may be moderated to preserve a long accepted name in its accustomed meaning." Article 23 deals specifically with priority and its exceptions.

"When stability of nomenclature is threatened in an individual case, the strict application of the Code may under specified conditions be suspended by the International Commission on Zoological Nomenclature." Of particular importance are the limitations to priority specified in Arts. 23$b$, 23$d$(ii), 41, and 79.

### 4.  FIRST-REVISER PRINCIPLE

A basic Linnaean principle which was widely followed by his disciples was that the action of the first reviser should be adopted in all equivocal nomenclatural situations. This principle, although sanctioned by usage, was almost entirely ignored in the drafting of the 1901 Code. It has been incorporated, however, in numerous provisions of the 1961 Code.

Linnaeus himself was an inveterate first reviser. The reviser principle to him was quite clearly more important than the principle of priority. In later publications he again and again changed what he had done in his own earlier publications, and his students and followers invariably accepted his last action rather than that which had priority. This is why the generations following Linnaeus accepted the 12th edition (1766) of the *Systema Naturae* as the standard of reference rather than the 10th edition (1758). This is why, for at least the first 100 years after Linnaeus, the type-species of a genus was determined by the method of elimination rather than that of fixation. It was the almost complete disregard of the first-reviser principle in the original Code of Zoological Nomenclature (1901) that was responsible for much of the name changing in the first half of the century.

There are numerous ways by which a first reviser can help to stabilize nomenclature. In the case of simultaneously published names (Art. 24$a$ and Recommendation 24$a$) he can select the name which is better known rather than that which has line or page precedence (which is not priority!). If a new name is spelled in more than one way in the original publication, Art. 32$b$ permits the first reviser to accept that which is most commonly used. If no type-species is fixed in the publication in which a new genus was described prior to 1931, the first reviser can select the species whose designation is in the best interest of stability (Art. 69). In the case of a composite name, based on a series of syntypes consisting of several species, the first reviser can help stability by making either of two choices (Art. 74$a$). He can designate as *lectotype* a specimen of the species to which the name has always been applied in the past, or if it is a name that is best suppressed, for the sake of stability, he can designate a *lectotype* belonging to a species for which a senior synonym is available. (See 13.56 for all first-reviser actions concerning types.) Finally, the first-reviser principle is important in the designation and restriction of type-localities (see 13.57).

## 5. RANGE OF AUTHORITY OF THE CODE

The Code applies to both living and extinct animals (Art. 1). If a still-living species was first named on the basis of fossil material, that name is also valid for the living species. If a generic name has been used for a fossil animal, it cannot be used again for a different genus of living animals and vice versa. Zoological nomenclators therefore contain the names given to both living and extinct animals so that the possible occurrence of homonymy can be determined.

There are, however, separate codes for botanical and bacteriological nomenclature, and Art. 2 of the Zoological Code spells out the relation between the codes and the status of names transferred between the kingdoms. Familiarity with the botanical rules is important for authors working with protists.

## 6. APPLICATION OF NAMES

Names are given only to taxa, and all taxa are populations or sets of populations. Consequently only populations are named. Names are given to individuals only as representatives of populations. Names given to individuals as such, or to phena within populations, have no official status. "Name given to . . . infrasubspecific forms as such . . . are excluded" from zoological nomenclature (Art. 1; see also 45c). Any "variety" name published after 1960 is an unavailable name [Art. 45e(ii)]. Likewise unavailable are all names given to hypothetical concepts, to teratological specimens as such, or names proposed for other than taxonomic use (Art. 1).

The absence of a convenient non-Linnaean nomenclature for single specimens is a considerable handicap in paleontology, particularly in the study of fossil man. To have a reference name many anthropologists created a generic and specific name for every new specimen they found. Simpson (1963) has exposed the fallacy of this custom. When dealing with individuals one must adopt some sort of vernacular nomenclature. Let us use "Olduwai L1" or "Trinil D7" rather than a set of scientific names which imply a nonexistent zoological status. The same is true for infrasubspecific names (13.41).

## 7. RECOMMENDATIONS

Recommendations indicate the best procedure in cases not covered by the strict application of the rules. They are designated by the number of the article with which they are associated, followed by an appropriate letter. Although compliance with the recommendations is not mandatory, it is nevertheless highly advisable as representing good practice.

## 8. IMPORTANT DATES

The most important dates in zoological nomenclature are summarized on page xviii of the 1961 Code (xx in the 1964 edition). The starting date for zoological nomenclature is January 1, 1758, and the tenth edition of the *Systema Naturae* has been conveniently decreed to have been published on that date (Art. 3). The only exception is that spider nomenclature is considered to have started in 1757 [Clerck, C. 1757. Aranei suecici (Svenska Spindler). Stockholmiae]. See *BZN,* 4:319.

## 9. RETROACTIVITY

It is one of the basic concepts of good law that no new law should be applied retroactively. An article stating this principle explicitly was adopted by the London congress but was eliminated by the editorial committee with approval of the majority of the Commission. This has led to some difficulties, particularly with respect to the emendation of family names (Art. 29), to the universal adoption of unjustified emendations (Art. 33), to the types of replacement names (Art. 72*d*), and the revival of secondary homonyms (Art. 59*c*). See *BZN,* **18**:323 and elsewhere.

## 10. VERNACULAR NOMENCLATURE

Vernacular names for animals and plants exist in all languages. A primitive tribe of Papuans in the mountains of northwestern New Guinea has 137 different names for the 138 species of local birds. Hunting peoples usually have a better knowledge of nature and consequently a richer taxonomic nomenclature than agricultural or particularly urban peoples. The more conspicuous species of mammals, birds, fishes, and insects have names in all the languages in Europe. In English many of our so-called "common names" are of Anglo-Saxon origin. For better-known animals they are usually uninominal. For example, among British birds, raven, rook, jay, (mag)-pie, (jack)daw, robin, redwing, thrush, linnet, nightingale, partridge, and many others; and among the butterflies, monarch, greyling, ringlet, peacock, comma, etc. In other cases group names exist, like bear, frog, woodpecker, and species names are formed by modifying these group names with a descriptive noun or adjective, thus polar bear, brown bear, etc. Still other names are polynominal, such as small pearl-bordered fritillary, dark green fritillary, etc. These more complex polynominals, consisting of a generic name and a qualifying epithet, occur in many native languages all over the world.

## 11.  SCIENTIFIC NOMENCLATURE

Latin was the international language of European scholars in the Middle Ages, and up to the eighteenth century the majority of scientific treatises were written in that language. The Swede Linnaeus, to make his work more widely available, wrote mostly in Latin. In the *Systema naturae* he introduced a system of scientific names for animals and plants, based on Aristotelian logic, which is still the official system of zoological nomenclature, 200 years later. The Linnaean method of naming (see Chap. 4) has been excellently explained by Svenson (1945) and by Cain (1958). The generic name was the important element in the Linnaean nomenclature, always consisting of a single word. The *differentia* which followed the generic name specified the characters of the given species within the genus. It could be up to 12 words long and often had to be changed when new species or characters were discovered. To cope with this complexity and instability Linnaeus invented the *trivial* epithet for each species—a single specific term, now called the specific name and, as it turned out, of immense practical value. Subsequent workers abandoned the *differentia* together with the Aristotelian logic but retained as the scientific designation of a species a combination of a generic and a specific name, a *binomen* (see Arts. 5, 6, 11c). Linnaeus consistently applied this system to animals for the first time in 1758 in the tenth edition of his *Systema naturae*.

There are slight differences of meaning between the terms binominal, binomial, and binary. Binary refers to designations consisting of two kinds of names. The Linnaean combination of generic name and differentia was polynominal, yet binary. The nomenclature of many early post-Linnaean authors was binary, having a uninominal generic name, as for instance that of Brisson. The 1961 Code uses only the term binominal nomenclature even though admitting that the name of a subspecies is a trinomen. Binominal is an amendment of the formerly used term binomial.

## 12.  ADVANTAGES AND DISADVANTAGES OF BINOMINAL NOMENCLATURE

The Linnaean binominal nomenclature is a compromise between Aristotelian logic and a simple information retrieval system. The genera adopted by Linnaeus were extremely wide in scope and truly an aid to the memory. Not only a zoologist but even an educated lay person could memorize the 312 generic names which Linnaeus used for all animals. The scientific name thus consisted of two components, a group-designating generic name as an aid to the memory, and a specific name emphasizing distinctness and uniqueness.

Since Linnaeus' time this system has lost much of its usefulness. There are far more kinds of organisms than Linnaeus ever imagined. The total number of genera of animals now recognized may well exceed 50,000. A generic name no longer tells much to a zoologist except in a few popular groups of animals; it has become a tool for the specialist. The degrading of the usefulness of the generic name has been made worse by the extreme generic splitting so prevalent in many areas of systematic zoology (see 10.5).

## 13.  NAME CHANGING

The worst aspect of the binominal system is its extreme instability. Every change in the delimitation of a genus or in the assignment of a species to a genus will lead to a change in one or several scientific names. Genera are split, genera are lumped, and species are frequently shifted from one genus to another. For instance, most newts were at one time included in the genus *Triturus*. Subsequently it was discovered that this so-called genus was actually a highly artificial aggregate of very different kinds of newts. What was once *Triturus viridescens* became *Notophthalmus viridescens,* and *Triturus rivulosus* became *Taricha rivulosa.* Zoologists who had long used these species for embryological studies were understandably upset when their experimental material suddenly had a different name. In a group of bees of the tribe Paracolletini, a single revision resulted in a change of 288 out of a total of 332 binomina and, even worse, in the need for renaming 16 species owing to secondary homonymy (Michener, 1964). In these examples we are not dealing with the results of excessive splitting or any sort of arbitrariness, but with a serious weakness of the entire system of binominal nomenclature. Increasingly often the question is therefore asked whether the entire system of binominal nomenclature has not outlived its usefulness.

## 14.  ALTERNATE SYSTEMS OF NOMENCLATURE

Two sets of suggestions have been made to meet the difficulties mentioned in 13.13. Herrera (1899) and other authors in the nineteenth and early twentieth century suggested various prefixes and suffixes to the generic name in order to indicate to what order and class of animals the genus belongs. For instance, it was suggested modifying the generic name with initial letters indicating class and order, and terminations indicating subkingdom (*us* equals Vertebrata, *a* equals Invertebrata, *um* equals Protozoa). Thus *Papilio* would become *Ylpapilia* (*Y* equals Insecta, *l* equals Lepidoptera). A number of other schemes to facilitate the ready placing of a generic name in the animal system were published.

More radical proposals suggest replacement of binominalism by uninominalism. Michener (1964), for instance, proposes freezing the original scientific name of a new species for all time by connecting generic and specific names with a hyphen. The disadvantage of this proposal is a perpetuation of all early errors of judgment. Thus *Leioproctus jenseni* of the bee family Colletidae would continue to be listed under its original name, *Nomia-jenseni*, even though *Nomia* is a genus of a different family (Halictidae). A rather high percentage of the names proposed in a poorly studied group would probably be misleading when the group became better understood. The obvious advantage of this system is that any name can be retained for all time regardless of any taxonomic shifts. Michener very perceptively also points out that many improvements in classification are quietly resisted by taxonomists when they realize the nomenclatural havoc the change would produce. Among North American ornithologists, for instance, it has long been realized that the genera *Dendroica* (1842) and *Parula* (1838) are not separable. Yet *Dendroica* with over 25 species is so well known that no one dares to list these species under the older name, *Parula,* particularly since this, in turn, may be a synonym of the still older name *Vermivora.*

A scientific name serves two functions in the current zoological nomenclature—that of an identification tag (like a social security number) and, being a binominal, that of a classificatory device. There was no conflict between the two functions, as far as Linnaeus was concerned. According to his essentialist concept of the genus, once a species was correctly identified as to genus, there was little danger of it ever changing its name. Also the number of taxonomists was so small that Linnaeus and his associates did not foresee a major problem with synonymy. Both assumptions proved to be wrong, and the result has been a disastrous instability in the Linnaean system of nomenclature.

More and more voices are now heard recommending either that the two nomenclatural functions be separated or that they be expressed numerically, or both. Instead of a binomen there might be a running identification number (for example, 968-1, 968-2) for all species described in 1968, the numbering to be done in a central international office. To facilitate information retrieval, it would also be desirable to have a numerical scheme for the hierarchy of higher taxa in which a given species is classified. Various authors have made such proposals, for instance Hull (1966) and Bullis and Roe (1967). With a steadily increasing need for easy information retrieval and ease of programming for computers, there is little doubt that taxonomists will have to adopt some uninominal system sooner or later. This will presumably exist, at least at first, side by side with the Linnaean nomenclature, but it may eventually replace it. Obviously we need a system of identifying symbols for species that can be used by nonspecialists and is suited for mechanical information-retrieval devices. Yet all numerical systems so far proposed have failed to cope fully with the difficulties produced by the discovery of new taxa and the inevitably continuing shifts in classification.

## 15.  CRITERIA OF PUBLICATION

No nomenclatural action is valid until it has been published. What constitutes publication is spelled out in Art. 8, and what does not in Art. 9. In the eighteenth and early nineteenth century many authors thought they could make a name available by labeling specimens in their collection with that name or by demonstrating such specimens at scientific meetings. The Code is quite specific in requiring a printing method of publication by which numerous identical copies can be reproduced. Although not explicitly rejected by the Code, replication methods such as mimeographing are considered by most zoologists as unsuitable for nomenclatural publications.

Editors of scientific journals must be familiar with the provisions of the Code if they permit the publication of new names of taxa. Corliss (1962b) stated perceptively:

> It seems to me that exercise of just a little more editorial care . . . would have reduced impressively the number of improper names now in need of mandatory emendation, the hundreds of inappropriate and undesirable names in the literature which cannot be changed, the many incorrectly cited dates and authorships, erroneous combinations, double publications of a new name as new, and myriads of inaccurate and incomplete bibliographic citations of taxonomic and nomenclatural importance.

No author should ever submit the description of a new taxon simultaneously to several journals. This inevitably causes difficulties in dating and synonymy.

## 16.  CONCEPT OF AVAILABILITY

In order to be available, a name must have been published in a way to satisfy the requirements detailed in Arts. 8–20. The term *available,* as used in the Code, means "legitimate" or, as the botanists put it, "effectively published." A name may be "available" in this technical sense even though, as a junior primary homonym, it is automatically invalid at the moment of publication (Arts. 53, 59*a*). It is most important to understand clearly the difference between "available" and "valid." To have this distinction made clear is one of the major advances of the 1961 Code. To employ the name of a taxon one must make sure that it is available (properly published) as well as valid (not preoccupied by a senior synonym or homonym).

A name published without satisfying the conditions of availability is generally called a *nomen nudum,* particularly if it fails to satisfy the conditions of Arts. 12–16. A *nomen nudum* has no standing in zoological nomenclature and is best never recorded, not even in synonymy. There is always the danger that such a listing provides an indication in the sense of Art. 16 and thus inadvertently makes the manuscript name available.

## 17.  KINDS OF NAMES

In addition to available names and *nomina nuda* the Code recognizes directly or by implication several other kinds of names. *Pre-Linnaean* names, that is, names published prior to January 1, 1758, have no status. A *nomen dubium* is an available name which cannot be assigned to a definite taxon owing to shortcomings in the original diagnosis or the type material. Since most *nomina dubia* are a potential threat to the validity of universally accepted names, zoologists frequently apply to the Commission to place such names on the official index of rejected names. For *nomina oblita* see 13.24 (statute of limitation). Names to be excluded from zoological nomenclature are specified in Art. 1. Finally, *rejected names* are included in the official indexes of rejected names published by the Commission (see 13.63).

## 18.  PUBLICATION IN SYNONYMY

Article 11*d* states that a name first published as a synonym is not thereby made available. This is a wise and important provision. Many compilers of name lists had the habit of introducing manuscript names into synonymy. Their taxonomic judgment was frequently faulty, so that they listed the manuscript name in the synonymy of a species to which it did not belong. Most of these names were ultimately made available by proper publication and were then universally adopted as names of the taxa to which they really belong. To consider such names as preoccupied by the erroneous synonymation would be a violation of the guarantee of taxonomic freedom (13.2). Any available name can be placed into and withdrawn from synonymy without permanently affecting the availability or validity of other names (Art. 17(1)). Placing a name into synonymy is a taxonomic action; it is a nomenclatural one only as long as the taxonomic decision is considered correct. However, if a name, first published as a synonym, has been (universally) treated prior to 1961 as an available name, such a name shall be considered as available, with its original date and authorship, and be either adopted as the name of a taxon or be used as a senior homonym (Art. 11*d*, 1964 ed.).

## 19. DIFFERENTIAL DIAGNOSIS

The 1901 Code specified that in order to qualify for availability a name had to be accompanied by a description, definition, or indication. So many flimsy and meaningless so-called descriptions were, however, published that the International Congress at Budapest in 1927 adopted a provision that in the future (after 1930) names would have to be accompanied by a differential diagnosis. Since it is impossible to prove that a description qualifies as a differential diagnosis, this was modified in the 1961 Code [Art. 13*a*(i)] to state that the name must be "accompanied by a statement that purports to give characters differentiating the taxon," etc. However, a direct comparison with another specified taxon (differential diagnosis) is not required.

Respectable taxonomists go well beyond this minimal requirement and describe and illustrate carefully the diagnostic differences between a newly proposed taxon and its closest relative(s). If this relative is a little-known species, the characters which separate the new species from a well-known or common species should also be stated. The same provision is true for new genera and new subspecies. Thus, direct comparison is not a mandatory provision of the Code but one which characterizes the conscientious worker, the worker who has the respect of his peers.

## 20. INDICATION

The various conditions which satisfy the requirement of an "indication" in the sense of Art. 12 are listed in Art. 16*a*, and what does not constitute an indication in Art. 16*b*. It is important to emphasize that reference to a vernacular name or to a type-locality is not sufficient to make a name available. Many of the indications sufficient to ensure availability prior to 1931 (Art. 12) are now no longer sufficient (Arts. 13–15).

## 21.  NAMES GIVEN TO HYBRIDS

Hybrids are normally individuals, not populations, and hence not taxa. Article 1 states clearly that a name given to a hybrid as such is not available. Article 17 (2) specifies that a name given to an animal later found to be a hybrid remains available only for the purposes of homonymy, not (as was clarified by the International Congress at Washington) for the purposes of synonymy. (A Declaration is in preparation.) A name given to a species hybrid cannot be applied to either of the parental species.

The Code does not specify what to do with names given to specimens taken from hybrid populations. It is here recommended that Art. 17 (2) be applied to cases of hybridization between two species, but that such a name be considered available for the purposes of homonymy *and* synonymy in the case of intergrading populations between subspecies. Such populations are usually highly variable, and if the intermediate population is not recognized as a separate subspecies, a name given to an individual in this population may, if appropriate, be applied to the more similar of the two adjacent subspecies.

## 22.  DATE OF PUBLICATION

The relevant provisions are clearly stated in Art. 21. The date of publication is the date on which the publication was mailed to subscribers, placed on sale, or, where the whole edition is distributed free of charge, mailed to institutions and individuals to whom such free copies are normally distributed. Journals are sometimes mailed weeks or even many months after the date printed on the covers. In these cases the mailing date is the correct date, not the printed date. (For publication, see also Arts. 8 and 9.)

Even more confusing are serial publications, parts of which are sometimes issued over a period of 20 or 30 years. Each part of such publications has a separate publication date, the date on which it was actually mailed. In tracing the actual publication date of historical periodicals and series, the *Journal of the Society for the Bibliography of Natural History* (starting with vol. 1, 1936) is invaluable.

### 23. VALIDITY

Validity is a term that refers to the rights of names in relation to homonyms and synonyms. *Synonyms* are different names for the same thing. The earliest published synonym is referred to as the senior synonym; all later synonyms are junior synonyms. A name becomes a senior synonym at the time when a second name for the same taxon is made available.

Two kinds of synonyms can be distinguished. One consists of names that objectively refer to the same thing, such as a new name for a supposedly preoccupied name, or names based on the same specimen or illustration. These are called *objective synonyms*. The other consists of names based on different type-material. Taxonomists may nevertheless decide that they refer to the same taxon. These are called *subjective synonyms*. A generic lumper, for instance, may consider as synonyms certain generic names which a splitter would consider valid.

Incorrect spellings of names (Art. 33*b*) do not qualify as synonyms. At a given time only one name can be the valid name of a taxon. This is normally the oldest available name that is not preoccupied by a senior synonym or homonym and has not been suppressed by the Commission for the sake of nomenclatural stability. The provisions are clearly stated in Art. 23. Suppression by the Commission converts a previously valid name into an invalid name.

One provision which is occasionally misunderstood concerns the seniority ranking of coordinate taxa [Arts. 23*e*(iii) and 47*b*]. If it is found that the name of a polytypic species is invalid, it is replaced by the next oldest name given to any subspecies of this species and not necessarily by the replacement name for the formerly nominate subspecies. The same is true for the replacement name of a genus consisting of several subgenera (Art. 44*b*). The situation with respect to family names is different (13.36).

### 24. STATUTE OF LIMITATION

In order to give concreteness to the statement in the Preamble "when stability of nomenclature is threatened in an individual case," a statute of limitation was adopted by the Copenhagen and London congresses, as stated in Art. 23*b*. Statutes of limitation are almost universal in law. Demands that a limit be set on priority had been made early in the history of taxonomy. When the ornithologist Coues found in 1880 that the well-known name *Ectopistes migratorius* of the Passenger Pigeon was threatened, he wrote: "Is it worthwhile to make this change? Cases like this make one wish that there were in our nomenclature some law of limitation by which a name which has not been challenged for, say, 50 years or a century, might then acquire an inalienable right to recognition." Article 23b is intended to achieve this.

The article gives an automatic protection to names that have been in unchallenged use for a period of at least 50 years. When after the end of such a 50-year period a senior synonym is discovered that has never been used during that period, it is to be considered a forgotten name (*nomen oblitum*).

Article 23b is not well worded, which is one of the reasons why it has been widely criticized. Instead of stressing the basic intent of the rule, the automatic protection of well-established and universally used junior names, it concentrates on the suppression of unused older names (which is merely a method for achieving the basic aim). It does not clearly define "unused" and "widely used," and a few zoologists have even misinterpreted the definition of the 50-year period, which actually is the 50-year period immediately preceding the rediscovery of the senior synonym.

In view of these difficulties the Washington Zoological Congress instructed the Commission "to prepare an interpretative Declaration on Article 23b." The President of the Commission has appointed a committee which is now scrutinizing each of the individual provisions in order to determine (1) whether they are workable, (2) whether they should be replaced by a preferable alternative, and (3) what wording to adopt in order to exclude all conceivable misunderstandings. The report of this committee will eventually be voted on by the Commission as a whole, so that it can discharge its task. For instance, the Declaration may specify, as was suggested at the Washington congress, that in order to deserve protection the junior name must have been used at least ten times and by at least two authors during the stated 50 years. Such a provision would greatly help zoologists who work in inactive groups, for whom the benefits of Art. 23b are questionable.

It has been quite rightly suggested that the term "used" should be construed rigidly, so that only use in the primary zoological literature qualifies. This is a term for literature in which the name is actually applied to a definite zoological object. A report on a collection, a faunal list, any zoological or physiological paper, any textbook—all such publications are primary. Anything that does not refer to zoological objects but is merely a catalog of names without diagnostic or distributional information does not qualify as primary zoological literature. Accordingly, the mere listing of a name in synonymy or in a bibliographic publication such as the *Zoological Record*, or *Biological Abstracts*, or any index of names does not qualify as primary zoological literature. Catalogs and checklists cannot be designated as "primary or not" on an overall basis. Some checklists like the American Ornithologists' Union Checklist are critical surveys of zoological objects. Other so-called checklists are nothing but uncritical compilations of names from the *Zoological Record* and other bibliographic sources. If an author is in doubt, he should be guided by the spirit of the Preamble. Whenever stability is not endangered, it is advisable to abide by strict priority.

The primary purpose of Art. 23*b* is to prevent the revival of a long-forgotten name which would replace a well-known one, a common practice prior to the existence of such a statute. There are many groups of organisms in which species names are referred to in several hundred publications annually. Furthermore, many of these publications, such as field guides, zoological textbooks, and literature on embryology and public health, have a circulation of thousands of copies. In the past there was no automatic protection for such names. When a bibliographer exhumed a long-forgotten name, the entire burden of saving the well-known name was placed on the general zoologist. By the time the Commission was able to get to a ruling—and in periods of disturbed international communication there was sometimes a lag of ten or more years—the replacement name had gained sufficient currency to make the situation virtually insoluble. Article 23*b* provides for the automatic protection of all names that have been in current use for more than 50 years without depriving names in inactive taxa of their rights.

The greatest difficulty the Commission faces in the preparation of its interpretative Declaration is the need to come up with a formulation that is equally acceptable to zoologists working on animals with a very active literature, like mammals, birds, butterflies, and disease vectors, and also to zoologists working in groups in which there is only a single specialist every 30 to 50 years and names are only rarely cited in the primary literature. Hopefully, the principles laid down in the Preamble will guide the Commission in arriving at a workable compromise.


### 25.  COMBINED AND DIVIDED TAXA

A taxon formed by the union of two or more taxa takes the oldest valid name among those of its components [Arts. 23*d*(i), 23*e*] (see also Art. 67*k*). In the case of family names, the strict application of priority may be set aside [Art. 23*d*(ii)] (see 13.33).

If a taxon is subdivided, its valid name must be retained for one of the components. This rule applies equally on the family (Art. 37), generic (Art. 44), and species levels (Art. 47). For instance, when a species is subdivided into several subspecies, the subspecies which contains the topotypical population becomes the nominate subspecies, its subspecific name being the same as that of the species.

Linnaeus described the red-winged blackbird (*Agelaius phoeniceus*) on the basis of Catesby's drawings and descriptions from South Carolina. South Carolina has therefore been designated as the type-locality. When this species was subdivided into several subspecies, the subspecies of eastern North America (including the region of South Carolina) became automatically the nominate subspecies, namely, *Agelaius phoeniceus phoeniceus* (Linnaeus).

Occasionally authors create synonyms by ignoring this rule. For instance, Thienemann (1938) found that the well-known turbellarian worm *Planaria alpina* consisted of two subspecies. The northern one (northern Germany, Scandinavia) he called *septentrionalis:* the southern one (Alps) he called *meridionalis.* Since the species had been described originally from Scandinavia and the Alps, it is obvious that either *meridionalis* or *septentrionalis* is a synonym of the nominate subspecies *P. alpina alpina,* the type-locality of which needs to be restricted to one of the two areas.

If a genus is divided into several genera, one of the new genera must bear the name of the originally undivided genus. Traditionally this has also been considered the preferred method for higher taxa (see 13.34).

## 26.  PAGE AND LINE PRECEDENCE

If two names for the same taxon are published in the same publication, they are considered as "published simultaneously" (Art. 24*a*). One of these names may have line or page precedence, but their priority is determined by the action of the first reviser (13.4). The first reviser must give serious consideration to Recommendation 24A and "select the name that will best ensure stability and universality of nomenclature." Almost invariably one of the two competing names is better known, is based on a better description, is based on better type-material, is based on an adult phenotype rather than on an immature stage, or has some other nomenclatural advantage. This is the name the first reviser should select. Chronological priority is not involved in the case of simultaneous publication, it is replaced by *designated priority.* The same is true for simultaneous publication in different works (also Art. 24*a*).

## 27.  WORK OF AN ANIMAL

Prior to 1931 the description of the work of an animal, for instance its gall, was considered a valid indication [Arts. 16*a*(viii), 24*b*(iii)]. After December 31, 1930, a name does not become available unless it also satisfies Art. 13*a* by being accompanied by a statement attempting "to give characters differentiating the taxon" from close relatives.

## 28.  FORMATION OF NAMES

Mandatory provisions governing the formation of zoological names are given in Arts. 26–30. Since such names must be either Latin or latinized or at least so constructed that they can be treated as a Latin word (Art. 11*b*), it is most important for the zoologist to familiarize himself with the rules that govern the correct transliteration and latinization of words. Three appendices (B, C, D) to the Code are specifically devoted to these matters. They are not mandatory provisions but have the same status as recommendations. Some older treatments are still useful.[1] See also Brown (1954).

Appendix B (Code, pp. 95–101) deals with the transliteration and latinization of Greek words. Appendix C (Code, pp. 101–103) deals with the latinization of geographical and proper names, while appendix D (Code, pp. 105–141) deals quite generally with recommendations on the formation of names. Every zoologist who plans to propose a new name for a taxon should carefully read these recommendations.

A few points might be stressed particularly. Complex, long, and difficult-to-pronounce names should be avoided. It is good practice to propose only the sort of new generic name that differs from an old, already existing, generic name in more than a single letter or its termination. It is very unwise in describing a new species to choose a specific name already in use in a related genus. Such a name would become a homonym if a subsequent author should combine these genera. For instance, names like *africanus, robustus, capensis,* etc. were given repeatedly to closely related hominids which subsequent authors might well place in the genus *Homo.* Names for protists should be such as would also be valid under the botanical rules.

A noun in opposition cannot be treated as an adjective. Words like *longicauda, melanogaster,* or *albipectus* in which the ending is a noun cannot be changed when transferred to a genus with different gender.

Compound names are to be united into a single word without a hyphen (Art. 26), with one exception (Art. 26*c*).

## 29.  DIACRITIC MARKS

Diacritic marks, apostrophes, or diaereses are not to be used in a zoological name and are to be deleted from names originally published with such marks (Art. 27). The only exception is the German umlaut sign, which is to be deleted from a vowel, the letter "*e*" to be inserted after the vowel [Art. 32*c*(i)]. Thus *ä, ö, ü* become *œ, æ, ue*.

---

[1] W. Miller, 1897. *Calif. Acad. Sci.,* (3) vol. 1, no. 3, and F. E. Clements, 1902. Univ. Studies, Nebraska, vol. 3, no. 1, are two examples.

## 30.  STEM OF A FAMILY NAME

Names of the family group are formed by the addition of prescribed endings (-idae, -inae) to the stem of the name of the type-genus (Art. 29).

The 1901 Code did not specify how the stem was to be determined, and many family names were formed in such a way as to give euphonious combinations. The 1961 Code specifies (Art. 29*a*) that "the stem is found by deleting the case-ending of the appropriate genitive singular." When retroactively applied, this provision will lead to a change in the names of some well-known families. For instance, the family of stick-insects or phasmids, almost universally called Phasmidae in zoology textbooks, will have to be called Phasmatidae.

## 31.  GENDER OF GENERIC NAMES

If the specific name is an adjective in the nominative singular, it must agree in gender with the generic name. To apply this rule correctly the zoologist must understand the rules governing the gender of Latin nouns, whether of Latin, Greek, or nonclassical origin. These are given in Art. 30. Various handbooks on scientific names are helpful, and standard Greek and Latin dictionaries are indispensable (see also Appendices B, C, D).

Latin and Greek grammar are full of pitfalls. A Latin noun ending in -*us* is by no means necessarily a noun of the second declension and thus automatically masculine: it may be feminine like *domus* or neuter like *pectus*. A latinized Greek word takes the gender appropriate to its Latin termination [see examples under Art. 30*a*(i)(3)].

A zoologist who proposes a new name in the genus group should give the etymology and the gender of the new name (Appendix E, 16).

## 32.  ORIGINAL SPELLING AND EMENDATION

One of the curses of nomenclature has always been the zoologists' habit of "correcting" names proposed by other authors. Article 32 states clearly that the original spelling must be retained unless some exceptional condition exists. There must be evidence in the original publication of an inadvertent error such as a *lapsus calami* or a copyist's or printer's error. Other kinds of errors such as incorrect transliteration, improper latinization, and use of an inappropriate connecting vowel are not to be considered inadvertent errors.

This provision has been severely criticized by those with a classical education. But it must be realized that fewer and fewer zoologists have such an education and that the meaning of zoological nomenclature has changed since the time of Linnaeus. At that time it was a Latin language, now scientific names are essentially arbitrary combinations of letters that serve as labels for taxa. Many names for species are misleading, but this is irrelevant considering the now accepted meaning of nomenclature. It is contrary to this meaning and to the Code to change *Australopithecus* to *Australanthropus* because one considers this fossil hominid to be closer to man than to the anthropoids.

Although the article prohibits any corrections of grammatical mistakes in generic and specific names, except for matters of form (Arts. 26–28, 30), it makes mandatory in Art. 32a(iii) emendations in the stem of family names, which seems quite inconsistent.

An incorrect original spelling, as defined in Art. 32c, is to be corrected wherever it is found. The incorrect spelling has no separate status in nomenclature. The same is true for incorrect subsequent spellings (Art. 33b), that is, any unintentional error in the spelling of a name.

Any demonstrably intentional change in an original spelling of a name is an "emendation." If this emendation is not justified by the provisions of Art. 32, it is to be considered an "unjustified emendation" [Art. 33a(ii)]. An unjustified emendation has the status of a new name with its own date and author and is a junior objective synonym of the name in its original form. It is available as a replacement name in case the name as originally spelled is found to be invalid owing to homonymy.

### 33.  FAMILY NAMES

The names of taxa above genus rank are always uninominal and in the plural. Provisions relating to names in the family group are found in Arts. 11e, 23d, 29, 35–41, 63–65. To avoid grammatical mistakes, remember that these names are in the plural. One can say, "The family Fringillidae is the largest family of songbirds" but must say, "The Fringillidae are the largest. . . ." The same is true for the names of orders, classes, and other higher taxa.

Linnaeus did not employ the family category, but after it was introduced by French zoologists in the 1780s and 1790s, it was soon universally adopted. The early codes, however, had relatively few provisions concerning family names.

An author proposing a new taxon of family rank has the taxonomic freedom to select the genus which he considers central ("most typical") for the new taxon. Seniority of the name of the type-genus is irrelevant. Provisions for the formation of family names are given in Art. 29.

The 1961 Code introduces limited priority for family names as the result of a decision made by the Copenhagen Congress. However, since family names are widely used in zoology, even by nonspecialists, stability is even more important on the family level than on those of genus and species. The Code therefore provides for two important stabilizing devices. Article 23$d$(ii) decrees that "if a zoologist observes that the strict application of the Law of Priority to two or more synonymous family-group names would upset general usage," he is to request the Commission to decide which name is to be accepted.

There are no satisfactory nomenclators that list the dates and authors of names in the family category. Prior to 1900 (Art. 11$e$) no family name needed to have standardized endings, but they are now mandatory (Art. 29). In groups with little history, little published literature, and few specialists it is usually not difficult to establish authorship and original date of family-group names. In groups with a rich literature and long history, such as mammals and birds, it is virtually impossible to establish the original date of publication for the older, traditional names. This has created countless difficulties. Most specialists will agree with Myers and Levitan (1962, p. 290), who state: "An extension of priority to family names will take zoologists into a maze of old group names which often cannot be clearly recognized as of familial (or any other) hierarchical grade." Some mitigating provisions will have to be found, perhaps along the lines suggested by Bradley (1962, $SZ$, 11:178–179), to facilitate the placing of family names on the Official List.

The other provision favoring stability relates to name changes in the type-genus. The name of a taxon in the family group must be formed from the stem (see 13.30) of the name of the type-genus (Art. 29). The type of a family taxon is the zoological object identified by the name of the type-genus. There is, however, no need, for the sake of stability, to change the name of a family each time the currently valid name of the type-genus is changed. The type of the family is, as stated, a zoological object and not a name. The validities of the name of the family and of the name of the type-genus are independent (Art. 40). Family names are formed by tautonymy from the name of the type-genus, and Art. 40 merely codifies what has always been true for generic names formed by tautonymy from specific names. Likewise, they are never changed when the specific name which served in the tautonymy loses its validity.

When a zoologist finds that the name of the type-genus of a family is threatened owing to the availability of a senior synonym, he may well want to bring this situation to the attention of the Commission. It is sometimes advisable to suppress the senior name in order to retain the convenient tautonymous relation between the name of the family and that of its type-genus.

## 34. NAMES FOR HIGHER TAXA (ABOVE THE FAMILY GROUP RANK)

The 1961 Code does not contain rules dealing with names of taxa in categories above the family group. Attempts were made at the Copenhagen and London congresses to draft some rules, but these were unsuccessful. There is still much uncertainty about the basic subdivision of some of the phyla, like the Porifera (sponges) and Turbellaria. A premature freezing of the names of classes and orders might stultify subsequent efforts for a complete reclassification on the basis of new character complexes.

The worst threat to stability has been the misconception of a few zoologists that removal of some genera from a higher taxon requires a complete renaming (13.25). There is no excuse whatsoever for renaming the well-known Bryozoa with more than 3,000 species as Ectoprocta because a few genera with 75 species were removed into the separate class or phylum Entoprocta (Mayr 1968b). Even though the names of higher taxa are not based on the type method, the removal of some taxa of lower rank does not justify the renaming of a higher taxon. Fortunately this principle was firmly adhered to when heterogeneous elements were removed from the Mollusca and Insecta as defined by Linnaeus.

It is frequently proposed that uniform endings for the names of the higher taxa should be employed. This proposal, on first sight, seems so logical and sensible that only mature consideration brings out its grave shortcomings. First of all, it violates the concept of priority as well as that of stability. For instance, six of the seven orders of insects recognized by Linnaeus, i.e. Coleoptera, Hemiptera, Lepidoptera, Neuroptera, Hymenoptera, and Diptera are still recognized essentially in the same sense as 200 years ago. They all use the suffix -ptera (wings). The major orders of mammals, such as Rodentia, Carnivora, Insectivora, and Primates are equally old and well-known. Many ordinal names in birds, such as Impennes, Tubinares, Oscines, are likewise not based on a type-genus. Simpson 1952, SZ, 1:20–23) well presents the argument against uniform endings.

Furthermore, little would be gained by having uniform endings because the ranking of higher taxa is a particularly disputed area of zoology. Different authors may rank the same taxon as superfamily, suborder, order, or even subclass. Standardized endings would invest a given ranking with far greater authority and definiteness than is justified by the facts. Finally, there are several competing systems of uniform endings.

It would seem important, however, for the zoological congresses or specialists in the groups to stabilize the nomenclature of higher taxa in all cases where several names compete for the same taxon, as for instance in the case of the bivalve mollusks (Bivalvia, Pelecypoda, Lamellibranchia).

There are a few general principles concerning the names of higher taxa above family rank. They are single words in the plural. Although they are not based on a type-genus, there is generally a consensus as to the central ("most typical") component of a higher taxon. Zoologists maintain well-established names of higher taxa even though they may find some other word "more appropriate." When a composite higher taxon is divided, the original name should be retained for the "more typical" or larger group and a new name applied to the newly recognized group. See also MLU, 276–278. The history of the names of many higher taxa is uncertain. Corliss (1957) has established it in exemplary fashion for the ciliates.

## 35.  COORDINATE STATUS OF CATEGORIES

Article 36 specifies that the proposal of a new taxon in the family-group implicitly constitutes permanent authorship for that taxon, even when changed in rank. When a given taxon, first described as a new subfamily, is raised to family rank, "it is thereupon available with its original date and author" for use, based on the same type-genus, in the higher ranking category. Article 37 specifies that describing a taxon in the family-group automatically makes the describer also the implicit author of any nominate subtaxon. The given example of the family Tipulidae well illustrates this principle. Article 43 confirms the same principle for the genus and Art. 46 for the species.

The creation of superfamilies creates difficulties not properly taken care of in Art. 36 (Mayr, 1954). If a family is raised to rank of a super-family, the provisions of Art. 36 must be applied. On the other hand, if an author makes an entirely new superfamily by bringing together various families not previously associated, he is not giving a new rank and name to a previously established taxon. It would violate the principle of taxonomic freedom (13.2) if he had to choose the type of the new superfamily on the basis of the oldest included family name. Quite correctly, the principle is specifically rejected in Art. 64 for the choice of the type-genus of a family and implicitly in the provisions for selection of type-species for genera (Arts. 66–70).

## 36.  SUBORDINATE TAXA

The provisions for subordinate taxa are essentially the same for the family group (Art. 37), for the genus group (Art. 44), and for the species group (Art. 47). The nominate subtaxon consists in each case of that subtaxon which contains the type. One of the included subtaxa, which therefore is called the nominate subtaxon, must have the same name as the taxon. The only exception is on the family level in the case of renaming the type-genus (Art. 40). Also, in the family group the subtaxon has a different ending.

In the case of genus and species (Arts. 44*b*, 47*b*) it is provided that the subtaxon with the next oldest name becomes the nominate subtaxon when it is found that the name of the original nominate subtaxon is invalid. This provision is a slight deviation from a rigid application of the type method (see below). For this reason it is not applied (in Art. 37) to subtaxa on the family level. Changes in the names of families have become rarer under the new Code owing to the provisions of Arts. 23*d* and 40.

## 37.  HOMONYMY BETWEEN FAMILY NAMES

Homonymy between such names occurs only infrequently. It is of course irreconcilable with the principle of uniqueness of names, laid down in the Preamble, for a family name to be based on a generic name that is a junior homonym. Should an author discover such a situation, it is advisable that he report it to the Commission. If the senior homonym should be either a forgotten name (Art. 23*b*) or a never-used junior synonym, the Commission can exercise its plenary powers to suppress such a name. This would result in preservation of the family name based on the junior homonym.

Article 55 advises on situations where identical family names are based on nonidentical generic names. Such cases must always be referred to the Commission.

## 38.  GENUS GROUP NAMES

Provisions dealing with generic names are found in Arts. 11*f*, 13*b*, and 42–44. A genus group name must be a noun in the nominative singular or be treated as such. A noun in the plural when first published takes its date and authorship from its first publication in the singular. An author proposing a new generic name should make certain that his proposal does not omit any of the following five essential points:

1. That it is clearly indicated as a new genus: *X-us*, new genus.
2. That the proposed generic name does not violate the rules and the recommendations concerning the formation of names (consult Code Appendix D).
3. That the newly proposed name is neither a homonym nor a synonym.
4. That the description contains a clear statement of the characters in

which the new genus differs from previously described genera (Art. 13*a*).
It is desirable that a differential diagnosis is added in which a direct
comparison is made with that genus or those genera which are believed
to be most closely related to the newly described one.

5. That the type-species is clearly and definitely stated (Art. 13*b*). The
   generic limits may be interpreted differently by succeeding authors. But
   the type-species will forever anchor the concept of the stated genus to
   a clearly defined zoological object.

For additional comments on generic names see MLU, pp. 261–270.

## 39.  COLLECTIVE GROUPS

A collective group of identifiable species of which the generic position
is uncertain may be treated, for the sake of taxonomic convenience, as
a genus (Art. 42*c*). Such a collective group does not require a type-species.
Assignment to a collective group is a temporary identifying label which
is particularly convenient in parasitology and paleontology.

## 40.  NAMES OF TAXA IN THE SPECIES GROUP

The rules concerning these names are given in Arts. 5, 11*g*, 23*e*(ii),
34*b*, and 45–49. There are only two categories in the species group, the
species and the subspecies. The *specific name* is the second word in the
binomen (Art. 5), and the *subspecific name,* when employed, is the third
word of the trinomen. Linnaeus used the term *nomen triviale* for the specific
name, but for various reasons this usage became essentially obsolete in
zoology after 1800. The initial letter of these names is lowercase (Art.
28). Since a scientific name in the modern interpretation is considered
a recognition symbol formed by a sequence of letters (13.32), it is not
advisable (indeed leading to confusion with generic names) to capitalize
specific names at the beginning of a sentence.

Articles 11 to 19 list the prerequisites for making a specific name
*available.* Particularly important is Art. 13*a*(i), according to which names
published after 1930 must be "accompanied by a statement that purports
to give characters differentiating the taxon." It is good taxonomic practice
to include in the original description not only such diagnostic information
but also a differential diagnosis that consists of an actual comparison with
closely related species (see 13.19). Advice on correct endings of species-
group names formed from modern personal names is given in Recommendation 31A of the Code. A vote of the Washington congress in 1963 deleted
Art. 31 (of 1961) and replaced it with this recommendation, thereby vali-

dating many of the practices current prior to 1961 and converting some of the mandatory provisions of Art. 31 (of 1961) into recommendations. Three widespread usages in particular were permitted, namely (1) treating modern personal names as nouns in apposition like *Calypte anna,* (2) employing the correct grammatical genitive of a latinized family name, e.g. *fabricii* instead of *fabriciusi,* and (3) employing the correct masculine genitive ending, *-ae,* in words derived from names like Costa, Molina, Kuroda. When used as masculine nouns, Latin words like *agricola* also have the genitive -ae, e.g. *agricolae.*

The frequently made suggestion "to correct" the latinized dedication genitive -ii to -i is implicitly rejected by the wording of Art. 32*a* that the original spelling of a name is to be retained as the "correct original spelling." The ending *-ii* is not listed as contravening this provision.

## 41.  INFRASUBSPECIFIC NAMES

Linnaeus did not recognize the subspecies category. The term *variety* used by him and his followers referred to a medley of deviations from the type of the species (3.3.1; also, see Mayr, 1963, pp. 334–346). No distinction was made by him between different kinds of infraspecific variants. As a result of a long and tortuous history we now clearly distinguish between two kinds of infraspecific variants. The subspecies designates a genuine taxonomic category based on populations; variety names are names for phena and aberrant individuals and have no standing in nomenclature (Art. 1).

For more that 150 years zoologists did not make a clear distinction between varieties that are genuine taxa (based on populations) and varieties that designate merely groups of individuals or phena. It would be ritualistic to determine the availability of infraspecific names on the basis of the original terminology chosen by the author. Although the Code clearly states that names given to "infrasubspecific forms as such" are excluded (Art. 1), it deals with other aspects of infrasubspecific names in Arts. 10*b* and 45*c, d, e.* Any name proposed after 1960 is unavailable if not clearly given to a subspecies taxon. In the case of infrasubspecific names given prior to 1961, corroborating evidence is to be used, such as the statement that a taxon is "characteristic of a particular geographical area or geological horizon" [Art. 45*d*(ii)]. A name first established with infrasubspecific rank becomes an available name (Art. 10*b*) when used for a taxon of the species group and takes the date and authorship from the time of such change in usage. It is taxonomic practice to give the benefit of the doubt to authors having introduced "varieties" prior to 1961 (Art. 45*e*).

The indiscriminate naming of infrasubspecific forms has brought much discredit to taxonomy. The Latin language is traditionally used for the names of taxa, and if one wants to designate infrasubspecific forms, it is far better to use vernacular words from modern languages. The names for the mutant forms of *Drosophila* are a good example. Linsley (1944) has listed other kinds of infrasubspecific variants that should not be given Latin designations. This includes sexual dimorphs, castes in social insects, alternate generations, all kinds of morphs and polymorphs, seasonal forms, pathological forms, and various kinds of aberrations and teratologies. Now that zoologists clearly understand the difference between populations and individual variants (phena), there is no longer any excuse for terminological sloppiness.


## 42. AUTHORSHIP

The meaning of authorship and the rules concerning the citation of names of authors are given in Arts. 50 and 51. Contrary to the statement of the Code [Art. 51*b*(i)], almost all zoologists customarily separate the name of a taxon from that of a subsequent user (not its original author) by a comma.

In view of the fact that an extraordinarily large number of species were superficially and hurriedly described so that some vain author would "get his name in print," the whole institution of citing authors' names has been frequently attacked, not without reason. If zoologists have not given it up, it is for the single reason that the name of an author is an identifying label, particularly useful in the case of homonymy. The name of the author does not form part of the name of the taxon, and its citation is optional (Art. 51*a*).

The author of a name is the person who is alone responsible both for the name and for "the conditions that make it available" (Art. 50), that is, for the diagnostic description. Consequently, if someone other than the author of the paper is responsible for a name and its availability, it is this other contributor who shall be considered the author of the name (Art. 51*c*). It is sometimes difficult to determine "responsibility" when the unfinished manuscript of a zoologist is published after his death by another zoologist. If the manuscript contains only the names, and the editor supplied the diagnostic descriptions that "make the names available," it is the editor who is to be regarded as the author of these names, according to Art. 50.

When there were only a few zoologists, *abbreviations* for the names of authors were customary, like L. for Linnaeus or F. for Fabricius. So little is gained by these abbreviations that they are no longer considered good practice. Special problems arise when an author changes his name during the period when he is actively publishing, or assumes a title (e.g. Laporte to Comte de Castelnau). In these cases it is customary to cite as author's name the name under which the new taxon originally was published. Names published anonymously after 1950 are no longer available (Art. 14).

### 43.   HOMONYMY

Homonyms are identical names for two or more different taxa. The earliest of such names is the senior homonym; later ones are junior homonyms. Articles 52–60 of the Code deal with the validity of homonyms and with replacement names for invalid homonyms. This area is one of the most difficult in zoological nomenclature, and it is probable that future congresses may modify some of the provisions, particularly where they seem to be in conflict with either taxonomic freedom or stability.

A junior homonym in the genus group is always invalid (Arts. 53, 56). However even a single-letter difference prevents homonymy of generic names.

There are some real conflicts with respect to the application of the law of homonymy to species-group names. Homonymous specific names originally published in the same genus (primary homonyms) necessitate the renaming of the junior homonym. However, when two identical specific names are secondarily brought together under one generic name, such homonymy may be a matter of purely subjective taxonomic judgment. For an author who rejects the lumping of the two genera into one, no homonymy exists, and there is no need whatsoever to reject the junior name. The 1901 Code said that a junior homonym was to be rejected. Some zoologists interpreted this to mean only primary junior homonyms, others both primary and secondary junior homonyms. Article 59c copes in part with this situation by stating very clearly that "a secondary homonym is to be restored as the valid name whenever a zoologist believes that the two species-group taxa in question are not congeneric." Unfortunately, the editorial committee added after the London congress that this provision was applicable only to "a name rejected after 1960." It did not make clear what to do with subjective junior homonyms rejected prior to 1961 only by a minority of workers in a field. This again is an area where different practices prevail in groups where there is only one specialist every 50 years and in those where there are simultaneously 10, 20, or 50 specialists. In ornithology, for example, it has never been the practice to suppress secondary homonyms permanently. For instance, in the 1950s one ornithologist adopted an exceedingly broad generic concept for the Old World flycatchers and lumped many well-known genera under the name *Muscicapa*. As a result numerous specific names, some of them very well known and of long standing, became for him junior homonyms, and he renamed them. Nearly all other ornithologists rejected the sweeping lumping of the first author and refused to accept his replacement names. Several recent applications in *BZN* indicate that the custom of not accepting taxonomically unjustified secondary homonymies is not limited to ornithology. Indeed it would be an intolerable infringement of the taxonomic freedom guaranteed in the Preamble if Art. 59c were rigidly interpreted.

On the other hand it would be in the interest of stability to adopt a statute of limitation for the revival of names long and universally rejected as secondary junior homonyms. Otherwise every generic split may disturb stability on the species level.


## 44. ACTUALITY PRINCIPLE

The classification of all homonyms into primary and secondary homonyms is an oversimplification. Blackwelder (1948) and MLU (p. 226) present various examples of homonymy being complicated by changes in the generic association of two potentially homonymous specific names. There is a strong trend to get away from all ritualism and to sanction the renaming of a specific name only when an author thinks that two specific names are truly congeneric at the time of the discovery of the potential homonymy. The principle of basing nomenclatural decisions on the actual situation existing at the time when a potential conflict of names is discovered may be designated as the *actuality principle*.

For instance, an African weaver bird was described by Cretzschmar in 1827 as *Ploceus superciliosus*. Another weaver was described by Shelley in 1873 as *Hyphantornis superciliosus*. Eventually the latter species was transferred to the genus *Ploceus,* but long after the original *P. superciliosus* of Cretzschmar had been transferred to the genus *Plocepasser*. No homonymy ever existed in the sense that any ornithologist ever thought the two species were congeneric. As a result even the most recent monographs and catalogs retain *superciliosus* as valid specific name in both respective genera.

A further brake to needless renaming is applied in Art. 57c, which states that homonymy does not exist when two identical species-group names are used in entirely different genera that merely happen to have homonymous names, such as when the name *Noctua* at one time was used both for some birds and for a group of Lepidoptera.

## 45.  REPLACEMENT NAME (FIRST REVISER)

Articles 56*c* and 57*e* deal with the rare eventuality of the simultaneous publication of two homonyms now placed in different categories. In order to serve best the interests of stability, the mandatory provision is made that if one of the two names is for a subcategory (e.g. subgenus versus genus, or subspecies versus species), the name proposed for the full category takes precedence over the one proposed for the subcategory. As in other cases of renaming of homonyms, the "actuality principle" should be invoked. When at the time of the discovery of homonymy one of the homonyms is the name of a subgenus or of a subspecies, this is the homonym to be renamed, in order to preserve the well-established older name for the full genus or species.

## 46.  HOMONYMY AMONG SUBSPECIES

All specific and subspecific names in a genus are potentially homonymous (Art. 57). As the example given in this article illustrates, it is not permissible to use the same subspecies name in different species of the same genus.

## 47.  REPLACEMENT OF REJECTED HOMONYMS

Provisions governing the publication of replacement names are given in Art. 60. Before proposing a new name as a replacement name (*nomen novum*) for one which is preoccupied, an author must be certain that it meets these conditions:

1. That there is no other name available for the species (or genus). In the past there have been a few nomenclaturists who published replacement names for all junior homonyms whenever a catalog or nomenclator was published. Since in most of these cases replacement names had already been provided by specialists of the respective groups, such wholesale renaming resulted in nothing but an added burden to synonymies and a loss of respect for the author of these unnecessary names.
2. That the original author of the preoccupied name is no longer alive. According to paragraph 3 of the Code of Ethics (Code, Appendix A, p. 93) a zoologist should communicate with the author of a junior homonym, if he is still alive, and give him a reasonable opportunity to publish a replacement name.
3. That the new name is proposed in the form recommended in the Code. As stated in Art. 13*a*(ii) and as recommended by several zoological congresses, there must be "a full bibliographic reference" (not merely

"Smith, 1907") to the original citation of the preoccupied name, and, in the case of preoccupied generic names, to the name of the type-species (Appendix E, 15, p. 145).

4. That it is desirable to propose a replacement name. For instance, there is no excuse for renaming preoccupied names that are invalid synonyms. More importantly, a replacement name automatically takes the same type and the same type-locality (in case of a species) of the preoccupied name. If the type of the preoccupied name is no longer in existence, or if there is even the slightest doubt as to the identity of the species with the preoccupied name, it is sometimes preferable simply to redescribe the taxon in question, as if entirely new, and to provide it thus with an unambiguous type and type-locality.

Under exceptional circumstances a homonym may provide an opportunity for shifting an originally ill-chosen type-locality of a subspecies. For example, let us assume that there is a species with a northern and southern subspecies, meeting in a narrow zone of intergradation. The type-locality of the southern subspecies is far in the south, but the type-locality of the population whose name had always been applied to the northern subspecies is actually located in the zone of intergradation. If it is found that this name is preoccupied, it is better not to replace it by a *nomen novum* but to redescribe the northern subspecies and select a new type-locality in the middle of its range. The number of cases where such a shift of type-locality would be desirable is undoubtedly very small; in most cases it would only be confusing. A genus (or species) divided into subtaxa takes the name of the next oldest available subtaxon if the name of the nominate subtaxon is found to be a junior homonym (see 13.36 and Arts. 44*b*, 47*b*).

## 48. THE TYPE METHOD

In taxonomic practice doubt often arises as to the identity of taxa to which names are attached. The descriptions are frequently insufficient for establishing such identity, particularly the rather short descriptions of earlier authors. Sometimes a description applies equally well to several species since the species-specific diagnostic characters were not mentioned in the early description. In the case of higher taxa, the contents of these taxa change as additional species are discovered. When such a higher taxon is split, the question arises as to which of the components the name should be attached to. It is obvious that a secure standard of reference is needed to tie taxonomic names unequivocally to definite, objectively recognizable taxa. These standards are the types, and the method of using types in order to tie names to taxa is called the type method.

Few current workers in taxonomy realize how comparatively modern the type method is. Indeed its introduction during the last 100 years was one of the major conceptual changes in the theory of taxonomy and nomenclature. Early taxonomy was dominated by the Aristotelian concept of types. All specimens that conformed to the taxonomist's concept of the type of a taxon were considered typical. The types—and there were as many types as there were typical specimens—formed the basis of the description of the species. This was the concept of type held by Linnaeus and his contemporaries. As is well described by Cain (1958) and Svenson (1945), Linnaeus never hesitated to replace specimens in his collections which we would now consider as types with "better ones."

"Linnaeus never designated any specimens as type. Whether his description was based on one single or on several specimens, it cannot even be taken for granted that [these] were preserved in his collection. Or, the original specimen may later have been substituted by another, in better condition, by Linneaus himself" (Lindroth, 1957). Macan (*BZN*, 18:328) remarks, "The Linnaean collection is notoriously unreliable since many of the original specimens . . . were replaced later by other specimens," an observation confirmed by Lindroth and other Linnaean scholars. No nomenclatural decision should ever be made by relying on a "Linnaean type." There is no such thing. Consequently, "If description and authentic specimen disagree, the former is decisive" (Lindroth, *loc. cit.*). This point must be made emphatically because there have been a few ill-advised attempts in recent years to change well-known Linnaean names because imaginary types were found to belong to species different from those to which the name had always been applied.

Linnaeus was not alone in these practices. It was customary in several European museums in the first half of the nineteenth century to substitute "new" type-specimens whenever the old ones became faded or were damaged by insect pests. This was quite legitimate under the Aristotelian type-concept. In other instances, the inadvertent transfer of labels from one specimen to another has caused an obscuring of the identity of type-specimens. Evidence derived from old types must be treated with extreme care and discrimination and never be used to upset stable nomenclature. In such cases, as with Linnaean "types," if description and so-called authentic specimen disagree, the former is decisive.

Consistent with the Aristotelian type-concept, typification of higher taxa was done by the process of "elimination" of atypical elements. This was the standard method, practiced until far into the nineteenth century, and it was one of the most upsetting decisions of the 1901 Code not to ratify this traditional procedure in the case of old type-fixations but to apply retroactively the modern type method.

The history of the gradual adoption of the type method after about 1850 has not yet been written. How little the importance of this change was realized is indicated by the fact that the original 1901 Code does not include any directives concerning types. Provisions for generic types were adopted at the Boston congress in 1907, and for the types of species even later. Some authors who speak of allotypes, plesiotypes, metatypes, homotypes, hypotypes, etc., still do not fully understand the true nomenclatural function of the type-specimen. Such pseudotypes have no nomenclatural significance, but may be helpful in identification. Even in the new Code there are still relics of the Aristotelian type-concept, such as the provision (Art. 72b): "The type-series of a species consists of all the specimens on which its author bases the species," as if it were the function of the types to form the basis of a species.

Species consist of variable populations, and no single specimen can represent this variablity. No single specimen can be typical in the Aristotelian sense. As Simpson (1961) has discussed perceptively, the only function of the type-specimen is to be a "name bearer." Indeed he even suggested that one might drop the misleading term "type" and call the name-bearing specimen the *onomatophore* (Greek for name-bearer). The term type, however, is too firmly fixed in the taxonomic tradition for such a change to be practical. The younger generation of taxonomists now understands clearly that a type is nothing more and nothing less than a specimen (or taxon) which tells us to which taxon a given name should be attached. The term type-series is an anachronism with respect to the current concept.

A type is always a zoological object, never a name. The type of a genus is a species, the type of family is a genus (Art. 61). This is important in the case of "misidentification of the type," that is, in a case where an author, when designating the type of a new taxon, refers to it under a wrong name (13.49). Once designated the type cannot be changed, not even by the author of the taxon, except—by exercise of the plenary powers of the Commission (Art. 79)—through the designation of a neotype (Art. 75).

Description of a new species is based on the entire material available to the zoologist, including the type-specimen. It is *not* the function of the type to serve as the exclusive or primary basis of the description. Simpson (1961, pp. 183–186) discusses this aspect fully. He also introduces the term *hypodigm* for the entire sample of specimens personally known to a given taxonomist at a given time and considered by him to consist of unequivocal members of the taxon. Disagreements among taxonomists are often caused by the fact that they have studied different hypodigms.

## 49.  CORRECTING THE MISIDENTIFICATION OF TYPES

It is unfortunately true that the type-species of new genera or the type-genera of new taxa of the family group have sometimes been misidentified (= misnamed) by the original authors of the new taxa. Articles 41, 49, 65b, 67j, and 70a contain provisions about how to correct the error of the original author. The principle on which such corrections are based is that the type of a taxon is not a name but a zoological object. The type (-species or -genus) is then the zoological object which the original author had before him (when making the type designation) and not the name which he may have erroneously attached to this object.

However, it would be an intolerable burden on the working taxonomist if he had to make sure in each case that the original type was correctly identified and named. It is therefore provided in the Code that a zoologist must assume that the author correctly identified this type (Arts. 65*a*, 70).

If there is strong or clear evidence that a misidentification is involved, the case is to be referred to the Commission. The Commission in such cases tends to make a ruling that will maintain stability and continuity.

If reexamination of the type of a well-known species proves that the type actually belongs to a different species, the Commission can by its plenary powers (Art. 79) suppress the original type and designate a neotype (see 13.55) which conforms to the accepted concept of the species.

Wisely administered, the various provisions on misidentification can do much to preserve stability and continuity in nomenclature.


## 50.  TYPE OF REPLACEMENT NAME

Articles 67*i* and 72*d* unequivocally provide that if the name of a taxon is expressly proposed as a replacement name for a prior name, it retains the type of the prior name.

Former generations of zoologists did not necessarily follow this rule, and many names, seemingly replacement names, clearly have a composite basis. They are based on the description of specimens (including a new type) but also refer to a preoccupied name. A retroactive application of the new Art. 72*d* to such composite names, when the name has been universally applied to the newly designated type-material for 50 years or more, would surely be in conflict with the stability concept expressed in the Preamble.


## 51.  KINDS OF TYPE-SPECIMENS

Since the type-specimen is the official standard of reference for the name of a species, it can have full authority only if it is unique (Art. 45*b*). When there are two or more type-specimens, it has all too often been found that they actually belong to different species. Which of these species then shall have the name? The new Code clearly specifies (Art. 72*a*) that a taxon of the species group (species or subspecies) can have only a single type, either the specimen designated or indicated as the type by the original author at the time of publication of the original description (holotype) or one designated from the type-series (lectotype) [misprinted in the Code], or a neotype.

Specimens before the original describer that are neither the holotype nor the lectotype (of a subsequent author) are conventionally referred to as *paratypes*. Paratypes have no special standing under the Code and do not qualify as types, by the exclusion clearly formulated in Art. 72*a*. Recognition of a "type-series" in Art. 72*b* introduces a contradiction.

## 52. TYPE DESIGNATION

Since the information concerning the type-specimen is nomenclaturally the most important information given on the occasion of naming a new taxon, it has become customary in taxonomic literature to record this information immediately following the new name. The information concerning the type takes the same place in the sequence as does the synonymy in a redescription (see 11B.1.7–11B.2).

The definite fixation of a type-species is, after 1930, a prerequisite for making a name in the genus group available (Art. 13*b*). Designation of a type-specimen is not a prerequisite for making a name in the species group available. In certain groups, for instance certain genera of protozoans, it is exceedingly difficult to preserve specimens in such a way that they can serve as a permanent standard of reference. Yet even in protozoology the designation of type-specimens has increasingly become standard practice (Corliss, 1963). For rules concerning types in the species group, see Arts. 71–75.

Even though the publication of information on the type-specimen of a new species-name is not mandatory, it is nevertheless expected from the zoologist that he should clearly designate a single specimen as the holotype of the new species and that he should supply in the original description the following information, in addition to measurements and other descriptive data, characteristic for the type-specimen:

1. Precise collecting locality and other relevant data on the labels of the specimen
2. Sex
3. Developmental stage or form (if significant) to which the type is referable
4. In the case of parasites, name of the host species
5. Name of the collector
6. Collection in which the holotype is deposited and, when specimens are numbered, the number assigned to it

7. Altitude of the type-locality or depth in meters below sea level at which the holotype was taken

8. In the case of fossil species, geological horizon

Great care should be exercised by the author, if a large hypodigm is available, to select as the type a specimen which—owing to its state of preservation, sex, age, or locality data—is most suitable as a name-bearer. The diagnostic characters relevant in a given genus are sometimes better accessible in some specimens than in others. The type should have maximal usefulness for taxonomic discrimination.

In the case of fossil material, if the hypodigm consists of many individual pieces (e.g. bones), it is advisable to designate the most diagnostic of them as type, particularly if there is even the slightest doubt as to whether the pieces actually belong to a single individual. Many "types" of formerly described fossil species have, on reexamination, turned out to be composites of several different species.

The following additional advice may be offered with regard to types:

1. Type designation or fixation should always be completed before publication.
2. Type designation should be clear and unambiguous; deposition (and museum number) of type should always be recorded.
3. Types of undescribed species should not be distributed prior to publication.
4. Type labels should never be changed or removed.
5. Type fixation for species of older authors should be attempted only by a specialist during revisionary work.

### 53.  TYPE-MATERIAL

Type-specimens, as the official standards for the names of species, "are to be regarded as the property of science," according to Art. 72*f*. Recommendations 72A, 72B, 72C, and 72D specify the responsibilities of the caretakers of such type-material. All types should be transferred as quickly as possible to public institutions where their safety is guaranteed and where they are accessible to other research workers. In most museums distinctive labels are used to indicate clearly that a given specimen is a type.

Types should not be handled in connection with ordinary identification work. However, monographers should have free access to types, particularly when there is doubt as to the zoological identity of a type-specimen. Most museums are increasingly ready to mail types to specialists. In case of loss there is always the possibility of replacing a type by a neotype (Art. 75). For further comments on type collections see also 6.2.5.

## 54.  LECTOTYPES

If the name of a species was based on a series of syntypes, any zoologist may designate one of these syntypes as the lectotype (Art. 74 and Recommendations 74A, 74B, 74C, and 74D). Selection of a lectotype should be undertaken only by a specialist during revisionary work and ordinarily only if it contributes to the unambiguous affixation of a name to a given taxon. It should never be done merely in order to add a type-specimen to the collection. If the description of a species is clearly based on a particular specimen, that specimen should be made the lectotype. If one of the syntypes was illustrated, it should be selected as lectotype, other things being equal.

Syntypes are often widely scattered as the result of exchanges. This requires special considerations in lectotype selection. When possible, a lectotype should be chosen from syntypes in the collection of a public institution, preferably of the institution containing the largest number of syntypes of the species, or containing the collection on which the author of the nominal species worked, or containing the majority of his other types.

Many classical authors clearly designated one specimen as the type in their collections without specifically citing such a specimen as "the type" in the published description. The present wording of Art. 73b would imply that such a specimen does not qualify as the holotype. Such specimens have traditionally been accepted as holotypes, in contrast to cases where an author labeled numerous syntypes as "type." It would seem that Art. 73b could well be modified to legitimize the traditional practice (see Fennah, 1957, and Young, 1958, SZ, 7: 120–122).

In cases where the syntypes are from several localities and a previous reviser has already restricted the type-locality of the species, a responsible zoologist will give due consideration to this fact in the selection of the lectotype.

## 55.  NEOTYPES

In Art. 75 the Code regulates the designation of neotypes: "If through loss or destruction no holotype, lectotype, or syntype exists." The provision specifically forbids the manufacturing of neotypes as a matter of curatorial routine or for the sake of having a type for every species. Most of the older species of birds, for instance, have no types but are such clearly understood taxa that no type is needed. Neotypes ought to be designated "only in connection with revisory work, and then only in exceptional circumstances, when a neotype is necessary in the interest of stability of nomenclature" (Art. 75*a*). Even when the original type is damaged or even lost, it is in most cases quite unnecessary to designate a neotype.

Least desirable is for a neotype to be designated for a species whose name is not in general use either as a valid name or as a synonym, or for any name, not in use, that is a *nomen dubium*. No author should designate a neotype until he has carefully checked that his action conforms entirely with all the provisions of Art. 75. (See 75*e* for neotypes designated prior to 1961.)

The Commission has the power (Art. 79) to suppress an existing type if this is in the interest of stability of nomenclature and to designate a neotype to conform with the traditional usage of a name. For instance, some years ago it was discovered that the type of the Hottentot Teal (*Anas punctata*) in the Oxford University Museum actually belonged to the Maccoa Duck (*Oxyura maccoa*) and that a most confusing switching of names was inevitable if this type-specimen were not suppressed. The Commission designated a neotype for the species, and the stability of names that had existed for the previous 125 years was preserved. If a zoologist finds that the type of a well-known species has been misidentified, he can, if stability is seriously threatened, apply to the Commission for a suppression of the type and the designation of a neotype conforming to the traditional usage of the name. Such action should be confined to exceptional cases.

## 56.  TYPES AND FIRST REVISERS

There are a number of situations in which a zoologist may have to undertake an action in order to clarify the status of a type.

*A. Type of a Species*

1. If a series of syntypes belongs to two or more species, a zoologist must determine whether or not a previous first reviser restricted the name to one of the components. If not, he has to select a lectotype from among the syntypes in such a way that it best serves stability of nomenclature and is in accordance with the provisions of Art. 74 and the comments given in 13.54.

2. If the type is found not to belong to the species to which the name is traditionally applied, two avenues are open. In inactive groups the zoologist may prefer simply to shift the name to the correct species. However, if such action would cause a serious disturbance of stability, particularly in the case of names in active and universal use for more than 50 years, the zoologist may request the International Commission to use its plenary powers to suppress the original type and designate a neotype conforming to existing usage (see 13.55).

3. The Code contains no provisions regarding what to do when the type

lacks all diagnostic characters. In the case of a forgotten name or *nomen dubium,* such a name is best transferred to the Official Index of Rejected Names. In the case of names in general use, a neotype may be designated by the Commission provided such an action is "essential for solving a complex zoological problem" (Art. 75). In most cases no such neotype designation will be needed.

### B. *Type of a Genus*

1. If no species is included in the original naming of a genus, published before 1931, follow the provisions of Art. 69*a* (ii).
2. If none of several included species has been previously designated as type-species, the first reviser will again act according to the provisions of Art. 69.
3. If the evidence indicates that the name used for the previously designated type-species resulted from misidentification, the provisions of Art. 70 are to be followed. In such cases the Commission is charged "to designate as the type-species whichever species will in its judgment best serve stability and uniformity of nomenclature."

## 57.  TYPE-LOCALITIES

The type-locality is the place where the population occurs from which the type-specimen was taken. Specimens collected at the type-locality are called *topotypes,* and the population that occurs at the type-locality is called the topotypical population. The 1961 Code does not contain any mandatory provisions concerning type-localities, but Recommendation 72E contains advice on their designation or restriction. The comments contained in the ensuing paragraphs represent the best current practices, but they are not part of the Code. As in the case of lectotypes or neotypes, the designation or restriction of type-localities should not be done routinely but only by a specialist in connection with revisionary work.

Species can ordinarily be identified by single specimens, subspecies often only be adequate population samples. The type-locality, consequently, is unimportant at the species level but frequently decisive for the determination of the validity of subspecies. In view of the not infrequent overlap in the characteristics of subspecies, a single specimen can be the name-bearer ("type") for a subspecies only to the extent that it helps to identify the population from which it was sampled. Where it fails to do this, a knowledge of the type-locality becomes a necessity.

*Original Designation of a Type-locality.* When describing a new species or subspecies a worker often has before him material from many localities within the range of the new taxon. It is his duty to make as prudent a choice of the type-locality as possible. In this he should be guided by the following considerations, among others:

1. To choose a type-locality from which are available many topotypes which constitute a fair sample of the population and illustrate its variation.
2. In the case of a variable species or subspecies, to place the type-locality in the area inhabited by the population that the describer considers most typical for the new taxon.
3. In the case of a new subspecies that is part of a cline, to place the type-locality at the other end of the cline from the location of a previously described subspecies.
4. Never to place a type-locality in an area of intergradation or hybridization.
5. To record, in the description of a new taxon, the location of the type-locality as precisely as possible. Taxonomically distinct populations of a species sometimes occur only 100 m apart from each other. In paleontology a few centimeters vertically may mark a change from one bed to another. This is why type-localities must be fixed with extreme accuracy. It is also the reason why it is so important that the collector give precise locality data on the labels of the specimens (see Chap. 6). When such data are lacking, it is sometimes possible to obtain them from the collector if he is still alive. In other cases the information may be found in published or unpublished journals or fieldbooks.

*Restriction of a Designated Type-locality.* Earlier authors, not appreciating the need for exact type-localities, often described new species from "California" or "Brazil" or "Africa." When later collections indicate that the species from "Brazil" is geographically variable and consists of two or more subspecies, it becomes necessary to determine the exact locality from which the type of the nominate subspecies came.

Most workers accept the principle that the "first reviser," the person who first realizes the geographical variability of such a species, has the right to designate arbitrarily a more restricted type-locality, provided that evidence derived from a study of the type itself does not contradict his designation. Such a fixation is followed unless it can be shown that the action of the first reviser is erroneous. Obviously, if the first reviser restricts to Rio de Janeiro the type-locality of a species from "Brazil," his restriction should not be binding if the type is still in existence and belongs to a subspecies which is confined to the neighborhood of Cayenne. To avoid such mistakes, the first reviser should make a careful investigation of the probable route of the collector. Even in the absence of exact information certain conclusions may be obvious; a type collected in China in 1775 most likely came from Canton or some part of Fukien, not from Szechuan, Kansu, or some other place far in the interior.

In the case of a "Voyage," it is often possible to determine the exact locality by a study of the course of the voyage. For example, a small owl, *Ninox ocellata,* collected by the *Voyage au Pôle Sud,* was described by Hombron and Jacquinot as having come from Chile, South America. This is an obvious error, since the genus does not occur in the Americas. Later on, Mathews, believing *ocellata* to be an earlier name of *N. roseoaxillaris* Hartert (1929, San Cristobal, Solomon Islands), restricted the type-locality of *ocellata* to San Cristobal. However, it is stated in the report of the above-mentioned *voyage* that the expedition landed in the Solomon Islands only on Ysabel Island (and adjacent St. George), where no owl resembling *ocellata* occurs. Mathews' restriction of the type-locality is therefore untenable. Subsequently it was shown by Peters that the Coburg Peninsula, Northern Territory, Australia, is the only locality touched by the *voyage* where an owl occurs that agrees with the description of *N. ocellata*. Peters therefore restricted the type-locality to Coburg Peninsula, and this restriction has been universally accepted. This case clearly shows how much care must be exercised in the correction or restriction of type-localities.

*Correction of a Wrong Type-locality.* There are two sets of circumstances under which an error in the originally stated designation of the type-locality can be corrected:

1. Exact type-locality given in the original description. If the author or some subsequent worker can prove beyond doubt that the type(s) did not come from the locality given in the original description (owing to some error or misinformation), he can shift the type-locality to the place from which the type really came, or, at least, where the taxon is known to occur. Actually this is not a shift of the type-locality but only of the "stated" type-locality, since the type never came from the originally designated locality.

   A type-locality cannot be altered because an author finds that the population at a different locality is "more typical" or because he has received "better material" from a new locality. Proposals for the shift of type-localities for these and similar reasons must be rejected.

2. Exact type-locality not given in the original description. If no type-locality is given, or if only a vague one ("India"), the first reviser may designate a restricted type-locality. Such a restriction may later be set aside if it conflicts with the available evidence, but only if the case is unequivocal. The fixation of a type-locality cannot be set aside because that locality, at the time of the collection, was "less accessible" than some other locality, or because the species is "rather rare" at that locality. It should be changed, however, if it is clearly outside the range of the species.

Designation of a lectotype is equivalent to a restriction of the type-locality if the original syntypes came from several localities. If there is a conflict between lectotype selection and restriction of type-locality, lectotypes take precedence. A type-locality should never be changed or restricted in such a way that it upsets the stability of nomenclature.

When designating a lectotype an author should take into careful consideration the possibility of a prior restriction of type-locality (see 13.54).

## 58. NOMINAL TAXON

The Code frequently refers to "nominal species" or to "nominal" combined with some other category (rank) designation. The expression *nominal species* means a species for nomenclatural purposes without reference to its taxonomic status. Such a species is indicated exclusively by its name, and the name, in turn, is defined by its name-bearer ( = type). The term is therefore correctly defined in the glossary of the Code as "a named species, objectively defined by its type-specimen." Equivalent definitions pertain to nominal genus and nominal family.

## 59. INTERNATIONAL COMMISSION ON ZOOLOGICAL NOMENCLATURE

The duties, powers, organization, and operation of the Commission are regulated by Arts. 76–82 of the Code, the Constitution of the Commission (*BZN, 21*(3) : 181–185), and the By-laws (*BZN 22*(1) : 3–8).

Since this is often misunderstood, it must be pointed out that the rules of nomenclature are phrased and adopted by the Zoological Congresses. These are the legislative bodies responsible for the rules. The Commission is authorized by these International Congresses, as stated in Arts. 76–79, to interpret or suspend provisions of the Code in individual cases and to submit to the Congresses recommendations for the clarification or modification of the Code.

In particular the Commission is empowered to suspend the application of any provision of the Code "if such application to a particular case would in its judgment disturb stability or universality or cause confusion" (Preamble and Art. 79). Under these plenary powers it may "annul or invalidate any name, type-designation, or other published nomenclatural act, or any publication, and validate or establish replacements," by a two-thirds majority.

## 60. APPLICATIONS TO THE COMMISSION

Any zoologist may submit cases involving nomenclatural problems to the Commission (Art. 78). It must be remembered, however, that the Commission is a judicial, not a fact-finding board. It is under no obligation to supplement or verify information contained in applications. A zoologist planning to submit an application will do well to study the style of such applications in recent volumes of *BZN* and also to circulate his application first informally among cospecialists. The processing and publication of the applications is described in Art. 17 of the Constitution, and the By-laws, Sec. IIIC [*BZN* **21** (3) and **22**(1)] respectively.

## 61.  CASES UNDER CONSIDERATION

For the sake of stability in nomenclature it is of extreme importance that zoologists obey the provisions of Art. 80. This states that, while an application to the Commission is pending, "existing usage is to be maintained, until the decision of the Commission is published."

## 62.  AMENDMENTS TO THE CODE

As stated in Art 87, amendments to the Code can be made only by an international congress of zoology. All such amendments are first processed by the Commission and are submitted to the Congress only if approved by a majority of the section of nomenclature of that Congress (Arts. 77 and 78*a*). In accordance with its legislative authority the Congress can vote to add or delete articles and to review any decision of the Commission (Art. 78*g*).

## 63.  OFFICIAL LISTS AND INDEXES

Article 14 of the Constitution provides that "the Commission shall compile and maintain" the following lists and indexes:

Official List of Family-group Names in Zoology
Official List of Generic Names in Zoology
Official List of Specific Names in Zoology
Official Index of Rejected and Invalid Family-group Names in Zoology
Official Index of Rejected and Invalid Generic Names in Zoology
Official Index of Rejected and Invalid Specific Names in Zoology
Official List of Works Approved as Available for Zoological Nomenclature
Official Index of Rejected and Invalid Works in Zoological Nomenclature

Section III(A) of the By-laws delegates to the secretariat of the Commission responsibility for "the preparation and editing for publication of official lists and indexes." Provisions in the Code pertaining to these official lists and indexes are found in Arts. 23*b*, 23*d*, 77, and 78*f*.

### 64.  APPENDICES TO THE CODE

Attached to the Code are five appendices to serve as a guide to good usage in nomenclature. They are not mandatory, as are Arts. 1–87 of the Code proper, but rather have the same status as Recommendations in the Code.

Appendix A (Code, p. 93) is a Code of Ethics. Every zoologist should carefully study the eight paragraphs of this appendix in order not to violate well-established conventions and risk losing the respect of his colleagues. For instance, no zoologist should publish a replacement name for a junior homonym during the lifetime of its author without following the procedure of Paragraph 3 of the Code of Ethics. A new name proposed in violation of these provisions is, however, available if it otherwise satisfies the provisions of the Code.

Appendix B deals with the transliteration and latinization of Greek words (Code, pp. 95–101).

Appendix C deals with the latinization of geographical and proper names (Code, pp. 101–103).

Appendix D contains extraordinarily detailed and helpful recommendations on the formation of names (Code, pp. 105–141). Since every zoologist must have his own copy of the Code handy at all times, we have not included a corresponding section in this text.

### 65.  GENERAL RECOMMENDATIONS

Appendix E (Code, pp. 143–147) contains a useful summary of recommendations to the working taxonomist. The beginner, in particular, should carefully study the 24 recommendations, because this will surely help him to avoid making mistakes or at least to improve the quality of his publications. Numbers 15, 16, 22, and 24 are of special importance.

### 66.  GLOSSARY AND INDEX

If a zoologist is in doubt as to the meaning of a word used in the Code, he should consult the Glossary to be found on pp. 148–154 (English) and pp. 155–161 (French).

Even the experienced worker sometimes has difficulties in finding the exact article containing provisions concerning a given case. Even though there are only 86 articles, there are over 600 individual provisions in the Code and in the appendices (exclusive of the tables). The index of the Code (pp. 163–176) is an invaluable guide to these provisions. Items printed in boldface in the index refer to the glossary.

# Bibliography

AGASSIZ, L., and H. E. STRICKLAND. 1848. Bibliographia Zoologiae et Geologiae. London, Ray Society, vols. 1–4.

ALBRECHT, F. O. 1962. Physiologie, comportement et écologie des acridiens, etc. *Colloq. Int. Centre Nat. Rech. Sci.* No. **114**:283–297.

ALEXANDER, R. D. 1962. The role of behavioral study in cricket classification. *Syst. Zool.,* **11**:53–72.

———. 1967. Acoustical communication in arthropods. *Ann. Rev. Ent.,* **12**:495–526.

AMADON, D. 1967. The superspecies concept. *Syst. Zool.,* **15**:245–249.

ANDERSON, E. 1936. Hybridization in American Tradescantias. *Ann. Mo. Bot. Garden,* **23**:511–525.

———. 1949. Introgressive hybridization. John Wiley & Sons, Inc., New York, 199 pp.

———. 1954. Efficient and inefficient methods of measuring specific differences. *In* O. Kempthorne, et al. (eds.), Statistics and mathematics in biology. Iowa State College Press, Ames, Iowa, pp. 93–106.

ANDERSON, R. M. 1965. Methods of collecting and preserving vertebrate animals, 4th ed. *Bull. Natl. Mus. Canada,* Dept. Mines, no. 69, Biol. Ser. 18, Ottawa, 199 pp.

ANTHONY, H. E. 1945. The capture and preservation of small mammals for study. *Amer. Mus. Nat. Hist. Sci. Guide* 61, 54 pp.

ARKELL, W. J., and J. A. MOY-THOMAS. 1940. Paleontology and the taxonomic problem. *In* J. S. Huxley (ed.), The new systematics. Clarendon Press, Oxford, pp. 395–410.

BADER, R. S. 1958. Similarity and recency of common ancestry. *Syst. Zool.,* **7**:184–187.

BAER, J. G. (ed.). 1957. First symposium on host specificity among parasites of vertebrates. *Zool. Inst. Univ. Neuchatel,* 324 pp.

BAKER, W. W., et al. 1958. Guide to the families of mites. Contribution no. 3, Institute of Acarology, University of Maryland, College Park, 242 pp.

BALAZUC, J. 1948. La Tératologie des Coléoptères et expériences de transplantation sur *Tenebrio molitor* L. *Mem. Mus. Natl. d'Hist. Nat.* (n.s.), **25**:1–293.

BATES, M. 1940. The nomenclature and taxonomic status of the mosquitoes of the *Anopheles maculipennis* complex. *Ann. Ent. Soc. Amer.,* **33**:343–356.

BECKNER, M. 1959. The biological way of thought. Columbia University Press, New York, 200 pp.

BEER, J. R. DE, and E. F. COOK. 1957. A method for collecting ectoparasites from birds. *J. Parasitol.,* **43**:445.

BESSEY, C. E. 1908. The taxonomic aspect of the species. *Amer. Nat.,* **42**:218–224.

BIANCO, S. L. 1899. The methods employed at the Naples zoological station for the preservation of marine animals (translated from Italian by E. O. Hovey). *U.S. Nat. Mus. Bull.,* no. 39, part M:3–42.

BISHOP, S. 1952. Handbook of salamanders. Comstock Publishing Associates, Cornell University Press, Ithaca, N.Y., 555 pp.

BLACKWELDER, R. E. 1948. The principle of priority in biological nomenclature. *J. Wash. Acad. Sci.,* **38**:306–309.

————. 1967. Taxonomy. John Wiley & Sons, Inc., New York, 698 pp.

————, and R. M. BLACKWELDER. 1961. Directory of zoological taxonomists of the world. Southern Illinois University Press, Carbondale, Ill., 404 pp.

————, and A. BOYDEN. 1952. The nature of systematics. *Syst. Zool.,* **1**:26–33.

BLAIR, W. F. 1962. Non-morphological data in anuran classification. *Syst. Zool.,* **11**:72–84.

————. 1963. Evolutionary relationships of North American toads of the genus *Bufo. Evolution,* **17**:1–16.

BLAKER, A. A. 1965. Photography for scientific publication. A handbook. W. H. Freeman and Company, San Francisco, 158 pp.

BLOCH, K. 1956. Zur Theorie der naturwissenschaftlichen Systematik (unter besonderer Berücksichtigung der Biologie). *Acta Biotheoretica,* Leiden, **7**:1–138.

BOCK, W. J. 1963. Evolution and phylogeny in morphologically uniform groups. *Amer. Nat.,* **97**:265–285.

BORGMEIER, T. 1957. Basic question of systematics. *Syst. Zool.,* **6**:53–69.

BOTTLE, R. T. and H. V. WYATT. 1966. The usage of biological literature. Butterworth & Co. (Publishers), Ltd., London, 286 pp.

BOWMAN, R. I. 1961. Morphological differentiation and adaptation in the Galapagos finches. *Univ. Calif. Publ. in Zool.,* **58**:1–326.

BOYCE, A. J. 1964. The value of some methods of numerical taxonomy with reference to hominoid classification. The Systematics Assoc. (London) Publ. no. 6, pp. 47–65.

BRINKMANN, R. 1929. Statistisch-biostratigraphische Untersuchungen an mitteljurassischen Ammoniten über Artbegriff und Stammesentwicklung. *Abhandl. Ges. Wiss. Göttingen,* Math. Nat. Kl. (N.F.), **13**:1–249.

British Museum (Nat. Hist.). 1936ff. Instructions for collectors.

BROWN, R. W. 1954. Composition of scientific words. A manual of methods and a lexicon of materials for the practice of logotechnics. Publ. by the author, Washington (U.S. Nat. Mus.), 882 pp.

BROWN, W. L., and E. O. WILSON. 1956. Character displacement. *Syst. Zool.,* **5**:49–64.

BRUES, C. H., A. L. MELANDER, and F. M. CARPENTER. 1954. Classification of insects. *Bull. Mus. Comp. Zool.,* **108**:1–917.

BRYSON, V., and H. J. VOGEL (eds.). 1965. Evolving genes and proteins. Academic Press, New York and London, 629 pp.

BUCHNER, P. 1966a. Endosymbiosis of animals with plant microorganisms. (Interscience) John Wiley & Sons, Inc., New York.

————. 1966b. Die Symbiosen der Palaeococcoidea. *Z. Morph. Ökol. Tiere,* **56**:275–362.

BUCK, R. C., and D. L. HULL. 1966. The logical structure of the Linnaean hierarchy. *Syst. Zool.,* **15**:97–111.

BÜHLER, P. 1964. Zur Gattungs- und Artbestimmung von Neomys-Schädeln. *Ztschr. Säugetierkunde,* **29**:65–93.

BULLIS, H. R., JR., and R. B. ROE. 1967. A bionumeric code application in handling complex and massive faunal data. *Syst. Zool.,* **16**:52–55.

BURMA, B. H. 1948. Studies in quantitative paleontology. I. Some aspects of the theory and practice of quantitative invertebrate paleontology. *J. Paleontol.,* **22**:725–761.

BUSNEL, R. G. 1963. Acoustic behavior of animals. Elsevier Publishing Company, Amsterdam and New York.

CAILLEUX, A. 1954. How many species? *Evolution,* **8**:83–84.

CAIN, A. J. 1954. The superspecies. *Syst. Zool.,* **3**:145–146.

———. 1956. The genus in evolutionary taxonomy. *Syst. Zool.,* **5**:97–109.

———. 1958. Logic and memory in Linnaeus' system of taxonomy. *Proc. Linn. Soc. London,* **169**:144–163.

———. 1959a. Deductive and inductive methods in post-Linnaean taxonomy. *Proc. Linn. Soc. London,* **170**:185–217.

———. 1959b. Taxonomic concepts. *Ibis,* **101**:302–318.

———. 1962. The evolution of taxonomic principles. *In* G. C. Aimsworth and P. H. A. Sneath (eds.), Microbial classification. 12th Symp. Soc. Gen. Microbiology, Cambridge University Press, London, pp. 1–13.

———, and G. A. HARRISON. 1958. An analysis of the taxonomist's judgment of affinity. *Proc. Zool. Soc. London,* **131**:85–98.

———, and ———. 1960. Phyletic weighting. *Proc. Zool. Soc. London,* **135**:1–31.

CAMIN, J. H., and R. R. SOKAL. 1965. A method for deducing branching sequences in phylogeny. *Evolution,* **19**:311–326.

CAMP, W. H., and C. L. GILLY. 1943. The structure and origin of species. *Brittonia,* **4**:323–385.

DE CANDOLLE, AUGUSTIN, P. 1813. Théorie élémentaire de la botanique. Chez Deterville, Paris, viii + 500 + 27 pp.

CAPPE DE BAILLON, P. 1927. Recherches sur la tératologie des insectes. Encyclopédie Entomologie, **8**:5–291.

CARPENTER, G. D. H. 1949. *Pseudacrea eurytus* (L.) (Lep. Nymphalidae): a study of a polymorphic mimic in various degrees of speciation. *Trans. Roy. Entomol. Soc. London,* **100**:71–133.

CARSON, H. L., F. E. CLAYTON, and H. D. STALKER. 1967. Karyotypic stability and speciation in Hawaiian Drosophila. *Proc. Nat. Acad. Sci.,* **57**:1280–1285.

CAVALLI-SFORZA, L. L., and A. W. F. EDWARDS. 1964. Analysis of human evolution. *In* Genetics today (*Proc. XI Intern Congr. Genetics, The Hague,* 1963), **3**:923–933, Pergamon Press, New York.

CHIARELLI, B. 1966. Caryology and taxonomy of the catarrhine monkeys. *Amer. J. Phys. Anthropol.,* **24**:155–169.

CLARCK, G. L., (ed.). 1961. The encyclopedia of microscopy. Reinhold Publishing Corporation, New York, 693 pp.

CLAUSEN, C. D. 1942. The relation of taxonomy to biological control. *J. Econ. Entomol.,* **35**:744–748.

CLAY, T. 1949. Some problems in the evolution of a group of ectoparasites. *Evolution,* **3**:279–299.

———. 1958. Revisions of Mallophaga genera. *Bull. Brit. Mus. (Nat. Hist.) Ent.,* **7**:123–208.

COCHRAN, W. G. 1959. Sampling techniques. John Wiley & Sons, Inc., New York.

COLEMAN, W. 1964. Georges Cuvier, zoologist. Harvard University Press, Cambridge, Mass., 212 pp.

CONKLIN, H. C. 1962. Lexicographical treatment of folk taxonomies. *Intern. J. Amer. Linguistics,* **28**:119–141.

CONSTANCE, L. 1964. Systematic botany—an unending synthesis. *Taxon,* **13**:251–273.

CORLISS, J. O. 1957. Nomenclatural history of the higher taxa in the subphylum Ciliophora. *Arch. Protistenk.,* **102**:113–146.

———. 1959. An illustrated key to the higher groups of the ciliated protozoa, with definition of terms. *J. Protozool.,* **6**:265–284.

———. 1962a. Taxonomic procedures in classification of protozoa. *Symp. Soc. Gen. Microbiol.,* **12**:37–67.

———. 1962b. Taxonomic-nomenclatural practices in protozoology and the new International Code of Zoological Nomenclature. *J. Protozool.,* **9**:307–324.

———. 1963. Establishment of an international type-slide collection for the ciliate protozoa. *J. Protozool.,* **10**:247–249.

CROWSON, R. A. 1958. Darwin and classification. *In* S. A. Barnett (ed.), A Century of Darwin. Harvard University Press, Cambridge, Mass., pp. 102–129.

———. 1965. Classification, statistics, and phylogeny. *Syst. Zool.,* **14**:144–148.

CULLEN, J. M. 1959. Behaviour as a help in taxonomy. *Syst. Assoc. Publ.,* **3**:131–140.

DAHL, F. 1925 *et seq.* Die Tierwelt Deutschlands. Gustav Fischer, Jena, 55 pts. *et seq.*

DALL, W. H. 1898. Contributions to the tertiary fauna of Florida. *Trans. Wagner Free Inst. Sci. Phila.* **3**:675–676.

DARWIN, C. 1859. On the origin of species by means of natural selection, or the preservation of favoured races in the struggle for life. John Murray, London, ix + 502 pp.

———. [1964] Facsimile edition ed. by E. Mayr. Harvard University Press, Cambridge, Mass.

DETHIER, V. G. 1947. Chemical insect attractants and repellents. McGraw-Hill Book Company, New York.

DIAMOND, J. M. 1965. Zoological classification system of a primitive people. *Science,* **151**:1102–1104.

DOWNEY, J. C. 1962. Host-plant relations as data for butterfly classification. *Syst. Zool.,* **11**:150–159.

DREISBACH, R. R. 1952. Preparing and photographing slides of insect genitalia. *Syst. Zool.,* **1**:134–136.

DRIVER, E. C. 1950. Name that animal. A guide to the identification of the common land and fresh-water animals of the U.S. with special reference to the area east of the Rockies. Kraushar Press, Northampton, Mass., 558 pp.

EADES, D. C. 1965. The inappropriateness of the correlation coefficient as a measure of taxonomic resemblance. *Syst. Zool.,* **14**:98–100.

EDMONDSON, W. T. 1949. A formula key to the rotatorian genus Ptygura. *Trans. Amer. Microsc. Soc.,* **68**:127–135.

——— (ed.). 1959. Ward & Whipple's fresh-water biology. John Wiley & Sons, Inc. New York.

EDMUNDS, G. F. 1962. The principles applied in determining the hierarchical level of the higher categories of Ephemeroptera. *Syst. Zool.,* **11**:22–31.

EHRLICH, P. R. 1958. The comparative morphology, phylogeny and higher classification of the butterflies (Lepidoptera: Papilionoidea). *Univ. Kansas Science Bull.,* **39**:305–370.

———. 1964. Some axioms of taxonomy. *Syst. Zool.,* **13**:109–123.

ELTON, C. 1947. Animal ecology. Sidgwick & Jackson, Ltd., London, xx + 209 pp., 13 figs.

EMDEN, F. I. VAN. 1957. The taxonomic significance of the characters of immature insects. *Ann. Rev. Entomol.,* **2**:91–106.

ENGELMANN, W. 1846. Bibliotheca Historico-Naturalis. Verzeichnis der Bücher über Naturgeschichte 1700–1846. W. Engelmann, Leipzig, 786 pp.

EVANS, H. E. 1957. Studies on the comparative ethology of digger wasps of the genus *Bembix*. Cornell Studies in Entomology, Cornell University Press, Ithaca, N.Y., 248 pp.

―――. 1964. The classification and evolution of digger wasps as suggested by larval characters (Hymenoptera: Sphecoidea). *Entomol. News,* **75**:225–237.

―――. 1966. The comparative ethology and evolution of the sand wasps. Harvard University Press, Cambridge, Mass., 526 pp.

FARRIS, J. S. 1966. Estimation of conservatism of characters by consistency within biological populations. *Evolution,* **20**:587–591.

FAURE, J. C. 1943*a*. The phases of the lesser army worm, Laphygma exigua (Hübn.) *Farming in S. Africa,* **18**:69–78.

―――. 1943*b*. Phase variation in the army worm, Laphygma exempta (Walk.). *Union S. Africa Dept. Agr. Farm. Bull.,* **243**:1–17.

FELL, H. B. 1965. The early evolution of the Echinozoa. *Breviora,* no. 219:1–17.

FENNAH, R. G. 1957. A guiding principle for lectotype selection. *Syst. Zool.,* **6**:47–48.

FERRIS, G. F. 1928. The principles of systematic entomology. Stanford University Press, Stanford, Calif.

FINGERMAN, M. 1963. The control of chromatophores. Pergamon Press, New York, 184 pp.

FISHER, R. A. 1936. The use of multiple measurements in taxonomic problems. *Ann. Eugenics,* **7**:179–188.

―――. 1938. The statistical use of multiple measurements. *Ann. Eugenics,* **8**:376–386.

FITCH, W. M., and E. MARGOLIASH. 1967. Construction of phylogenetic trees. *Science,* **155**:279–284.

FORD, E. B. 1945. Polymorphism. *Biol. Rev.* **20**:73–88.

―――. 1965. Genetic polymorphism. The M.I.T. Press, Cambridge, Mass., 101 pp.

FOSTER, J. B. 1965. The evolution of the mammals of the Queen Charlotte Islands, British Columbia. *Occ. Papers Brit. Columbia Prov. Mus.,* no. 14:1–130 (esp. 78–86).

FRANCON, M. 1961. Progress in microscopy. Pergamon Press, New York, 295 pp.

GARN, S. M., and R. H. HELMRICH. 1967. Next step in automated anthropometry. *Amer. J. Phys. Anthropol.,* **26**:97–99.

GATENSBY, J. B., and H. W. BEANS (eds.). 1950. Lee's microtomist's vademecum. Eleventh edition. McGraw-Hill Book Company, 753 pp.

GEROULD, J. J. 1921. Blue-green caterpillars: the origin and ecology of a mutation in hemolymph color in *Colias (Eurymus) philodice. J. Exp. Zool.,* **34**:385–415.

GERSCH, M. 1964. Vergleichende Endokrinologie der wirbellosen Tiere. Akad. Verlagsges., Geest & Portig, Leipzig.

GÉRY, J. 1962. Le problème de la sous-espèce et de sa définition statistique (à propos du coefficient de Mayr-Linsley-Usinger). Vie et Milieu, **13**:521–541.

GHISELIN, M. T. 1966*a*. An application of the theory of definitions to systematic principles. *Syst. Zool.,* **15**:127–130.

―――. 1966*b*. On psychologism in the logic of taxonomic controversies. *Syst. Zool.,* **15**:207–215.

GILMOUR, J. S. L. 1940. Taxonomy and philosophy. *In* J. S. Huxley (ed.), The new systematics. Clarendon Press, Oxford, pp. 461–474.

―――. 1961. Taxonomy. *In* A. M. MacLeod and L. S. Cobley, Contemporary botanical thought. Oliver & Boyd Ltd., Edinburgh, pp. 27–45.

GINSBURG, I. 1938. Arithmetical definition of the species, subspecies and race concept, with a proposal for a modified nomenclature. *Zoologica,* **23**:253–286.

GISIN, H. 1964. Synthetische Theorie der Systematik. *Z. Zool. Syst. Evol. Forsch.*, 2:1–17.

GOLDSCHMIDT, R. 1933. Lymantria. *Bibliog. Genetica*, 11:1–185.

GOSLINE, W. A. 1965. Thoughts on systematic work in outlying areas. *Syst. Zool.*, 14:59–61.

GOULD, S. J. 1966. Allometry and size in ontogeny and phylogeny. *Biol. Rev. (Cambridge Phil. Soc.)*, 41:587–640.

GRAY, P. 1954. The microtomist's formulary and guide. McGraw-Hill Book Company, New York.

————. 1958. Handbook of basic microtechnique. McGraw-Hill Book Company, New York.

GREGG, J. R. 1954. The language of taxonomy. Columbia University Press, New York, 70 pp.

GRIMPE, G., and E. WAGLER. 1925 *et seq*. Die Tierwelt der Nord- und Ostsee, Bd. 1 *et seq*.

GÜNTHER, K. 1956. Systematik und Stammesgeschichte der Tiere, 1939–1953. *Fortschritte der Zoologie*, 10:33–278.

————. 1962. Systematik und Stammesgeschichte der Tiere, 1954–1959. *Fortschritte der Zoologie*, 14:268–547.

HACKETT, L. W. 1937. Malaria in Europe. Oxford University Press, London, xvi + 366 pp., 60 figs.

HAECKEL, E. 1866. Generelle Morphologie der Organismen, II. Georg Reiner, Berlin, vii–clx, 462 pp.

HAGMEIER, E. M. 1958. The inapplicability of the subspecies concept to the North American marten. *Syst. Zool.*, 7:1–7.

HANDLER, P. (ed.). 1964. Biochemistry symposium: Biochemical evolution. *Fed. Proc.*, 23:1229–1266.

HANDLIRSCH, A. 1926–1936. Orthoptera. *In* W. Kükenthal (ed.), Handbuch der Zoologie. De Gruyter, Berlin and Leipzig, 4:687–796.

HARRISON, G. A. 1959. Environmental determination of the phenotype. *Syst. Assoc.*, no. 3, 81–86.

HARTMAN, W. D. 1958. A re-examination of Bidder's classification of the Calcarea. *Syst. Zool.*, 7:97–110.

HATHEWAY, W. H. 1962. A weighted hybrid index. *Evolution*, 16:1-10.

HENNIG, W. 1950. Grundzüge einer Theorie der Phylogenetischen Systematik. Deutscher Zentralverlag, Berlin, 370 pp.

————. 1966. Phylogenetic systematics. University of Illinois Press, Urbana, Ill., 263 pp.

HERRERA, A. L. 1899. (See Opinion 72, Intern. Commission on Zool. Nomenclature. *Smithsonian Inst. Misc. Collect.*, 73:19, 1922.)

HESLOP-HARRISON, J. W. 1962. Purposes and procedures in the taxonomic treatment of higher organisms. *Symp. Soc. Gen. Microbiol.*, no. 12:14–36.

————. 1963. Species concepts: Theoretical and practical aspects. *In* T. Swain (ed.), Chemical plant taxonomy. Academic Press Inc., New York.

HIGHTON, R. 1962. Revision of North American salamanders of the genus *Plethodon*. *Bull. Fla. St. Mus.* (Biol. Sci.), 6:235–367.

HOLLAND, G. P. 1964. Evolution, classification and host relationships of Siphonaptera. *Ann. Rev. Entomol.*, 9:123–146.

HOPKINS, G. H. E. 1949. The host-association of the lice on mammals. *Proc. Zool. Soc. London*, 119:387–604.

HOPWOOD, A. T. 1950. Animal classification from the Greeks to Linnaeus. *In* T. A. Sprague et al. (q.v.). *Linn. Soc. London*.

HOYER, H. B., et al. 1964. A molecular approach in the systematics of higher organisms. *Science,* 144:959–967.

HULL, D. L. 1965. The effect of essentialism on taxonomy. *Brit. J. Phil. Sci.,* 15:314–326, 16:1–18.

———. 1966. Phylogenetic numericlature. *Syst. Zool.,* 15:14–17.

———. 1967. Certainty and circularity in evolutionary taxonomy. *Evolution,* 21:174–189.

HURT, P. 1949. Bibliography and footnotes. A style manual for college and university students, rev. ed. University of California Press, Berkeley, Calif., xii + 167 pp.

HUXLEY, J. S. 1939. Clines: an auxiliary method in taxonomy. *Bijdr. Dierk.* 27:491–520.

——— (ed.). 1940. The new systematics. Clarendon Press, Oxford, 583 pp.

———. 1958. Evolutionary processes and taxonomy, with special reference to grades. *Uppsala University Arsskr.,* 1958:6, 21–39.

IMBRIE, J. 1957. The species problem with fossil animals. *In* E. Mayr (ed.), The species problem. *Amer. Assoc. Adv. Sci.,* Publ. 50, Washington, D.C., pp. 125–153.

INGER, R. F. 1958. Comments on the definition of genera. *Evolution,* 12:370–384.

———. 1961. Problems in the application of the subspecies concept in vertebrate taxonomy. *In* W. F. Blair (ed), Vertebrate speciation. University of Texas Press, Austin, pp. 262–285.

———. 1967. The development of a phylogeny of frogs. *Evolution,* 21:369–384.

IRWIN, M. R. 1947. Immunogenetics. *Adv. in Genetics,* 1:133–159.

JEPSEN, G. L. 1944. Phylogenetic trees. *Trans. New York Acad. Sci.,* ser. 2:81–92.

JOHNSGARD, P. A. 1965. Handbook of waterfowl behavior. Cornell University Press, Ithaca, N.Y.

JOLICOEUR, P. 1959. Multivariate geographical variation in the wolf *Canis lupus* L. *Evolution,* 13:283–299.

JONES, R. McCLUNG. 1950. McClung's handbook of microscopical technique. Harper & Row, Publishers, Incorporated, New York.

JORDAN, K. 1905. Der Gegensatz zwischen geographischer und nichtgeographischer Variation. *Zeitschr. wissensch. Zool.,* 83:151–210.

KEAST, A. 1961. Bird speciation on the Australian continent. *Bull. Mus. Comp. Zool.,* 123:305–495.

KENDRICK, W. B., and L. K. WERESUB. 1966. Attempting Neo-Adansonian computer taxonomy at the ordinal level in the Basidiomycetes. *Syst. Zool.,* 15:307–329.

KENNEDY, J. S. 1956. Phase transformation in locust biology, *Biol. Rev.* 31:349–370.

——— (ed.). 1961. Insect polymorphism. *Symp. Roy. Entomol. Soc.* London, no. 1, pp. 1–115.

KEZER, J., T. SETO, and C. M. POMERAT. 1965. Cytological evidence against parallel evolution of *Necturus* and *Proteus. Amer. Nat.,* 99:153–158.

KIM, K. C., B. W. BROWN, and E. F. COOK. 1966. A quantitative taxonomic study of the *Hoplopleura hesperomydis* complex (Anoplura, Hoplopleuridae), with notes on *a posteriori* taxonomic characters. *Syst. Zool.,* 15:24–45.

KINSEY, A. C. 1930. The gallwasp genus *Cynips. Indiana Univ. Studies,* 16:1–577.

KIRBY, H. 1950a. Materials and methods in the study of Protozoa. University of California Press, Berkeley, Calif., 73 pp.

———. 1950b. Systematic differentiation and evolution of flagellates in termites. *Rev. Soc. Mex. Hist. Nat.,* 10:57–79.

KNUDSEN, J. W. 1966. Biological techniques. Collecting, preserving, and illustrating plants and animals. Harper & Row, Publishers, Incorporated, New York, 525 pp.

KOHN, A. J. 1959. The ecology of Conus in Hawaii. *Ecol. Monographs,* **29**:47–90.

———, and G. H. ORIANS. 1962. Ecological data in the classification of closely related species. *Syst. Zool.,* **11**:119–127.

KUHL, W. 1949. Das wissenschaftliche Zeichnen in der Biologie und Medizin. W. Kramer, Frankfurt a. M., 179 pp.

KUMMEL, B., and D. RAUP. 1965. Handbook of paleontological techniques. W. H. Freeman and Company, San Francisco, 852 pp.

LACK, D. 1947. Darwin's finches. Cambridge University Press, London, 208 pp.

LEONE, C. A. (ed.). 1964. Taxonomic biochemistry and serology. The Ronald Press Company, New York, 728 pp.

LÉVI, C. 1956. Étude des Halisarca de Roscoff. Embryologie et systematique des Dèmosponges. *Arch. Zool. Exptl. Gen.,* **93**:1–181.

———. 1957. Ontogeny and systematics in sponges. *Syst. Zool.,* **6**:174–183.

LEVI, H. W. 1959. Problems in the spider genus *Steatoda* (Theridiidae). *Syst. Zool.,* **8**:107–116.

———. 1966. The care of alcoholic collections of small invertebrates. *Syst. Zool.,* **15**:183–188.

LEWONTIN, R. C. 1966. On the measurement of relative variability. *Syst. Zool.,* **15**:141–142.

LIDICKER, W. Z., JR. 1962. The nature of subspecific boundaries in a desert rodent and its implications for subspecific taxonomy. *Syst. Zool.,* **11**:160–171.

LIGHT, S. F. ET AL., 1954. Intertidal invertebrates of the Central California coast. University of California Press, Berkeley, Calif., 446 pp.

LINDROTH, C. H. 1957. The Linnaean species of Carabid beetles. *J. Linn. Soc. Zool.,* **43**:325–341.

LINNAEUS, C. 1735. Systema naturae, sive regna tria naturae systematice proposita per classes, ordines, genera et species, Lugduni Batavorum, 12 pp.

———. 1737. Critica botanica. Lugduni Batavorum xiv + 270 pp.

———. 1753. Fundamenta botanica (In Alston Charles: Tirocinium botanicum edinburgense).

———. 1758. Systema naturae per regna tria naturae, secundum classes, ordines, genera, species cum characteribus, differentiis, synonymis, locis. Editio decima, reformata, Tom. I. Laurentii Salvii, Holmiae, 824 pp.

LINSLEY, E. G. 1937. The effect of stylopization on *Andrena porterae. Pan-Pacific Entomol.,* **13**:157.

———. 1944. The naming of infraspecific categories. *Ent. News,* **55**:225–232.

———, and R. L. USINGER. 1959. Linnaeus and the development of the international code of zoological nomenclature. *Syst. Zool.,* **8**:39–47.

LÖVE, A. 1964. The biological species concept and its evolutionary structure. *Taxon,* **13**:33–45.

LOVEJOY, A. O. 1936. The great chain of being. Harvard University Press, Cambridge, Mass., 382 pp.

LUBNOW, E., and G. NIETHAMMER. 1964. Zur Methodik von Farbmessungen für taxonomische Untersuchungen. *Verhandl. Deutsch. Zool. Gesellsch. München,* **1963**:646–663.

LYNES, H. 1930. Review of the genus Cisticola. *Ibis* (suppl.), 673 pp.

MACFADYEN, A. 1955. A comparison of methods for extracting soil arthropods. *Soil Zool.,* **1955**:315–332.

MACINTYRE, G. T. 1966. The Miacidae (Mammalia, Carnivora). *Bull. Amer. Mus. Nat. Hist.,* **131**:115–210.

MAERZ, A., and M. R. PAUL. 1950. A dictionary of color, 2d ed. McGraw-Hill Book Company, New York, 208 pp.

MANWELL, R. D., ET AL., 1957. Intraspecific variation in parasitic animals. *Syst. Zool.*, 6:2–28.

MASLIN, T. P. 1952. Morphological criteria of phylogenetic relationship. *Syst. Zool.*, 1:49–70.

MAYR, E. 1931. Birds collected during the Whitney South Sea expedition. 12. Notes on *Halcyon chloris* and some of its subspecies. *Amer. Mus. Novitates*, 469:1–10.

———. 1942. Systematics and the origin of species. Columbia University Press, New York, 334 pp.

———. 1943. *In* J. A. Oliver, Status of *Uta ornata lateralis.*, *Copeia*, p. 102.

———. 1954. Notes on nomenclature and classification. *Syst. Zool.*, 3:86–89.

——— (ed.). 1957a. The species problem. *Amer. Assoc. Adv. Sci. Publ.*, no. 50, 395 pp.

———. 1957b. Species concepts and definitions. See Mayr, 1957a, pp. 1–22.

———. 1958. Behavior and systematics. *In* A. Roe and G. G. Simpson (eds.), Behavior and evolution. Yale University Press, New Haven, pp. 341–362.

———. 1960. The emergence of evolutionary novelties. *In* S. Tax (ed.), The evolution of life. University of Chicago Press, Chicago, pp. 349–380.

———. 1963. Animal species and evolution. The Belknap Press, Harvard University Press, Cambridge, Mass., 797 pp.

———.1964a. Inferences concerning the Tertiary American bird faunas. *Proc. Natl. Acad. Sci.*, 51:280–288.

———. 1964b. From molecules to organic diversity. *Fed. Proc.*, 23:1231–1235.

———. 1964c. The new systematics. *In* C. A. Leone (ed.), Taxonomic biochemistry and serology. The Ronald Press Company, N.Y., pp., 13–32.

———. 1965a. Classification and phylogeny. *Amer. Zool.*, 5:165–174.

———. 1965b. Numerical phenetics and taxonomic theory. *Syst. Zool.*, 14:73–97.

———. 1965c. Classification, identification and sequence of genera and species. *L'Oiseau*, 35 Special No., pp. 90–95.

———. 1966. The proper spelling of taxonomy. *Syst. Zool.*, 15:88.

———. 1968a. The role of systematics in biology. *Science*, 159:595–599.

———. 1968b. Bryozoa vs. Ectoprocta. *Syst. Zool.*, 17:213–216.

———. and R. GOODWIN. 1956. Biological materials: Part I, Preserved materials and museum collections. Natl. Acad. Sci., Natl. Res. Counc. Publ., No. 399.

MCKEVAN, D. K. 1961. Current tendencies to increase the number of higher taxonomic units among insects. *Syst. Zool.*, 10:92–103.

MCKINNEY, F. 1965. The comfort movements of Anatidae. *Behaviour*, 25:120–220.

MEISE, W. 1936. Zur Systematik und Verbreitungsgeschichte der Haus- and Weiden-sperlinge, *Passer domesticus* (L.) und *hispaniolensis* (T.). *J. Ornithol.*, 84:634–672.

METCALF, Z. P. 1954. The construction of keys. *Syst. Zool.*, 3:38–45.

MICHENER, C. D. 1944. Comparative external morphology, phylogeny, and a classification of the bees (Hymenoptera). *Bull. Amer. Mus. Nat. Hist.*, 82:157–326.

———. 1949. Parallelism in the evolution of saturnid moths. *Evolution*, 3:129–141.

———. 1952. The Saturniidae (Lepidoptera) of the western hemisphere. *Bull. Amer. Mus. Nat. Hist.*, 98:335–502.

———. 1957. Some bases for higher categories in classification. *Syst. Zool.* 6:160–173.

———. 1963. Some future developments in taxonomy. *Syst. Zool.*, 12:151–172.

———. 1964. The possible use of uninominal nomenclature to increase the stability of names in biology. *Syst. Zool.*, 13:182–190.

———, and R. R. Sokal. 1957. A quantitative approach to a problem in classification. *Evolution,* 11:130–162.

———, and ———. 1966. Two tests of the hypothesis of nonspecificity in the *Hoplitis* complex. *Ann. Ent. Soc. Amer.,* 59:1211–1217.

Miller, R. L., and J. S. Kahn. 1962. Statistical analysis in the geological sciences. John Wiley & Sons, Inc., New York, 483 pp.

Minkoff, E. C. 1965. The effects on classification of slight alterations in numerical technique. *Syst. Zool.,* 14:196–213.

Moreau, R. E. 1959. The classification of the Musophagidae. *Publ. Syst. Assoc.,* no. 3:113–119.

Moynihan, M. 1959. A revision of the family Laridae (Aves). *Amer. Mus. Novitates,* no. 1928, 42 pp.

Munroe, E. 1960. An assessment of the contribution of experimental taxonomy to the classification of insects. *Rev. Canad. Biol.,* 19:293–319.

Myers, G. G., and A. E. Leviton. 1962. Generic classification of the high-altitude pelobatid toads of Asia (Scutiger, Aelurophryne, and Oreolalax). *Copeia* 1962 (2):287–291.

Needham, G. H. 1958. The practical use of the microscope. Charles C Thomas, Publisher, Springfield, Ill., 493 pp.

Newell, N. D. 1947. Infraspecific categories in invertebrate paleontology. *Evolution,* 1:163–171.

———. 1956. Fossil populations. *In* The species concept in paleontology. *Syst. Assoc. Publ.,* no. 2:63–82.

Nichols, D. (ed.). 1962. Taxonomy and geography. *Syst. Assoc. Publ.,* no. 4, 158 pp.

Oldroyd, H. 1958. Collecting, preserving and studying insects. Macmillan, New York, 327 pp.

Oman, P. W., and A. P. Cushman. 1946. Collection and preservation of insects. *U.S. Dept. Agr. Misc. Publ.,* 601:1–42.

Orton, G. L. 1957. The bearing of larval evolution on some problems in frog classification. *Syst. Zool.,* 6:79–86.

Osche, G. 1960. Aufgaben und Probleme der Systematik am Beispiel der Nematoden. *Verhandl. Deutsch. Zool. Gesell. Bonn,* 1960:329–384.

Papp, C. S. 1968. Scientific illustration: theory and practice. Wm. C. Brown, Dubuque, Iowa, xiv + 318 pp.

Park, O., W. C. Allee, and V. E. Shelford. 1939. A laboratory introduction to animal ecology and taxonomy with keys, etc. University of Chicago Press, Chicago x + 272 pp., 1–17 pls.

Parker, T. J., and W. A. Haswell. 1940. A textbook of zoology, 6th ed. Macmillan & Co., Ltd., London, 2 vols.

Pemberton, C. E. 1941. Contributions of the entomologist to Hawaii's welfare. *Hawaiian Planter's Rec.,* 45:107–119.

Pennak, R. W. 1953. Fresh-water invertebrates of the United States. The Ronald Press Company, New York, ix + 769 pp.

Perrier, E. 1893–1932. Traité de zoologie. Masson et Cie, Paris, vols. 1–10.

Peterson, A. 1934, 1937. A manual of entomological equipment and methods. Edwards Bros. Pt. 1, Ann Arbor, Mich., 1934, 21 pp. 138 pls. Pt. 2, St. Louis, Mo., 1937, 334 pp.

Popper, K. R. 1950. The open society and its enemies. Vol. 1, The spell of Plato. Routledge & Kegan Paul, Ltd., London, 351 pp.

Pratt, H. S. 1951. A manual of the common invertebrate animals (exclusive of insects). McGraw-Hill Book Company, New York, 854 pp.

PROSSER, C. L., and F. A. BROWN. 1961. Comparative animal physiology. 2d. ed. W. B. Saunders Co., Philadelphia, 688 pp.

PUBLICATIONS of the Systematics Association, London, 1953ff. (1) See Smart and Taylor, 1953. (2) The species concept in paleontology, 1956, ed. by P. C. Sylvester-Bradley. (3) Function and taxonomic importance, 1959, ed. by A. J. Cain. (4) Taxonomy and geography, 1962, ed. by D. Nichols. (5) Speciation in the sea, 1963, ed. by J. P. Harding and N. Tebble. (6) Phenetic and phylogenetic classification, 1964, ed. by V. H. Heywood and J. McNeill.

RAUP, D. M. 1962. Crystallographic data in echinoderm classification. Syst. Zool., 11:97–108.

REMANE, A. 1952. Die Grundlagen des natürlichen Systems, der vergleichenden Anatomie und der Phylogenetik. Akad. Verlagsges., Leipzig, 400 pp.

RENSCH, B. 1929. Das Prinzip geographischer Rassenkreise und das Problem der Artbildung. Borntraeger, Berlin, 206 pp.

———. 1934. Kurze Anweisung für zoologisch-systematische Studien. Akad. Verlagsges. Leipzig, 116 pp.

———. 1947. Neuere Probleme der Abstammungslehre. Die transspezifische Evolution. Ferdinand Enke Verlag, Stuttgart, 407 pp.

———. 1959. The laws of evolution. In S. Tax (ed.), The evolution of life. University of Chicago Press, Chicago, pp. 95–116.

———. 1960. Evolution above the species level. Columbia University Press, New York, 419 pp.

REYMENT, R. A. 1960. Studies on Nigerian Upper Cretaceous and Lower Tertiary Ostracoda, Part 1. Stockh. Contr. Geol., 7:1–238.

———. 1963. Part 2. Stockh. Contr. Geol., 10:1–286.

RIDGWAY, R. 1912. Color standards and color nomenclature. A. Hoen Co., Washington, D.C., iv + 43 pp.

RIECH, E. 1937. Systematische, anatomische, ökologische und tiergeographische Untersuchungen über die Süsswassermollusken Papuasiens und Melanesiens. Archiv Naturgesch. (N.S.) 6:37–153.

ROGERS, D. J. 1963. Taximetrics—new name, old concept. Brittonia, 15:285–290.

ROHLF, F. J. 1963. Classification of Aedes by numerical taxonomic methods (Diptera: Culicidae) Ann. Entomol. Soc. Amer., 56:798–804.

———. 1967. Correlated characters in numerical taxonomy. Syst. Zool., 16:109–126.

ROLLINS, R. C. 1965. On the bases of biological classification. Taxon, 14:1–6.

ROSEN, D. E., and R. M. BAILEY. 1963. The poeciliid fishes (Cyprinodontiformes), their structure, zoogeography and systematics. Bull. Amer. Mus. Nat. Hist., 126:1–176.

ROZEBOOM, L. E. 1962. Taxonomy concerning mosquito populations. J. Parasitol., 48:664–670.

RUSSELL, H. 1963. Notes on methods for the narcotization, killing, fixation and preservation of marine organisms. Systematics-Ecology Program, MBL, Woods Hole, Mass., 70 pp.

SABROSKY, C. W. 1950. Taxonomy and ecology. Ecology, 31:151–152.

———. 1955. The interrelations of biological control and taxonomy. J. Econ. Entomol., 48:710–714.

SAILER, R. I. 1961. Utilitarian aspects of supergeneric names. Syst. Zool., 10:154–156.

SAKAGAMI, S. F., and C. D. MICHENER. 1962. The nest architecture of the sweat bees (Halictinae). University of Kansas Press, Lawrence, 135 pp.

SALT, G. 1927. The effects of stylopization an aculeate Hymenoptera. J. Exp. Zool., 48:223–231.

———. 1941. The effects of host upon their insect parasites. Biol. Rev., 16:239–264.

SCHAEFFER, B., and M. K. HECHT (eds.). 1965. Symposium: The origin of higher levels of organization. *Syst. Zool.,* 14:245–342.

SCHMIDT, K. P. 1950. More on zoological nomenclature. *Science,* 111:235–236.

SCHNITTER, H. 1922. Die Najaden der Schweiz. *Rev. Hydrol.* (suppl.), 2:1–200.

SEAL, H. L. 1964. Multivariate statistical analysis for biologists. Methuen & Co., Ltd., London.

SELANDER, R. K., et al. 1965. Colorimetric methods in ornithology. *Condor,* 66:491–495.

———, and R. F. JOHNSTON, 1967. Evolution in the House Sparrow. 1. Intra-population variation in N. America. *Condor,* 69:217–258.

SHAROV, A. G. 1965. Evolution and taxonomy. *Z. Zool. Syst. Evol.,* 3:349–358.

SHEALS, J. G. 1965. The application of computer techniques to Acarine taxonomy: a preliminary examination with species of the *Hypoaspis-Androlaelaps* complex (Acarina). *Proc. Linn. Soc. London,* 176:11–21.

SHORT, L. L., JR. 1965. Hybridization in the flickers (*Colaptes*) of North America. *Bull. Amer. Mus. Nat. Hist.,* 129:307–428.

SIBLEY, C. G. 1957. The evolutionary and taxonomic significance of sexual dimorphism and hybridization in birds. *Condor,* 59:166–191.

———. 1960. The electrophoretic patterns of avian egg-white proteins as taxonomic characters. *Ibis,* 102:215–284.

———. 1961. Hybridization and isolating mechanisms. *In* W. F. Blair (ed.), Vertebrate speciation. University of Texas Press, Austin, pp. 69–88.

SIMPSON, G. G. 1945. The principles of classification and a classification of mammals. *Bull. Amer. Mus. Nat. Hist.,* 85:1–350.

———. 1953. The major features of evolution. Columbia University Press, New York, 434 pp.

———. 1959*a*. Anatomy and morphology: classification and evolution: 1859 and 1959. *Proc. Amer. Phil. Soc.,* 103:286–306.

———. 1959*b*. The nature and origin of supraspecific taxa. *Cold Spring Harbor Symp. Quant. Biol.,* 24:255–271.

———. 1961. Principles of animal taxonomy. Columbia University Press, New York, 247 pp.

———. 1962*a*. The status of the study of organisms. *Amer. Sci.,* 50:36–45.

———. 1962*b*. Primate taxonomy and recent studies of nonhuman primates. *Ann. New York Acad. Sci.* 102:497–514.

———. 1963. The meaning of taxonomic statements. *In* S. L. Washburn (ed.), Classification and human evolution. Viking Fund Publ. in Anthropology no. 37:1–31.

———. 1964. Organisms and molecules in evolution. *Science,* 146:1535–1538.

———. 1965. The geography of evolution. Chilton Company, Book Division, Philadelphia, 249 pp.

———, A. ROE, and R. C. LEWONTIN. 1960. Quantitative zoology, rev. ed. Harcourt, Brace & World, Inc., New York, 440 pp.

SIMS, R. 1966. The classification of the Megascolecoid earthworms: an investigation of oligochaete systematics by computer techniques. *Proc. Linn. Soc. London,* 177:125–141.

SMART, J., and G. TAYLOR. (eds.) 1953. Bibliography of key works for the identification of the British fauna and flora. The Syst. Assoc., c/o Brit. Mus. (Nat. Hist.). Publ. No. 1, xi + 126 pp.

SMITH, H. M. 1946. Handbook of lizards. Comstock Press Associates, Cornell University Press, Ithaca, N.Y., xxi + 557 pp.

————. 1965. More evolutionary terms. *Syst. Zool.,* **14**:57–58.

SMITH, H. S. 1942. A race of *Comperiella bifasciata* successfully parasitizes California red scale. *J. Econ. Entomol.,* **35**:809–812.

SMITH, R. C., and R. H. PAINTER. 1967. Guide to the literature of the zoological sciences, 7th ed. Burgess Publishing Company, Minneapolis, 238 pp.

SMITH, R. I. (ed.). 1964. Keys to marine invertebrates of the Woods Hole Region. Contr. no. 11, Syst. Ecol. Program, Marine Biol. Lab., Woods Hole, 208 pp.

SMITH, W. J. 1966. Communication and relationships in the genus *Tyrannus*. *Pub. Nuttall Ornith. Club* no. 6, Cambridge, Mass., 250 pp.

SNEATH, P. H. A. 1957. The application of computers to taxonomy. *J. Gen. Microbiol.,* **17**:201–226.

————. 1962. The construction of taxonomic groups. *In* Microbial classification. Symposia Soc. Gen. Microbiol., Cambridge University Press no. 12, pp. 289–332.

SOKAL, R. R. 1965. Statistical methods of systematics. *Biol. Rev.,* **40**:337–391.

————, and C. D. MICHENER. 1967. The effect of different numerical techniques on the phenetic classification of bees of the Hoplitis complex (Megachilidae). *Proc. Linn. Soc. London,* **178**:59–74.

————, and P. H. A. SNEATH. 1963. Principles of numerical taxonomy. W. H. Freeman and Company, San Francisco, 359 pp.

SOTAVALTA, O. 1964. Studies on the variation of the wing venation of certain tiger moths. *Ann. Acad. Sci. Fenn.* A IV (Biol.), **74**:1–41.

SPIETH, H. T. 1952. Mating behavior within the genus Drosophila (Diptera). *Bull. Amer. Mus. Nat. Hist.,* **99**:399–474.

SPRAGUE, T. A., et al. 1950. Lectures on the development of taxonomy delivered during the session 1948–1949. Linnaean Society of London in conjunction with the Systematics Association, London, 83 pp.

SQUIRES, D. F. 1966. Data processing and museum collections: a problem for the present. *Curator,* **9**:216–227.

STAFLEU, F. A. 1963. *In* M. Adanson, vol. 1, Adanson and his "Familles des plantes." The Hunt Botanical Library, Carnegie Inst. Tech., Pittsburgh, Pa., Monograph Ser., no. 1, pp. 123–264.

STANILAND, L. N. 1953. The principles of line illustration. Burke Publ. Co., London, 224 pp.

STEBBINS, R. C. 1954. Amphibians and reptiles of Western North America. McGraw-Hill Book Company, New York, ix + 539 pp.

STENZEL, H. B. 1963. A generic character, can it be lacking in individuals of the species in a given genus? *Syst. Zool.* **12**:118–121.

STOREY, M., and N. J. WILMOWSKY. 1955. Curatorial practices in zoological research collections. 1. Preliminary report on containers and closures for storing specimens preserved in liquid. *Circ. Nat. Hist. Mus. Stanford Univ.,* no. 3, pp. 1–22.

STRESEMANN, E. 1950. The development of theories which affected the taxonomy of birds. *Ibis,* **92**:123–131.

STRICKLAND, H. E. 1842. Rules for zoological nomenclature. Report of 12th meeting of British Association held at Manchester in 1842. *Brit. Assoc. Adv. Sci. Rept.,* **1842**:105–121.

STUNKARD, H. W. 1957. Intraspecific variation in parasitic flatworms. *Syst. Zool.,* **6**:7–18.

STURTEVANT, A. H. 1942. The classification of the genus Drosophila, with descriptions of nine new species. *Univ. Texas Publ.,* no. 4213, pp. 7–51.

Style Manual for Biological Journals (A.I.B.S.). 1964. 2d ed., 117 pp.

SVENSON, H. K. 1945. On the descriptive method of Linnaeus. *Rhodora,* **47**:273–302, 363–388.

SYLVESTER-BRADLEY, P. C. 1951. The subspecies in paleontology. *Geol. Mag.,* **88**:88–102.

——— (ed.). 1956. The species concept in paleontology. *Syst. Assoc. Publ.,* London, no. 2, p. 145.

———. 1958. Description of fossil populations. *J. Paleontol.,* **32**:214–235.

TAVOLGA, W. N., and W. N. LANYON (eds.). 1960. Animal sounds and communication. Amer. Inst. Biol. Sciences, Publ. no. 7, 443 pp.

TAXOMETRICS. 1962ff. A newsletter dealing with mathematical and statistical aspects of classification. Milano, Italy.

THIENEMANN, A. 1938. Rassenbildung bei *Planaria alpina.* Jub.-Festschr. Grig. Antipa, pp. 1–21 (from *Zool. Ber.* **49** (1940):84–85).

THOMPSON, W. R. 1952. The philosophical foundation of systematics. *Can. Entomol.,* **84**:1–16.

———. 1962. Evolution and taxonomy. *Studia Entomol.,* **5**:549–570.

THROCKMORTON, L. H. 1962. The problem of phylogeny in the genus Drosophila. *In* Studies in Genetics, 2. M. R. Wheeler (ed.), *Univ. Texas Publ.,* no. 6205, pp. 207–343.

———. 1965. Similarity versus relationship in Drosophila. *Syst. Zool.,* **14**:221–236.

———. 1969. Concordance and discordance of taxonomic characters in Drosophila classification. *Syst. Zool.* (in press).

TINBERGEN, N. 1959. Comparative studies of the behavior of gulls (Laridae): A progress report. *Behavior,* **15**:1–70.

TRELEASE, S. F. 1951. The scientific paper. How to prepare it. How to write it. The Williams & Wilkins Company, Baltimore, xii + 163 pp.

UVAROV, B. P. 1921. A revision of the genus *Locusta* L. (*Pachytylus* Fieb.), with a new theory as to the periodicity and migrations of locusts. *Bull. Entomol. Res.,* **12**:135–163.

———. 1928. Locusts and grasshoppers. Imperial Bureau of Entomology, London, xiii + 352 pp.

VACHON, M. 1952. Études sur les scorpions. Institut Pasteur d'Algérie, Algiers, 482 pp.

VAN TYNE, J. 1952. Principles and practices in collecting and taxonomic work. *Auk,* **69**:27–33.

VAURIE, C. 1949. A revision of the bird family Dicruridae. *Bull. Amer. Mus. Nat. Hist.* **93**:205–342.

———. 1955. Pseudo-subspecies. Acta XI Congr. Int. Orn., 1954, Basel, pp. 369–380.

VERHEYEN, R. 1958. Contribution à la systematique des Alciformes. *Inst. Roy. Sci. Nat. Belgique,* **34**:1–15.

VILLALOBOS-DOMINGUEZ, C., and J. VILLALOBOS. 1947. Atlas de los colores. El Ateneo, Buenos Aires.

VOSS, E. G. 1952. The history of keys and phylogenetic trees in systematic biology. *J. Sci. Labs. Denison Univ.,* **43**:1–25.

WAGNER, R. P. 1944. Nutritional differences in the *mulleri* group. *Univ. Texas Publ.,* no. 4920, pp. 39–41.

WAGNER, W. 1962. Dynamische Taxionomie angewandt auf die Delphaciden Mitteleuropas. *Mitt. Hamburg Zool. Mus. Inst.,* **60**:111–180.

WAGSTAFFE, R., and J. H. FIDLER. 1955. The preservation of natural history specimens. Vol. I: Invertebrates. H. F. and G. Witherby, London, 205 pp.

WALKER, T. J. 1964. Cryptic species among sound-producing ensiferan Orthoptera (Gryllidae and Tettigoniidae). *Quart. Rev. Biol.*, 39:345–355.

WARING, H. 1963. Color change mechanisms of cold-blooded vertebrates. Academic Press Inc., New York, 266 pp.

WATSON, L., W. T. WILLIAMS, and G. N. LANCE. 1967. A mixed-data numerical approach to angiosperm taxonomy: the classification of Ericales. *Proc. Linn. Soc. London*, 178:25–35.

WHITE, M. J. D. 1954. Animal cytology and evolution, 2d ed. Cambridge University Press, Cambridge, England, 454 pp.

————. 1957. Cytogenetics and systematic entomology. *Ann. Rev. Entomol.*, 2:71–90.

WICKLER, W. 1961. Ökologie und Stammesgeschichte von Verhaltensweisen. *Fortschr. Zool.*, 13:303–365.

————. 1967. Vergleichende Verhaltensforschung und Phylogenetik. *In* G. Heberer (ed.), Evolution Organismen, 3d ed. G. Fischer, Stuttgart, pp. 420–508.

WIEGMANN'S ARCHIV FÜR NATURGESCHICHTE. 1835. *et seq.* Bericht über die Leistungen im Gebiete der Naturgeschichte während des Jahres 1834 [. . . 1922], Berlin.

WILLIAMS, C. A., JR. 1964. Immunochemical analysis of serum proteins of the primates: a study in molecular evolution. *In* J. Buettner-Janusch (ed), Evolutionary and genetic biology of the primates. Academic Press Inc., New York, no. 2, pp. 25–74.

WILSON, E. O. 1965. A consistency test for phylogenies based on contemporaneous species. *Syst. Zool.*, 14:214–220.

————, and W. L. BROWN. 1953. The subspecies concept and its taxonomic application. *Syst. Zool.*, 2:97–111.

WIRTH, M., G. F. ESTABROOK, and D. J. ROGERS. 1966. A graph theory model for systematic biology with an example for the Oncidiinae (Orchidaceae). *Syst. Zool.*, 15:59–69.

WOOD, A. E. 1950. Porcupines, paleogeography, and parallelism. *Evolution*, 4:87–98.

WOOD, C. A. 1931. An introduction to the literature of vertebrate zoology. Oxford University Press, London, xii–xix + 643 pp.

WRIGHT, C. A. 1966. Experimental taxonomy: A review of some techniques and their applications. *Intern. Rev. Gen. Exptl. Zool.*, 2:1–42.

WRIGHT, A. H., and A. A. WRIGHT. 1949. Handbook of frogs and toads. Cornell University Press (Comstock), Ithaca, N.Y., xii + 640 pp.

———— and ————. 1957. Handbook of snakes. Comstock Press Associates, Cornell University Press, Ithaca, N.Y. (2 vols.), 1105 pp.

ZWEIFEL, F. W. 1961. A handbook of biological illustration. Phoenix Books, University of Chicago Press, vii +131 pp.

# Glossary

**Accessory sexual characters.** The structures and organs (except the gonads) of which the genital tract is composed, including accessory glands and external genitalia (cf. Secondary sexual characters).

**Adaptation.** The condition of showing fitness for a particular environment, as applied to characteristics of a structure, function, or entire organism; also the process by which such fitness is acquired (cf. Environment).

**Adaptive radiation.** Evolutionary divergence of members of a single phyletic line into a series of rather different niches or adaptive zones.

**Affinity.** Relationship. Sometimes misleadingly employed as synonym for phenetic similarity.

**Agamic.** A species or generation which is not reproducing sexually.

**Agamospecies.** A species without sexual reproduction, an asexual species.

**Albinism.** In zoology, the absence of pigmentation, and particularly of melanins, in an animal (cf. Melanism).

**Allele.** Any of the alternative expressions (states) of a gene (locus).

**Allochronic species.** Species which do not occur at the same time level (cf. Synchronic species).

**Allometric growth.** Growth in which the growth rate of one part of an organism is different from that of another part or of the body as a whole.

**Allopatric.** Of populations or species, occupying mutually exclusive (but usually adjacent) geographical areas.

**Allopatric hybridization.** Hybridization between two allopatric populations (species or subspecies) along a well-defined contact zone (cf. Sympatric hybridization).

**Allopatric speciation.** Species formation during geographical isolation (cf. Sympatric speciation).

**Allotype.** A paratype of the opposite sex to the holotype (cf. Paratype).

**Alpha taxonomy.** The level of taxonomy concerned with the characterization and naming of species (See 1.4.3).

**Alternation of generations.** The alternation of a bisexual with a unisexual (parthenogenetic) generation.

**Amphiploid.** A polyploid produced by the chromosome doubling of a species hybrid, that is, of an individual with two rather different chromosome sets.

**Analogous.** A similar feature in two or more taxa which cannot be traced back to the same feature in the common ancestor of these taxa (cf. Homologous, and 4.8.1).

**Anatomy.** The science of internal morphology, as revealed by dissection.

**Antibody.** A serum globulin which is produced in the blood of an immunized animal in response to the introduction of a foreign antigen (cf. Antigen, Antiserum, Serology).

**Antigen.** A substance capable of inducing the formation of antibodies when introduced into the bloodstream of an animal (cf. Antibody, Precipitin reaction, Serology).

**Antigenic.** With the properties of an antigen.

**Antiserum.** Blood serum containing specific antibodies (cf. Antibody, Precipitin reaction).

**Apomorph.** A derived character.

**A posteriori weighting.** The weighting of taxonomic characters on the basis of their proved contribution to the establishment of sound classifications, i.e. of monophyletic taxa.

**A priori weighting.** The weighting of taxonomic characters on the basis of preconceived criteria, e.g. their physiological importance.

**Archetype.** A hypothetical ancestral type arrived at by the elimination of specialized characters (cf. Phylogeny).

**Artenkreis** (Rensch). Superspecies (q.v.).

**Artificial classification.** Classification based on convenient and conspicuous diagnostic characters without attention to characters indicating relationship; often a classification based on a single arbitrarily chosen character instead of an evaluation of the totality of characters (cf. Classification, Phylogeny).

**Asexual reproduction.** Not involving the fusion of the nuclei of different gametes.

**Atlas.** In taxonomy, a method of presenting taxonomic materials primarily by means of comparative illustrations rather than by comparative descriptions (cf. Monograph).

**Authority citation.** The custom of citing the name of the author of a scientific name or name combination [e.g., *X-us* Jones, *X-us albus* Jones, *Y-us albus* (Jones)]. (See 13.42).

**Autopolyploid.** A polyploid originating through the doubling of a diploid chromosome set.

**Autosome.** One of the chromosomes other than a sex chromosome.

**Available name.** A name published in a manner satisfying the requirements specified in Arts. 8–20 of the Code (cf. Valid name, and 13.16).

**Baculum.** An ossification (bone) in the phallus of some mammals.

**Beta taxonomy.** The level of taxonomy concerned with the arranging of species into a natural system of lower and higher taxa (cf. Alpha taxonomy, Gamma taxonomy; see 1.4.3).

**Bibliographical reference.** For nomenclatural purposes, the citation of the name of the author and date of publication for a scientific name; a full bibliographical reference includes, in addition, the citation of the exact place of publication of a scientific name (i.e., title of book or journal, volume, page, etc.). (See 11.B.2.)

**Binary.** Refers to designations consisting of two kinds of names (see 13.11; cf. Binominal nomenclature).

**Binomen.** The scientific designation of a species, consisting of a generic and a specific name.

**Binomial nomenclature.** The system of nomenclature first standardized by Linnaeus and now generally referred to as binominal nomenclature.

**Binominal nomenclature.** The system of nomenclature adopted by the International Congress of Zoology, by which the scientific name of an animal is designated by both a generic and specific name (cf. Binary nomenclature).

**Biological classification.** The arranging of organisms into taxa on the basis of inferences concerning their genetic relationship.

**Biological races.** Noninterbreeding sympatric populations, which differ in biology but not, or scarcely, in morphology; supposedly prevented from interbreeding by preference for different food plants or other hosts (cf. Sibling species).

**Biological species concept.** A concept of the species category stressing reproductive isolation, and the possession of a genetic program effecting such isolation (cf. Species; see 2.2.3).

**Biota.** The flora and fauna of a region (cf. Fauna, Flora).

**Bisexual.** Of a population composed of functional males and females; sometimes also applied to an individual possessing functional male and female reproductive organ (= hermaphrodite).

**Catalog.** An index to taxonomic literature arranged by taxa so as to provide ready reference to at least the most important taxonomic and nomenclatural references to the taxon involved (cf. Checklist).

**Category.** *See* Taxonomic category.

**Character.** *See* Taxonomic character.

**Character displacement.** A divergence of equivalent characters in sympatric species resulting from the selective effects of competition.

**Character gradient.** *See* Cline.

**Character index.** A numerical value, compounded of the ratings of several characters, indicating a degree of difference of related taxa; also a rating of an individual, particularly a hybrid, in comparison with its most nearly related species (see 9A.3).

**Checklist.** Usually a skeleton classification of a group listed by taxa for quick reference and as an aid in the arrangement of collections (cf. Catalog).

**Cheironym.** A manuscript name (q.v.).

**Chorology.** The study of the geographical distribution of organisms.

**Chromatophore.** A pigment-bearing intracellular body.

**Chromosomal inversion.** Reversal of the linear order of the genes in a segment of a chromosome.

**Chromosome.** A deeply staining DNA-containing body in the nucleus of the cell, best seen during cell division.

**Circular overlap.** The phenomenon in which a chain of contiguous and intergrading populations curves back until the terminal links overlap geographically and behave like good species (noninterbreeding).

**Cladism.** A taxonomic theory by which organisms are ordered and ranked entirely on the basis of "recency of common descent," that is, on the basis of the most recent branching point of the inferred phylogeny.

**Cladistic.** Based on the principles of cladism.

**Cladogram.** A dendrogram based on the principles of cladism; a strictly genealogical dendrogram in which rates of evolutionary divergence are ignored (see 10.7.2).

**Classification.** The delimitation, ordering, and ranking of taxa (cf. Taxonomy, Systematics, Horizontal classification, Vertical classification, Artificial classification, Biological classification).

**Clinal.** Varying gradually, of characters.

**Cline.** A gradual and nearly continuous change of a character in a series of contiguous populations; a character gradient (cf. Subspecies).

**Clone.** All the offspring derived by asexual reproduction from a single sexually produced individual.

**Clustering methods.** Methods of grouping related or similar species into species groups or higher taxa (see 10.5.2).

**Coccids.** Scale insects.

**Code.** International Code of Zoological Nomenclature (see Chap. 12).

**Code of Ethics.** A set of recommendations on the propriety of taxonomic actions, to guide the taxonomist. Formulated in Appendix A of the International Code of Zoological Nomenclature (see 13.64).

**Coefficient of difference.** Difference of means divided by sum of standard deviations (see 9B1.1):

$$CD = \frac{M\,b - M\,a}{SD\,a + SD\,b}$$

**Coefficient of variability.** The standard deviation as percentage of the mean:

$$CV = \frac{SD \times 100}{M}$$

**Collective group.** An aggregate of related species of which the generic position is uncertain, used principally in paleontology and parasitology (see 13.39).

**Commission.** The International Commission on Zoological Nomenclature (see 13.59).

**Common name.** Colloquial name = vernacular name (q.v.).

**Competitive exclusion.** The principle that no two species can coexist at the same locality if they have identical ecological requirements.

**Complex.** A neutral term for a number of related taxonomic units, most commonly involving units in which the taxonomy is difficult or confusing (cf. Group, Neutral term).

**Congeneric.** A term applied to species of the same genus (cf. Genus).

**Conspecific.** A term applied to individuals or populations of the same species (cf. Species).

**Continuity.** In nomenclature, the principle that continuity of usage should take precedence over priority of publication in determining which of two or more competing scientific names should be adopted for a particular taxon (cf. Priority).

**Continuous variation.** Variation in which individuals differ from each other by infinitely small steps, as variation in quality of expression of a character or group of characters (cf. Discontinuous variation).

**Convergence.** Morphological similarity in but distantly related forms (see 10.1.3, 10.4.5).

**Cope's rule.** The generalization that there is a steady increase in size in phyletic series.

**Correlated characters.** Characters that are associated either as manifestations of a well-integrated ancestral gene complex (10.4.2) or because they are functionally correlated (10.4.3).

**Cotype.** Syntype (q.v.).

**Cryptic species.** Sibling species (q.v.).

**Cyclomorphosis.** A seasonal (and thus cyclic) nongenetic change of phenotype in species of planktonic freshwater organisms, particularly cladocerans and rotifers.

**Cytogenetics.** The comparative study of chromosomal mechanisms and behavior in populations and taxa, and their effect on inheritance and evolution.

**Cytology.** The study of the structure and physiology of the cell and its parts.

**Darwin principle.** The taxonomic importance of characters that are not the result of a specific ad hoc adaptation (10.4.2).

**Data matrix.** A tabulation of differences between species (or other taxa) in rows (characters) and columns (taxa) (10.2.2).

**Delimitation.** In taxonomy, a formal statement of the characters of a taxon which sets its limits (cf. Description, Diagnosis, Differential diagnosis).

**Deme.** A local population of a species; the community of potentially interbreeding individuals at a given locality.

**Dendrogram.** A diagrammatic drawing in the form of a tree designed to indicate degrees of relationship (see 10.7.2; cf. Phylogenetic tree).

**Derived character.** A character that differs materially from the ancestral condition (10.3.1).

**Description.** In taxonomy, a more or less complete formal statement of the characters of a taxon without special emphasis on those which set limits to the taxon or distinguish it from coordinate taxa (cf. Delimitation, Diagnosis, Differential diagnosis).

**Designated priority.** In cases of simultaneous publication of several names, the priority established by the first reviser (see 13.26).

**Diagnosis.** In taxonomy, a formal statement of the characters (or most important characters) which distinguish a taxon from other similar or closely related coordinate taxa (cf. Differential diagnosis, Description).

**Dichotomous.** Divided or dividing into two parts.

**Differentia.** Linnaeus' polynominal species diagnosis.

**Differential diagnosis.** A formal statement of the characters which distinguish a given taxon from other specifically mentioned equivalent taxa.

**Dimorphism.** Occurrence of two distinct morphological types (morphs, phena) in a single population (cf. Sexually dimorphic, Polymorphism).

**Diploid.** Having a double set of chromosomes ($2n$); the normal chromosome number of the cells (except for mature germ cells) of a particular organism derived from a fertilized egg (cf. Haploid, Polyploidy, Chromosome).

**Discontinuous variation.** Variation in which the individuals of a sample fall into definite classes which do not grade into each other (cf. Continuous variation).

**Discriminant function.** The sum of numerical values of certain diagnostic characters multiplied by calculated constants (9A.3).

**Dollo's rule.** The principle that evolution is irreversible to the extent that structures or functions once lost cannot be regained.

**Dominant.** An allele which determines the phenotype of a heterozygote (cf. Recessive, Homozygous, Heterozygous).

**Eclipse plumage.** Inconspicuous plumage of birds worn in alternation with a bright nuptial plumage.

**Ecological isolation.** A condition in which interbreeding between two or more otherwise sympatric populations is believed to be prevented by mating in different ecologic niches (cf. Reproductive isolation, Geographic isolation).

**Ecological race.** A local race that owes its most conspicuous attributes to the selective effect of a specific environment (cf. Ecotype).

**Ecology.** The study of the interactions between organisms and their environment.

**Ecophenotypic variation** (habitant variation). A nongenetic modification of the phenotype by specific ecological conditions, particularly those of a habitat.

**Ecospecies.** "A group of populations so related that they are able to exchange genes freely without loss of fertility or vigor in the offspring" (Turesson).

**Ecotype.** A descriptive term applied to plant races of varying degrees of distinctness which owe their most conspicuous characters to the selective effects of local environments (cf. Subspecies).

**Edaphic factor.** The influence of soil properties on organisms (especially plants).

**Electronic data processing (EDP).** The sorting and storage of data with the help of computers.

**Electrophoresis.** A process of separating different molecules, particularly polypeptides, owing to their differential rates of migration in an electric field.

**Emendation.** In nomenclature, an intentional modification of the spelling of a previously published scientific name (cf. Error, *Lapsus calami*) (see 13.32).

**Environment.** The totality of physical, chemical, and biotic conditions surrounding an organism.

**Equal weighting.** The method which treats all taxonomic characters as equally important, a key assumption of phenetics.

**Error.** In nomenclature, an unintentional misspelling of a scientific name, as a typographical error or an error of transcription (cf. Emendation, *Lapsus calami*).

**Essentialism.** A school of philosophers, originating with Plato and Aristotle, later maintained by the Thomists and so-called realists among the philosophers, who believed in the reality of underlying universals or essences; in taxonomy usually referred to as typology (see 4.3.2).

**Ethology.** The science of the comparative study of animal behavior.

**Eucaryotes.** Organisms with a well-defined nucleus and meiosis. All higher organisms above the level of procaryotes (q.v.).

**Eyepiece micrometer.** A linear scale in the field of vision of the eyepiece (or one of a pair of eyepieces) of a microscope for use as a measuring device.

**Family.** A taxonomic category including one genus or a group of genera or tribes of common phylogenetic origin, which is separated from related similar units (families) by a decided gap, the size of the gap being in inverse ratio to the size of the family (see 5.5).

**Family name.** The scientific designation of a taxon of family rank, recognized by the termination *idae,* which termination may not be used in names of other taxa.

**Fauna.** The animal life of a region (cf. Flora, Biota).

**Faunal work.** A publication in which taxa are included on the basis of their occurrence in a specified area rather than on the basis of relationship (cf. Local list; see 11A.6).

**First reviser.** The first author to publish a definite choice of one among two or more conflicting names or zoological interpretations which are equally available under the Code; in order to qualify as first reviser an author must give evidence of a choice between available alternatives (see 13.4).

**Flora.** The plant life of a region (cf. Fauna, Biota).

**Form.** A neutral term for a single individual, phenon, or taxon (cf. Group, Neutral term).

**Formenkreis.** A collective category of allopatric subspecies or species (Klein-schmidt); in paleontology, a group of related species or variants.

**Full bibliographical synonymy.** A reasonably complete list of references to a given taxon arranged so as simultaneously to serve the needs of nomenclature (chronology of names) and zoology (pertinent taxonomic and biological sources) (cf. Synonymy).

**Gamma taxonomy.** The level of taxonomy dealing with various biological aspects of taxa, ranging from the study of intraspecific populations to studies of speciation and of evolutionary rates and trends (see 1.4.3).

**Gause's rule.** The theory that no two species with identical ecological requirements can coexist in the same place (cf. Competitive exclusion).

**Gene.** A hereditary determiner; the unit of inheritance, carried in a chromosome, transmitted from generation to generation by the gametes, and controlling the development of the individual (cf. Chromosome).

**Gene flow.** The exchange of genetic factors between populations owing to dispersal of zygotes or gametes, e.g., pollen.

**Gene frequency.** The percentage of a given gene in a population (cf. Gene, Population).

**Gene pool.** The totality of the genes of a given population existing at a given time.

**Genetic drift.** Genetic changes in populations caused by random phenomena rather than by selection (cf. Local population).

**Genotype.** The genetic constitution of an individual or taxon (cf. Phenotype). Use of this term in nomenclature for the type-species of a genus is confusing and contrary to the terminology of the Code (see Art. 42*b,* Recommendation 67A).

**Genus.** A category for a taxon including one species or a group of species, presumably of common phylogenetic origin, which is separated from related similar units (genera) by a decided gap, the size of the gap being in inverse ratio to the size of the unit (genus) (see 5.4).

**Geographic isolate.** A population that is separated by geographic barriers from the main body of the species.

**Geographic isolation.** The separation of a gene pool by geographic barriers; the prevention of gene exchange between a population and others by geographic barriers.

**Geographical race.** Subspecies (q.v.).

**Grade.** A group of animals similar in level of organization; a level of anagenetic advance.

**Group.** A neutral term for a number of related taxa, especially an assemblage of closely related species within a genus (cf. Complex, Neutral term, Section).

**Gynandromorph.** An individual in which one part of the body is masculine, the other feminine; most frequent are bilateral gynandromorphs, in which the left and right halves are of different sex.

**Handbook.** In taxonomy, a publication designed primarily as an aid to field and laboratory identification rather than the presentation of new taxonomic conclusions (cf. Manual, Monograph).

**Haploid.** Having only a single set of chromosomes; gametes are usually haploid.

**Hermaphrodite.** An individual having both male and female reproductive organs (cf. Intersex).

**Heterozygous.** Having different alleles at homologous loci of the two parental chromosomes (cf. Allele, Locus, Homozygous).

**Hierarchy.** In classification, the system of ranks which indicates the categorical level of various taxa (i.e., kingdom to species) (cf. Taxonomic category).

**Higher category.** A taxonomic category of rank higher than the species (i.e., from subgenus to kingdom) (cf. Supraspecific).

**Higher taxon.** A taxon ranked in one of the higher categories.

**Histogram.** A set of rectangles in which the midpoints of class intervals are plotted on the abscissa and the frequencies (usually, number of specimens) on the ordinate (see 8C.3.1).

**Holistic.** Looking at wholes as more than the sums of their parts.

**Hollow curve.** For explanation see 10.5.3, Fig. 10-9. A curve demonstrating an excess over expectancy of very small (e.g. monotypic) and very large higher taxa.

**Holosteans.** A group of fishes ancestral to the teleost fishes.

**Holotype.** The single specimen designated or indicated as "the type" by the original author at the time of the publication of the original description (see 13.51).

**Homologous.** A feature in two or more taxa which can be traced back to the same feature in the common ancestor of these taxa (cf. Analogous).

**Homonym.** In nomenclature, one of two or more identical but independently proposed names for the same or different taxa (cf. Senior homonym, Junior homonym, Primary homonym, Secondary homonym; see 13.43).

**Homozygous.** Having identical alleles at the two homologous loci of a diploid chromosome set (cf. Allele, Locus, Heterozygous).

**Horizontal classification.** Classification which stresses grouping together species in a similar stage of evolution, rather than location on the same phyletic line (cf. Vertical classification).

**Host races.** Different genetic races of the same species in oligophagous food specialists or parasites occurring on different hosts.

**Hybrid belt.** A zone of interbreeding between two species, subspecies, or other unlike populations; zone of secondary intergradation (see 9B.2).

**Hybrid index.** *See* Character index (see 9A.3).

**Hybridization.** The crossing of individuals belonging to two unlike natural populations, principally species (cf. Allopatric hybridization).

**Hypodigm.** The entire material of a species that is available to a taxonomist.

**Identification.** The determination of the taxonomic identity of an individual (see 4.3.1 for theory and 7.3 for practices of identification).

**Indication.** In nomenclature the publication of certain types of evidence or cross references which establish the typification of a name and thus make it available (see Art. 16 and 13.20).

**Individual variation.** Variation within a population (see 8A, 8B).

**Industrial melanism.** The evolution of a darkened population owing to selection in the sooty surroundings of an industrial area (cf. Melanism).

**Infraorder.** An optional category below the suborder.

**Infraspecific.** Within the species; usually applied to categories (subspecies) and phena (varieties) (cf. Subspecies, Variety, Infrasubspecific form).

**Infrasubspecific form.** Individual and seasonal variants in a single interbreeding population.

**Infrasubspecific name.** A name given to an infrasubspecific form (see 13.41).

**Intergradation.** Merging gradually through a continuous series of intermediate forms or populations.

**International Code of Zoological Nomenclature.** The official set of regulations dealing with zoological nomenclature (see Chap. 12).

**Intersex.** An individual more or less intermediate in phenotype between male and female (cf. Hermaphrodite).

**Introgressive hybridization.** The spread of one or more genes of one species into the germ plasm of another species as a result of hybridization (cf. Hybridization).

**Irreversibility rule.** *See* Dollo's rule.

**Isolating mechanisms.** Properties of individuals that prevent successful interbreeding with individuals that belong to different populations.

**Isophene.** A line connecting points of equal expression of a character; lines at right angles to a cline on a map (cf. Cline).

**Junior homonym.** The more recently published of two or more identical names for the same or different taxa (cf. Homonym, Senior homonym).

**Junior synonym.** The more recently published of two or more available synonyms for the same taxon (cf. Synonym, Senior synonym).

**Karyological character.** A character involving chromosome structure or number.

**Key.** A tabulation of diagnostic characters of species (or genera, etc.) in dichotomous couplets facilitating rapid identification.

**Key character.** In taxonomy a character of special utility in a key.

**Lapsus calami.** In nomenclature, a slip of the pen, especially an error in spelling (cf. Error, Emendation; see 13.32).

**Lectotype.** One of a series of syntypes which, subsequent to the publication of the original description, is selected and designated through publication to serve as "the type" (see 13.54).

**Line precedence.** Occurrence of a name on an earlier line of the same page than another name for the same taxon (see 13.26).

**Linnaean hierarchy.** A structure of categorical ranks for taxa where each category except the lowest includes one or more subordinate categories (see 5.2).

**Local population.** The individuals of a given locality which potentially form a single interbreeding community (cf. Deme).

**Locus.** The position of a gene in a chromosome (cf. Gene, Chromosome).

**Lumper.** A taxonomist who emphasizes the demonstration of relationship in the delimitation of taxa and who tends to recognize large taxa (cf. Splitter; see 10.5.3).

**Macrotaxonomy.** The classification of higher taxa.

**Manuscript name.** In nomenclature, an unpublished scientific name (cf. *Nomen nudum*).

**Material.** In taxonomy, the sample available for taxonomic study (cf. Series, Hypodigm).

**Melanism.** An unusual darkening of color owing to increased amounts of black pigment; sometimes a racial character, sometimes, as in cases of polymorphism, restricted to a certain percentage of individuals within a population (cf. Industrial melanism, Albinism).

**Mendelian population.** A population with unrestricted interbreeding of individuals and free reassortment of genes.

**Meristic variation.** Variation in characters that can be counted, like number of vertebrae, scales, fin rays, and so forth.

**Metamorphosis.** A drastic change of form during development, as when a tadpole changes into a frog, or an insect larva into an imago.

**Metric system.** A decimal system of measures (with the meter as base) and weights (with the gram as base); the universal system in science for reporting measures and weights.

**Microgeographic race.** A local race, restricted to a very small area.

**Millimeter** (mm). 1/1,000 m, or 0.03937 in., approximately $\frac{1}{25}$ in. (cf. Metric system).

**Mimetic polymorphism.** Polymorphism in Lepidoptera in which the various morphs resemble other species distasteful or poisonous to a predator; often restricted to females.

**Mimicry.** Resemblance in color or structure to other species that are distasteful or poisonous to a predator.

**Monogenic.** Determined by a single gene (cf. Polygenic).

**Monograph.** In taxonomy, an exhaustive treatment of a higher taxon in terms of all available information pertinent to taxonomic interpretation; usually involving full systematic treatment of the comparative anatomy, biology, ecology, and detailed distributional analyses of all included taxa (cf. Revision, Synopsis).

**Monophyly.** The derivation of a taxon through one or more lineages from one immediately ancestral taxon of the same or lower rank (see 4.3.5).

**Monotypic.** A taxon containing but one immediately subordinate taxon, as a genus containing but one species, or a species containing but one (the nominate) subspecies.

**Morph.** Any of the genetic forms (individual variants) that account for polymorphism.

**Morphospecies.** A typological species recognized merely on the basis of morphological difference (cf. Phenon).

**Mosaic evolution.** Evolution involving unequal rates for different structures, organs, or other components of the phenotype.

**Muellerian mimicry.** Similarity (usually consisting of a similar warning coloration) of several species which are distasteful, poisonous, or otherwise harmful.

**Multivariate analysis.** The simultaneous analysis of several variable characters.

**Mutation.** In genetics, a discontinuous change of a genetic factor, usually the replacement or loss of one or several base pairs in the DNA.

**Natural selection.** The unequal contribution of genotypes to the gene pool of the next generation, through differential mortality and differences in reproductive success, caused by components of the environment.

**Neontology.** The science dealing with the life of Recent organisms (cf. Paleontology).

**Neoteny.** Attainment of sexual maturity in an immature or larval stage.

**Neotype.** A specimen selected as type subsequent to the original description in cases where the original types are known to be destroyed or were suppressed by the Commission (see 13.55).

**Neutral term.** A taxonomic term of convenience, such as *form* or *group,* which may be employed without reference to the formal taxonomic hierarchy of categories, and which has no nomenclatural significance.

**New name.** A replacement name for a preoccupied name (cf. Substitute name; see 13.45).

**Niche** (ecological). The precise constellation of environmental factors into which a species fits or which is required by a species.

**Nomenclator.** A book containing a list of scientific names assembled for nomenclatural, rather than taxonomic, purposes (cf. Catalog).

**Nomenclature.** A system of names (see Chaps. 12 and 13).

**Nomen conservandum.** A name preserved by action of the Commission and placed on the appropriate official list (see 13.63).

**Nomen dubium.** The name of a nominal species for which available evidence is insufficient to permit recognition of the zoological species to which it was applied.

**Nomen oblitum.** A name losing its validity under the statute of limitation (Art. 23*b*) (see 13.24).

**Nominal taxon** (species, genus, etc.). A named taxon, objectively defined by its type (see 13.58).

**Nominalism.** A school of philosophy, denying the existence of universals, and emphasizing the importance of man-given names for the grouping of individuals (see 4.3.3).

**Nominate.** Of a subordinate taxon (subspecies, subgenus, etc.), which contains the type of the subdivided higher taxon and bears the same name (see 13.36).

**Nondimensional species.** The species concept, represented by the noninterbreeding of species at a given place and time.

**Numerical phenetics.** The hypothesis that relationship of taxa can be determined by a calculation of an overall, unweighted similarity value.

**Objective synonym.** Each of two or more synonyms based on the same type.

**Official Index.** A list of names or works suppressed or declared invalid by the Commission (see 13.63).

**Official List.** A list of names or works which have been conserved or declared to be valid by the Commission (see 13.63).

**Oligogenic character.** A character determined by only few genes.

**Onomatophore.** "Name-bearer" = type (Simpson) (see 13.48).

**Ontogeny.** The developmental history of an individual organism from egg to adult.

**Original description.** A statement of characters accompanying the proposal of a name for a new taxon in conformance with Arts. 12 and 13 of the Code (see 11B.1.1).

**Orthogenesis.** Evolution of phyletic lines following a predetermined rectilinear pathway, the direction not being determined by natural selection.

**Overall similarity.** A (usually numerical) value of similarity calculated by the summation of similarities in numerous individual characters (see 10.2).

**Page precedence.** Occurrence of a name on an earlier page in the same publication than a synonym or homonym of it (see 13.26).

**Paleontology.** The science that deals with the life of past geological periods (cf. Neontology).

**Parallelism.** The independent acquisition of similar characters in related evolutionary lines (cf. Convergence).

**Parapatry.** Of populations or species, in nonoverlapping geographical contact without interbreeding.

**Parasitoid.** Wasps and flies, the larvae of which parasitize (and usually kill) individuals of the host species.

**Paratype.** A specimen other than the holotype which was before the author at the time of preparation of the original description and was so designated or indicated by the original author.

**Parthenogenesis.** The production of offspring from unfertilized eggs.

**Patronymic.** In nomenclature, a dedicatory name, a name based on that of a person or persons.

**Pheneticist.** *See* Numerical phenetics.

**Phenetic ranking.** Ranking into categories, strictly based on degree of overall similarity (see 10.2 and 10.5.2).

**Phenogram.** A diagram indicating degree of similarity among taxa (see 10.7.2).

**Phenon.** A sample of phenotypically similar specimens; a phenotypically reasonably uniform sample (see 1.2).

**Phenotype.** The totality of characteristics of an individual (its appearance) as a result of the interaction between genotype and environment.

**Phyletic.** Pertaining to a line of descent (cf. Phylogeny).

**Phyletic correlation.** Correlation of characters that are phenotypic manifestations of a well-integrated ancestral gene complex (see 10.4.2).

**Phyletic weighting.** Assessing the taxonomic importance of a character on the basis of its phyletic information content (see 10.4).

**Phylogenetic tree.** A diagrammatic presentation of inferred lines of descent, based on paleontological, morphological, or other evidence.

**Phylogeny.** The study of the history of the lines of evolution in a group of organisms; the origin and evolution of higher taxa (cf. Classification).

**Phylogram.** A tree-like diagram indicating degree of relationship among taxa (see 10.7.2).

**Pleiotropy.** The capacity of a gene to affect several characters, that is, several aspects of the phenotype.

**Plenary powers.** Special powers granted to the Commission (see 13.59 and Art. 79).

**Plesiomorph.** Of characters, primitive or as found in the ancestor (Hennig).

**Polygenic.** Of a character, controlled by several or numerous genes.

**Polymorphism.** The simultaneous occurrence of several discontinuous phenotypes or genes in a population, with the frequency even of the rarest type higher than can be maintained by recurrent mutation.

**Polynominal nomenclature.** A system of nomenclature consisting of a scientific designation of a species through more than two words; the antecedent of the Linnaean "binomial" system.

**Polyphyletic.** A term applied to a composite taxon derived from two or more ancestral sources; not of a single immediate line of descent (cf. Monophyletic).

**Polyploidy.** A condition in which the nuclear complement of chromosomes is an integral multiple (greater than 2) of the haploid number.

**Polythetic.** Of taxa, in which each member has a majority of a set of characters (see 4.7; also Simpson, 1961, p. 42, under polytypic).

**Polytopic.** Occurring in different places as, for instance, a subspecies composed of widely separated populations.

**Polytypic.** A taxon containing two or more taxa in the immediately subordinate category, as a genus with several species or a species with several subspecies (cf. Monotypic).

**Population.** *See* Local population.

**Precipitin reaction.** The formation of a visible precipitate at the interface when an antigen and the corresponding antiserum are brought together (cf. Antigen, Antiserum, Antibody).

**Predictive value.** The capacity of a classification to make predictions on newly employed characters or newly discovered taxa (see 4.5).

**Pre-Linnaean name.** A name published prior to Jan. 1, 1758, the starting point of zoological nomenclature (see 13.8).

**Primary homonym.** Each of two or more identical species-group names which, at the time of original publication, were proposed in combination with the same generic name (e.g., *X-us albus* Smith, 1910, and *X-us albus* Jones, (1920) (see 13.43).

**Primary intergradation.** A zone of intermediacy between two phenotypically different populations, having developed *in situ* as a result of selection (cf. Secondary intergradation; see 9B.2).

**Primary zoological literature.** Literature dealing with animals or zoological phenomena, not merely a listing of names (see 13.24).

**Priority.** The principle that of two competing names for the same taxon (below the rank of an infraorder) ordinarily that is valid which was published first (see 13.3).

**Procaryotes.** Those microorganisms (viruses, bacteria, blue-green algae) that lack well-defined nuclei and meiosis (cf. Eucaryotes).

**Pseudogamy** ( = **Gynogenesis**). Parthenogenetic development of the egg cell after the egg membrane has been penetrated by a male gamete.

**Q technique.** An analysis of association of pairs of taxa in a data matrix (see 10.2.2).

**Race.** Subspecies (q.v.).

**Ranking.** The placement of a taxon in the appropriate category in the hierarchy of categories (see 10.5.2).

**Rassenkreis** (Rensch). Synonym for polytypic species (cf. Subspecies, Polytypic; see 3.2).

**Recapitulation.** The theory that ontogeny recapitulates phylogeny (cf. Ontogeny, Phylogeny).

**Recent.** Of taxa which still exist, antonym of Fossil.

**Recessive.** Of a gene, not affecting the phenotype of the heterozygote.

**Reductionism.** The erroneous belief that complex phenomena can be entirely explained by reducing them to the smallest possible component parts and by explaining these.

**Redundant characters.** Characters so closely correlated with other, already used characters that they do not contribute new information to the analysis (see 10.4.3).

**Regression analysis.** A form of multivariate analysis (see 8C.2.5).

**Regressive character.** A character which is being reduced or lost in the course of phylogeny, sometimes independently in several related lines (see 10.4.3).

**Relationship.** For meaning in classification, see 10.1.

**Replacement name.** Substitute name (q.v.).

**Reproductive isolation.** A condition in which interbreeding between two or more populations is prevented by intrinsic factors (cf. Isolating mechanism).

**Reticulate evolution.** Evolution "dependent on repeated intercrossing between a number of lines, and thus both convergent and divergent at once" (Huxley).

**Revision.** In taxonomy, the presentation of new material or new interpretations integrated with previous knowledge through summary and reevaluation (cf. Synopsis, Monograph).

**R technique.** An analysis of association of characters in a data matrix (see 10.2).

**Saltation.** Discontinuous variation produced in a single step by major mutation (cf. Mutation).

**Sample.** That portion of a true population which is actually available to the taxonomist.

**Scatter diagram.** A bivariate or multivariate graphic method of population analysis (see 8C.3.3.).

**Scientific name.** The binominal or trinominal designation of an animal; the formal nomenclatural designation of a taxon (cf. Vernacular name).

**Secondary homonym.** Each of two or more identical specific names which, at the time of original publication, were proposed in combination with different generic names but which, through subsequent transference, reclassification, or combination of genera have come to bear the same (or an identical) combination of a generic and specific name (see 13.43).

**Secondary intergradation.** A zone of hybridization or strong steepening of character gradients where two separately differentiated populations have reestablished contact (see 9B.2).

**Secondary sexual characters.** Characters which distinguish the two sexes of the same species but which do not (like gonads or accessory sexual characters) function directly in reproduction (cf. Sexually dimorphism).

**Section.** A neutral term usually employed with reference to a subdivision of a taxon or a series of related elements in one portion of a higher taxon (cf. Higher category, Neutral term, Group).

**Selection.** See Natural selection.

**Semispecies.** The component species of superspecies (Mayr); also, populations that have acquired some, but not yet all, attributes of species rank; borderline cases between species and subspecies.

**Senior homonym.** The earliest published of two or more identical names for the same or different taxa (cf. Homonym, Junior homonym).

**Senior synonym.** The earliest published of two or more available synonyms for the same taxon (cf. Synonym, Junior synonym).

**Series.** In taxonomy, the sample which the collector takes in the field or the sample available for taxonomic study (cf. Material, Hypodigm).

**Serology.** The study of the nature and interactions of antigens and antibodies (cf. Antigen, Antibody).

**75-percent rule.** The rule that population $A$ can be considered subspecifically distinct from population $B$ if at least 75 percent of the individuals of $A$ are different from "all" the individuals of population $B$ (see 9B.1.2).

**Sex Chromosome.** A special chromosome, not occurring in identical number or structure in the two sexes and usually concerned with sex determination; the X chromosome or Y chromosome (cf. Chromosome, Autosome).

**Sex-limited character.** A character occurring in one sex only (cf. Secondary sexual character, Sex-linked character).

**Sex-linked character.** A character controlled by a gene located in the sex chromosome (cf. Sex chromosome).

**Sexual dimorphism.** The phenotypic difference between the two sexes of a species.

**Sexual reproduction.** Reproduction resulting in a diploid zygote with a maternal and a paternal chromosome set.

**Sibling species.** Pairs or groups of closely related species which are reproductively isolated but morphologically identical or nearly so (cf. Species; see 9A.2).

**Sonagram.** A graphic representation of the vocalization of an animal (see 7.4.10; Fig. 7-6).

**Speciation.** The splitting of a phyletic line; the process of the multiplication of species; the origin of discontinuities between populations caused by the development of reproductive isolating mechanisms (cf. Allopatric speciation, Sympatric speciation).

**Species.** Groups of actually (or potentially) interbreeding natural populations which are reproductively isolated from other such groups (cf. Subspecies, Population, Reproductive isolation).

**Specific name.** The second component of the binominal name of a species (see 13.40).

**Splitter.** In taxonomy, one who divides taxa very finely, to express every shade of difference and relationship, through the formal recognition of separate taxa and their elaborate categorical ranking (see 10.5.3).

**Standard deviation.** (SD) The square root of the sum $\Sigma$ of the squared deviations $d$ from the mean, divided by $N$:

$$SD = \sqrt{\frac{\Sigma d^2}{N}}$$

**Standard error** (of the mean). Standard deviation divided by the square root of the sample size, $N$:

$$SE = \frac{SD}{\sqrt{N}}$$

**Statute of limitation.** A provision in the Code (Art. 23$b$) to protect universally adopted junior names against the revival of forgotten senior synonyms (see 13.24).

**Strickland Code.** A code of nomenclature prepared by a committee of the British Association for the Advancement of Science under the secretaryship of H. E. Strickland and first published in 1842.

**Subfamily.** A category of the family-group subordinate to the family; an individual taxon ranked in the category subfamily (see 5.5).

**Subgeneric name.** The name of an optional category between the genus and the species, enclosed in parentheses when cited in connection with a binominal or trinominal combination and therefore excluded from consideration when determining the number of words of which a specific or subspecific name is composed [eg., *X-us* (*Y-us*) *albus rufus* is a trinominal].

**Subjective synonym.** Each of two or more synonyms based on different types, but regarded as referring to the same taxon by those zoologists who hold them to be synonyms.

**Subspecies.** A geographically defined aggregate of local populations which differs taxonomically from other such subdivisions of the species (see 3.3.2).

**Substitute name.** A name proposed to replace a preoccupied name and automatically taking the same type and type-locality (=New name; see 13.45).

**Superfamily.** The taxonomic category immediately above the family and below the order; an individual taxon ranked in this category.

**Superspecies.** A monophyletic group of entirely or largely allopatric species (cf. Allopatric, Semispecies; see 3.7).

**Supraspecific.** A term applied to a category or evolutionary phenomenon above the species level.

**Sympatric hybridization.** The occasional production of hybrid individuals between two otherwise well-defined sympatric species.

**Sympatric speciation.** Speciation without geographic isolation; the acquisition of isolating mechanisms within a deme.

**Sympatry.** The occurrence of two or more populations in the same area; more precisely, the existence of a population in breeding condition within the cruising range of individuals of another population.

**Symplesiomorphy.** The sharing of ancestral characters by different species (Hennig) (see 10.1.3).

**Synapomorphy.** The sharing of derived characters by several species (see 10.1.3).

**Synchronic species.** Species which occur at the same time level (cf. Allochronic species).

**Synonym.** In nomenclature, each of two or more different names for the same taxon (cf. Senior synonym, Junior synonym, Objective synonym, Subjective synonym).

**Synonymy.** A chronological list of the scientific names which have been applied to a given taxon, including the dates of publication and the authors of the names.

**Synopsis.** In taxonomy, a brief summary of current knowledge of a group (see 11A.2).

**Syntype.** Every specimen in a type-series in which no holotype was designated (see 13.54).

**Systematics.** The science dealing with the diversity of organisms (see 1.1).

**Taxon** (pl. taxa). A taxonomic group that is sufficiently distinct to be worthy of being distinguished by name and to be ranked in a definite category (see 1.2).

**Taxonomic category.** Designates rank or level in a hierarchic classification. It is a class, the members of which are all taxa assigned a given rank (see 1.2).

**Taxonomic character.** Any attribute of a member of a taxon by which it differs or may differ from a member of a different taxon (see 7.1).

**Taxonomy.** The theory and practice of classifying organisms (cf. Classification, Systematics).

**Teratology.** The study of structural abnormalities, especially monstrosities and malformations.

**Therapsid reptiles.** The reptilian group from which the mammals evolved.

**Topotype.** A specimen collected at the type-locality (see 13.57).

**Tribe.** A taxonomic category intermediate between the genus and the subfamily.

**Trinominal nomenclature.** An extension of the binominal system of nomenclature to permit the designation of subspecies by a three-word name (see 13.40).

**Triploid.** A cell or individual with three haploid chromosome sets, one of the forms of polyploidy.

**Trivial name.** An obsolete designation by Linnaeus for the specific name; also a synonym for "vernacular name."

**Type.** A zoological object which serves as the base for the name of a taxon (see 13.48–13.56).

**Type designation.** Determination of the type of a genus under Arts. 67–69 of the Code (see 13.52).

**Type-locality.** The locality at which a holotype, lectotype, or neotype was collected (cf. Topotype) (see 13.57).

**Type method.** The method by which the name for a taxon is unambiguously associated with a definite zoological object belonging to the taxon (see 13.48).

**Type selection.** Type designation (q.v.).

**Type-species.** The species which was designated as type of a nominal genus.

**Typological thinking.** A concept in which variation is disregarded and the members of a population are considered as replicas of the "type," the Platonic *eidos*.

**Uninominal nomenclature.** The designation of a taxon by a scientific name consisting of a single word; required for taxa above species rank.

**Univariate analysis.** A biometric analysis of a single character.

**Valid name.** An available name that is not preoccupied by a valid senior synonym or homonym (see 13.23).

**Variance.** The square of the standard deviation.

**Variation, ecophenotypic.** Variation caused by nongenetic responses of the phenotype to local conditions of habitat, climate, etc.

**Variety.** An ambiguous term of classical (Linnaean) taxonomy for a heterogeneous group of phenomena including nongenetic variations of the phenotype, morphs, domestic breeds, and geographic races.

**Vernacular name.** The colloquial designation of a taxon (cf. Scientific name).

**Vertical classification.** Classification which stresses common descent and tends to unite ancestral and descendant groups of a phyletic line in a single higher taxon, separating them from contemporaneous taxa having reached a similar grade of evolutionary change (cf. Horizontal classification, 10.5.4).

**Weighting.** A method for determining the phyletic information content of a character; the evaluation of the probable contribution of a character to a sound classification (see 10.4.).

*Index*

# Index

Page references in *italics* indicate Figures.